Blockchain Technology for Data Privacy Management

Advances in Intelligent Decision-Making, Systems Engineering, and Project Management

This new book series will report the latest research and developments in the field of information technology, engineering and manufacturing, construction, consulting, healthcare, military applications, production, networks, traffic management, crisis response, human interfaces, and other related and applied fields. It will cover all project types, such as organizational development, strategy, product development, engineer-to-order manufacturing, infrastructure and systems delivery, and industries and industry-sectors where projects take place, such as professional services, and the public sector including international development and cooperation etc. This new series will publish research on all fields of information technology, engineering, and manufacturing including the growth and testing of new computational methods, the management and analysis of different types of data, and the implementation of novel engineering applications in all areas of information technology and engineering. It will also publish on inventive treatment methodologies, diagnosis tools and techniques, and the best practices for managers, practitioners, and consultants in a wide range of organizations and fields including police, defense, procurement, communications, transport, management, electrical, electronic, aerospace, requirements.

Blockchain Technology for Data Privacy Management
Edited by Sudhir Kumar Sharma, Bharat Bhushan, Aditya Khamparia, Parma Nand Astya, and Narayan C. Debnath

For more information about this series, please visit: https://www.routledge.com/Advances-in-Intelligent-Decision-Making-Systems-Engineering-and-Project-Management/book-series/CRCAIDMSEPM

Blockchain Technology for Data Privacy Management

Edited by
Sudhir Kumar Sharma
Bharat Bhushan
Aditya Khamparia
Parma Nand Astya
Narayan C. Debnath

CRC Press is an imprint of the
Taylor & Francis Group, an **informa** business

First edition published 2021
by CRC Press
6000 Broken Sound Parkway NW, Suite 300, Boca Raton, FL 33487-2742

and by CRC Press
2 Park Square, Milton Park, Abingdon, Oxon, OX14 4RN

CRC Press is an imprint of Taylor & Francis Group, LLC

© 2021 selection and editorial matter, Sudhir Kumar Sharma, Bharat Bhushan, Aditya Khamparia, Parma Nand Astya, and Narayan C. Debnath; individual chapters, the contributors

The right of Sudhir Kumar Sharma, Bharat Bhushan, Aditya Khamparia, Parma Nand Astya, and Narayan C. Debnath to be identified as the authors of the editorial material, and of the authors for their individual chapters, has been asserted in accordance with sections 77 and 78 of the Copyright, Designs and Patents Act 1988.

Reasonable efforts have been made to publish reliable data and information, but the author and publisher cannot assume responsibility for the validity of all materials or the consequences of their use. The authors and publishers have attempted to trace the copyright holders of all material reproduced in this publication and apologize to copyright holders if permission to publish in this form has not been obtained. If any copyright material has not been acknowledged please write and let us know so we may rectify in any future reprint.

Except as permitted under U.S. Copyright Law, no part of this book may be reprinted, reproduced, transmitted, or utilized in any form by any electronic, mechanical, or other means, now known or hereafter invented, including photocopying, microfilming, and recording, or in any information storage or retrieval system, without written permission from the publishers.

For permission to photocopy or use material electronically from this work, access www.copyright.com or contact the Copyright Clearance Center, Inc. (CCC), 222 Rosewood Drive, Danvers, MA 01923, 978-750-8400. For works that are not available on CCC please contact mpkbookspermissions@tandf.co.uk

Trademark notice: Product or corporate names may be trademarks or registered trademarks and are used only for identification and explanation without intent to infringe.

Library of Congress Cataloging-in-Publication Data
Names: Sharma, Sudhir Kumar, editor.
Title: Blockchain technology for data privacy management / edited by Sudhir Kumar Sharma, Bharat Bhushan, Aditya Khamparia, Parma Nand Astya and Narayan C. Debnath.
Description: First edition. I Boca Raton : CRC Press, [2021] I Series: Advances in intelligent decision-making, systems engineering, and project management I Includes bibliographical references and index.
Identifiers: LCCN 2020044200 (print) I LCCN 2020044201 (ebook) I ISBN 9780367679200 (hbk) I ISBN 9781003133391 (ebk)
Subjects: LCSH: Blockchains (Databases) I Personal information management. I Privacy, Right of.
Classification: LCC QA76.9.B56 B574 2021 (print) I LCC QA76.9.B56 (ebook) I DDC 005.74–dc23
LC record available at https://lccn.loc.gov/2020044200
LC ebook record available at https://lccn.loc.gov/2020044201

ISBN: 978-0-367-67920-0 (hbk)
ISBN: 978-0-367-67923-1 (pbk)
ISBN: 978-1-003-13339-1 (ebk)

Typeset in Times
by SPi Global, India

Contents

Preface......vii

Editors......viii

Contributors......xi

Chapter 1 Overview of the Internet of Things and Ubiquitous Computing........1

Shashi Mehrotra, Shweta Sinha, and Sudhir Kumar Sharma

Chapter 2 Case Study on Future Generation IoT: Secure Model for IoT-Based Systems—Health Care Example......21

Mais Tawalbeh, Muhannad Quwaider, and Lo'ai A. Tawalbeh

Chapter 3 Spy-Bot: Controlling and Monitoring a Wi-Fi Controlled Surveillance Robotic Car Using Windows 10 IoT and Cloud Computing......37

A. Diana Andrushia, J. John Paul, K. Martin Sagayam, S. Rebecca Princy Grace, and Lalit Garg

Chapter 4 Multi-Antenna Communication Security with Deep Learning Network and Internet of Things......61

Garima Jain and Rajeev Ranjan Prasad

Chapter 5 Security Vulnerabilities, Challenges, and Schemes in IoT-Enabled Technologies......81

Siddhant Banyal, Amartya, and Deepak Kumar Sharma

Chapter 6 Advanced Security Using Blockchain and Distributed Ledger Technology......109

M. Al-Rawy and A. Elci

Chapter 7 Blockchain Technology and Its Emerging Applications......133

N. Rahimi, I. Roy, B. Gupta, P. Bhandari, and Narayan C. Debnath

Chapter 8 Emergence of Blockchain Technology: A Reliable and Secure Solution for IoT Systems......159

A.K.M. Bahalul Haque and Bharat Bhushan

vi   Contents

Chapter 9   Internet of Things and Blockchain: Amalgamation, Requirements,
                Issues, and Practices 185

   Ansh Riyal, Parth Sarthi Prasad, and Deepak Kumar Sharma

Chapter 10 Edge-Based Blockchain Design for IoT Security........................... 209

   *Pao Ann Hsiung, Wei-Shan Lee, Thi Thanh Dao, I. Chien,
   and Yong-Hong Liu*

Chapter 11 Blockchain for the Security and Privacy of IoT-Based
                Smart Homes 239

   Somya Goyal, Sudhir Kumar Sharma, and Pradeep Kumar Bhatia

Chapter 12 A Framework for a Secure e-Health Care System Using
                IoT-Based Blockchain Technology 253

   T. Sanjana, B. J. Sowmya, D. Pradeep Kumar, and K. G. Srinivasa

Chapter 13 Blockchain in EHR: A Comprehensive Review and Implementation
                Using Hyperledger Fabrics 275

   Ravinder Kumar

Chapter 14 Attacks, Vulnerabilities, and Blockchain-Based Countermeasures
                in Internet of Things (IoT) Systems... 295

   Nayanika Shukla and Bharat Bhushan

Index..317

Preface

With the emergence of numerous technologies, including embedded computing, sensors, actuators, cloud computing, and wireless devices, many objects in our routine life are becoming wirelessly interoperable. By enabling easy interaction with a wide range of objects (or physical devices), such as monitoring sensors, home appliances, surveillance cameras, actuators, vehicles, and so on, the Internet of Things (IoT) helps in the development of many different applications, including industrial automation, home automation, medical aids, intelligent energy management, mobile health care, and smart grids. The huge number of events generated by these objects, along with the heterogeneous technologies of the IoT, throw light on new challenges in application development, making ubiquitous computing even more difficult. The centralized IoT network architecture faces numerous challenges, such as single-point-of-failure, accountability, traceability, and security. These challenges make it necessary to rethink the structuring of the IoT. Currently, the most appropriate candidate technology that can support a distributed IoT ecosystem is "blockchain." Blockchain-based IoT systems have received enormous research interest, and researchers are making efforts to decentralize IoT communications using blockchain.

A blockchain is basically a chain of timestamped blocks linked by cryptographic hashes, which acts like a distributed ledger, whose data are shared among a network of peers. In general terms, blockchain is a continuously growing chain of blocks capable of storing all the committed transactions, with the help of a public ledger where every transaction is cryptographically verified and signed by all mining nodes. Blockchain operates in a decentralized environment supported by several core technologies, including distributed consensus algorithms, cryptographic hash, and digital signatures. Owing to its fault tolerance capabilities, decentralized architecture, and cryptographic security benefits such as authentication, data integrity, and pseudonymous identities, security analysts and researchers consider blockchain to resolve the privacy and security issues of the IoT. Blockchain also finds great application in various sectors, such as banking, health care, real estate, supply chains, government, cybersecurity, social media, and artificial intelligence. The maturity of blockchain technology and the establishment of its standard use cases indicate the need to gather all of the relevant research contributions into a single volume.

This book aims to provide a stepwise discussion, an exhaustive literature survey, rigorous experimental analysis, and discussions to demonstrate the usage of blockchain technology for securing communications. This book examines privacy issues and challenges related to data-intensive technologies in the IoT, embracing the role of blockchain in all possible facets of IoT security and other industrial applications. Furthermore, the book discusses the most promising state-of-the-art examples of ubiquitous computing, future IoT applications, blockchain technology, and other competitive technologies.

Editors

Sudhir Kumar Sharma is Professor of Computer Science at the Institute of Information Technology and Management, Guru Gobind Indraprastha University (GGSIPU), New Delhi, India. He has more than 21 years' experience in the field of computer science and engineering. He obtained his Ph.D. degree in Information Technology from University School of Information Communication and Technology USICT, GGSIPU, New Delhi, India. Dr. Sharma obtained his M. Tech degree in Computer Science and Engineering in 1999 from Guru Jambheshwar University, Hisar, India and an M.Sc. degree in Physics from the University of Roorkee (now IIT Roorkee), Roorkee, in 1997. His research interests include machine learning, data mining, and security. He has published a number of research papers in various prestigious international journals and international conferences. He is a lifetime member of Computer Society of India (CSI) and Institution of Electronics and Telecommunication Engineers (IETE). Dr. Sharma is Associate Editor of *the International Journal of End-User Computing and Development* (IJEUCD), IGI Global, USA. He was a convener of ICETIT-2019 and ICRIHE-2020.

Bharat Bhushan is Assistant Professor of Computer Science and Engineering (CSE) at the School of Engineering and Technology, Sharda University, Greater Noida, India. He is an alumnus as well as a Ph.D. scholar of Birla Institute of Technology, Mesra, India. He received his undergraduate degree (B.Tech in Computer Science and Engineering) with Distinction in 2012 and received his postgraduate degree (M.Tech in Information Security) with Distinction in 2015 from Birla Institute of Technology, Mesra. He has earned numerous international certifications, such as Cisco Certified Network Associate (CCNA), Cisco Certified Entry Networking Technician (CCENT), Microsoft Certified Technology Specialist (MCTS), Microsoft Certified IT Professional (MCITP) and Cisco Certified Network Professional Trained (CCNP). In the last three years, he has published more than 80 research papers at various renowned international conferences and in SCI indexed journals, including *Wireless Networks* (Springer), *Wireless Personal Communications* (Springer), *Sustainable Cities and Society* (Elsevier), and *Emerging Transactions on Telecommunications* (Wiley). He has contributed more than 20 chapters to various books and is currently in the process of editing seven books from publishers including Elsevier, IGI Global, and CRC Press. He has served as a reviewer/editorial board member for reputable international journals, including *IEEE Access, IEEE Communication Surveys and Tutorials*, and *Wireless Personal Communication* (Springer). He has also served as speaker and session chair at more than 15 national and international conferences. His current research interests include wireless sensor networks (WSNs), the Internet of Things (IoT), and blockchain technology. In the past, he has worked as an Assistant Professor at HMR Institute of Technology and Management, New Delhi, and Network Engineer at HCL Infosystems Ltd., Noida. He has qualified GATE exams for successive years, and he gained the highest percentile, 98.48, in GATE 2013.

Editors

Aditya Khamparia is Associate Professor of Computer Science and Engineering at Lovely Professional University, Punjab, India. His research areas are machine learning, soft computing, educational technologies, the IoT, semantic web and ontologies. An academic, researcher, author, consultant, community service provider, and Ph.D. supervisor, he focuses on rational and practical learning. With seven years' experience in teaching and two in industry, he has published more than 50 scientific research publications in reputable national and international journals and at conferences, which are indexed in various international databases. He has been invited as a faculty resource person, session chair, reviewer, and TPC member for different Faculty Deveopment programs (FDP)s, conferences, and journals. Dr. Aditya received research excellence awards in 2016, 2017, and 2018 at Lovely Professional University for his research contributions during the academic year. He is also acting as a reviewer and member of national and international conferences and journals.

Parma Nand Astya is Dean of the School of Engineering Technology at Sharda University, Greater Noida. He has more than 26 years of teaching, industry, and research experience. He has expertise in wireless and sensor networks, cryptography, algorithms, and computer graphics. He has a Ph.D. from IIT Roorkee, and an M. Tech and B. Tech in Computer Science and Engineering from IIT Delhi. He has been the head and a member of committees including the Board of Studies, faculty and staff recruitment committees, Academic Council, Advisory Committee, Monitoring and Planning Board, Research Advisory Committee, and Accreditation Committee. He was formerly President of the National Engineers Organization. He is a senior member of IEEE (USA). He is member of the Executive Council of IEEE UP section (R10), a member of the Executive Committee IEEE Computer and Signal Processing Society, a member of the Executive Council Computer Society of India, Noida section, and has acted as an observer at many IEEE conferences. He is also an active member of ACM, CSI, ACEEE, ISOC, IAENG, and IASCIT. He is a lifetime member of the Soft Computing Research Society (SCRS) and ISTE. He has delivered many invited and keynote talks at national and international conferences, workshops, and seminars in India and abroad. He has published more than 85 papers in peer-reviewed national and international journals and conferences. He has filed two patents. He is an active member of the advisory/technical program committee of reputable national and international conferences.

Narayan C. Debnath is Founding Dean of the School of Computing and Information Technology at Eastern International University, Vietnam. He is also Head of the Department of Software Engineering at Eastern International University, Vietnam. Dr. Debnath has been Director of the International Society for Computers and their Applications (ISCA) since 2014. Formerly, Dr. Debnath served as Full Professor of Computer Science at Winona State University, Minnesota, USA for 28 years (1989–2017). He was elected as Chairperson of the Computer Science Department at Winona State University for three consecutive terms and assumed the role of Chairperson of the Computer Science Department at Winona State University for seven years (2010–2017). Dr. Debnath earned a D.Sc. degree in Computer Science and also a Ph.D. in Physics. In the past, he served as the elected President for two separate terms, Vice President, and Conference Coordinator of the International Society for Computers and their Applications, and has been a member of the ISCA Board of Directors since 2001. Before being elected as Chairperson of the Department of Computer Science in 2010 at Winona State University, he served as Acting Chairman of the Department. In 1986–1989, Dr. Debnath served as Assistant Professor of the Department of Mathematics and Computer Systems at the University of Wisconsin–River Falls, USA, where he was nominated for the National Science Foundation (NSF) Presidential Young Investigator Award in 1989. Dr. Debnath is an author or coauthor of more than 425 publications in refereed journals and conference proceedings in computer science, information science, information technology, system sciences, mathematics, and electrical engineering. Dr. Debnath has been a visiting professor at universities in Argentina, China, India, Sudan, and Taiwan. He has been an active member of the ACM, IEEE Computer Society, and the Arab Computer Society, and is a senior member of the ISCA.

Contributors

M. Al-Rawy
Ark IT
Albania

Amartya
Netaji Subhas University of Technology
India

A. Diana Andrushia
Karunya Institute of Technology and
Sciences
India

Siddhant Banyal
Netaji Subhas University of Technology
India

A.K.M. Bahalul Haque
Department of Electrical and Computer
Engineering
School of Engineering and Physical
Sciences
North South University
Bangladesh

Bharat Bhushan
Department of Computer Science and
Engineering
School of Engineering and Technology
Sharda University
India

P. Bhandari
Southeast Missouri University
United States

Pradeep Kumar Bhatia
Guru Jambheshwar University of
Science and Technology
Hisar, India

I. Chien
National Chung Cheng University
Taiwan

Narayan C. Debnath
Eastern International University
Vietnam

A. Elci
Hasan Kalyoncu University
Turkey

Lalit Garg
L-Universita, Malta
Malta

Somya Goyal
Manipal University Jaipur
Jaipur, Rajasthan
and
Guru Jambheshwar University of
Science and Technology
Hisar, India

B. Gupta
Southern Illinois University
United States
bidyut@cs.siu.edu

Pao-Ann Hsiung
National Chung Cheng University
Taiwan

Garima Jain
Swami Vivekanand Subharti University
India

J. John Paul
Karunya Institute of Technology and
Sciences
India

D. Pradeep Kumar
Department of Computer Science and
Engineering
M S Ramaiah Institute of Technology
India

Ravinder Kumar
Shri Vishwakarma Skill University
India

Wei-Shan Lee
National Chung Cheng University
Taiwan

Yong-Hong, Liu
National Chung Cheng University
Taiwan

Shashi Mehrotra
Department of Computer Science and
Engineering
Koneru Lakshmaiah Education
Foundation
KL University, Vaddeswaram
India

K. Martin Sagayam
Karunya Institute of Technology and
Sciences
India

Parth Sarthi Prasad
Department of Computer Engineering
Netaji Subhas University of Technology
India

Rajeev Ranjan Prasad
Ericsson India Global Services Pvt. Ltd
India

Muhannad Quwaider
Jordan University of Science and
Technology
Jordan

S. Rebecca Princy Grace
Karunya Institute of Technology and
Sciences
India

N. Rahimi
Southern Illinois University
United States

Ansh Riyal
Department of Computer Engineering
Netaji Subhas University of Technology
India

I. Roy
Southern Illinois University
United States

T. Sanjana
Department of Electronics and
Communication Engineering

B. M. S. College of Engineering
India

Deepak Kumar Sharma
Department of Information Technology
Netaji Subhas University of Technology
India

Sudhir Kumar Sharma
Institute of Information Technology and
Management
Janakpuri, New Delhi, India

Nayanika Shukla
Department of Computer Science and
Engineering
HMR Institute of Technology and
Management
New Delhi, India

Shweta Sinha
Amity University
Haryana, India

K. G. Srinivasa
National Institute of Technical Teachers
Training and Research
Chandigarh
India

B. J. Sowmya
Department of Computer Science and
Engineering
M S Ramaiah Institute of Technology
India

Lo'ai A. Tawalbeh
Texas A&M University
San Antonio, United States

Mais Tawalbeh
Jordan University of Science and
Technology
Jordan

Thi Thanh Dao
National Chung Cheng University
Taiwan

1 Overview of the Internet of Things and Ubiquitous Computing

Shashi Mehrotra
Koneru Lakshmaiah Education Foundation, India

Shweta Sinha
Amity University, Haryana, India

Sudhir Kumar Sharma
Institute of Information Technology and Management, India

CONTENTS

1.1 Introduction 2
1.2 Overview of Ubiquitous Computing 3
    1.2.1 Technical Foundations 3
    1.2.2 Characteristics of a Ubiquitous Computing Environment 4
1.3 Challenges and Issues Associated with Ubiquitous Computing 5
1.4 Fundamentals of the Internet of Things 6
    1.4.1 Enabling Technologies of the IoT...7
1.5 Challenges and Issues of the IoT 8
    1.5.1 Security Issues of the IoT 8
    1.5.2 Security Requirements of the IoT..9
    1.5.3 Categories of Security Issues 9
        1.5.3.1 Low-Level Security Issues 9
        1.5.3.2 Intermediate-Level Security Issues 10
        1.5.3.3 High-Level Security Issues 10
1.6 The IoT's Impact on Ubiquitous Computing 10
1.7 Application Domains of Ubiquitous Computing 11
    1.7.1 Ubiquitous Computing in Health Care 11
        1.7.1.1 Ubiquitous Computing for Cardiac Patient Monitoring........12
        1.7.1.2 Ubiquitous Computing for Cognitive Training and Assessment...12
        1.7.1.3 Ubiquitous Computing for Rehabilitation Assessment.........12
        1.7.1.4 Behavior and Lifestyle Analysis 12
    1.7.2 Ubiquitous Computing in Education 13
    1.7.3 Ubiquitous Computing in Houses...13

1.7.4 Ubiquitous Computing for Transportation	13
1.7.5 Ubiquitous Computing in E-Commerce	13
1.7.6 Ubiquitous Computing in Environmental Control	14
1.8 Blockchain for IoT Security Enhancement	14
1.8.1 Blockchain Background	14
1.8.2 Potential Blockchain Solutions	15
1.9 Conclusion	15
References	16

1.1 INTRODUCTION

Ubiquitous computing is characterized by an explosion of small portable computer products in the form of smartphones, personal digital assistants (PDAs), etc. It refers to the idea of an infinite collection of almost any device with computational capability; the device is embedded with a chip in such a way that the connection is hidden and yet, available anytime, anywhere. Today, the smartphones to be seen in every hand show the most convincing evidence of ubiquitous computing. The main focus of ubiquitous computing has been toward human–things/machine interaction. With the shift in computing trends, from mainframe computers, to desktop computers for individuals, to multiple computers for one person, the demand for ubiquity could be achieved by providing mobile computing power. The evolution of the IoT has helped in this direction. With the advent of the IoT, thing–thing communication has become possible. The IoT enables devices to sense their surroundings and collect data, to provide usable information, i.e., allowing communication between the devices. The internet serves as the backbone of the IoT network. The power of the IoT has made computing genuinely ubiquitous, in the sense that now, the network size has been extended to cover humans, objects, machines, and the internet. The IoT enables more direct integration between computer-based structures and the physical world. Because ubiquitous computing deals with human–thing interactions, and the IoT deals with thing–thing businesses, both face similar challenges. However, undoubtedly, thing–thing interactions are the more challenging.

Today, technologies such as the IoT, mobile technology, and ubiquitous computing have permeated our lives on an incomparable scale, and the interaction of data, people, things, and computing is growing at an increasingly fast pace. The technology growth trend shows a future where augmented and virtual reality, mobility, and related network technology, including the IoT, will drive computing to an extent that would make existing human–machine interactions more omnipresent, i.e., more ubiquitous. But, the protection of networks and devices from threats and attacks is a major concern of the IoT, also extending up to the perception, transportation, and application layers [1]. The coupling of physical devices requires seamless security for data privacy and demands robustness against attack by means of a strong authentication process. Undoubtedly, the security, auditability, and decentralization features of blockchain technology can cater to the risks of the IoT associated with centralization and network attacks [2]. To achieve seamless, secure communication in the IoT world, blockchain needs to be remodeled to meet the IoT's specific needs.

Overview of the Internet of Things and Ubiquitous Computing

This chapter intends to introduce ubiquitous computing and present the IoT as a way toward achieving it. More specifically, the chapter guides the reader through the fundamental concepts of ubiquitous computing and its technical foundation, along with the fundamentals of the IoT and the enabling technology behind it. With the help of the literature available in these domains, the impact of the IoT on ubiquitous computing is analyzed, with a discussion of the way forward in strengthening ubiquitous computing.

The organization of the chapter is as follows: Section 1.1 covers the fundamentals of ubiquitous computing. An overview of ubiquitous computing, including its technical background and characteristics, is presented in Section 1.2. Section 1.3 focuses on challenges and issues associated with ubiquitous computing, followed by IoT fundamentals and technological aspects in Section 1.4. Section 1.5 discusses the challenges and problems of the IoT, with a focus on IoT security. The impact of the IoT on ubiquitous computing is presented in Section 1.6. Some of the applications of ubiquitous computing in synergy with the IoT are discussed in Section 1.7. Section 1.8 presents the potential of blockchain as an IoT security solution, followed by concluding remarks in Section 1.9.

1.2 OVERVIEW OF UBIQUITOUS COMPUTING

The computing wave started with mainframe computers, massive in dimension and capable of computing huge amounts of data. Due to their cost, only a few mainframe computers existed in the world. After that came the era of desktop computing, which brought one computer onto every desk, to help business-related events, coupled to a network through wired media. Then, in 1988, Marc Weiser coined the term "Ubiquitous Computing." Ubiquitous computing is a way of enhancing computer usage by making numerous computers available throughout the physical setting, but making them virtually unseen to the user [3]. He visualized that in the future, computing capabilities would be embedded in every object involved in our daily life, be it at the workplace or in the home [4]. It's a computing environment at the intersection of computing, networking, and embedded computing. With technological advancements, as devices are getting smaller and smaller, ubiquitous computing can be characterized as wireless connected computers embedded invisibly into any type of object in the surroundings.

1.2.1 TECHNICAL FOUNDATIONS

In 1965, Gordon Moore [5] stated that the computing power available on a microchip would double approximately every 18 months, and his prediction has proven accurate. Technological developments in the field of microelectronics have worked as a driving force behind ubiquitous computing. Recent progress in the fields of nanotechnology and microsystems has also influenced this growth. Figure 1.1 shows the three key technical advancements that have enabled ubiquitous computing.

Ubiquitous communication refers to the establishment of communication infrastructure and wireless technology to enable connectivity throughout. Ambient intelligence is a vision that applies to our everyday environment, especially the electronic

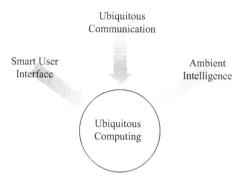

FIGURE 1.1 Key Technologies behind ubiquitous computing.

environment, making it sensitive to the presence of people. Intelligent user interface refers to an interface that can capture a broad range of inputs, such as human body motion, action, and preferences. It should be user friendly and provide cognitive flow, apart from being secure and productive. Factors that influence the existence of ubiquitous computing include:

- Processing Power: Cheaper, faster, smaller, more energy-efficient processing capability is desired.
- Storage: A large storage capacity with faster processing speed, small in dimension, is required.
- Networking: Low power, low latency, and high bandwidth to support local, global, and ad hoc networking.
- Sensors: Small in size, with high accuracy, precision, and less energy consumption.
- Displays: Flexible material, projection, and low power consumption.

1.2.2 Characteristics of a Ubiquitous Computing Environment

Based on the definition of ubiquitous computing, any ubiquitous computing environment should possess the following characteristics:

- Ubiquity: In such a system, information should be accessible from any location at any time.
- Embeddedness and Immediacy: Communication and computing both must exist together and should provide immediately useful information at any time.
- Nomadism: Mobility has become a vital communication feature. The ubiquitous computing environment must provide freedom of movement to both users and computing without following any fixed pattern.
- Adaptiveness, proactiveness, and personalization: The computing environment should be adaptable to the user's requirements and should provide flexibility

Overview of the Internet of Things and Ubiquitous Computing

and autonomy for communication and computing. It should be able to capture the correct information at the right time.

- Timelessness and seamlessness: There should be no need to restart the operation; the components should be easily upgradable and possess the ability to continue processing.

1.3 CHALLENGES AND ISSUES ASSOCIATED WITH UBIQUITOUS COMPUTING

Modern information and communication technologies (ICT) usage has become essential for economic growth in the global market. With the development and large-scale implementation of mobile telephone and internet technology, the economy, science, and private life all have witnessed enormous changes. Every evolving day, these objects transform in size and capability. The availability of smart devices and their interconnection throughout has influenced the economic and social perspective remarkably. But a few factors can directly or indirectly influence the growth of ubiquitous computing. These can be identified as:

1. Network dynamics: Nodes in the network can connect and depart at any time. The lack of a centralized system and the absence of fixed infrastructure create challenges for the self-configuration of the network.
2. Node behavior: The size of the communication network is significant, and a vast number of nodes are connected, in general. The nodes' behavior can be unpredictable, because some nodes can act maliciously or randomly. This type of behavior creates challenges for management and trust computation modules.
3. Availability and resource constraint: In general, resources are constrained. While nodes may be included in the network at any time, the resources needed to support them may not be available. Also, the nodes of the system are heterogeneous in terms of their capabilities, measured as their processing power, battery life, and communication capabilities. Due to the resource constraints, handling or managing the resource assignment of the device for complex computations is a challenge.
4. Data protection and authentication: Most smart devices today provide a locking mechanism, which prevents unauthorized access. But users who own more than one such connected device often keep the same password or store the password on the device itself, which poses a challenge for authentication/security units.
5. Standardization: One significant challenge in the implementation of ubiquitous computing is the lack of standardization. Every device connected to the network may belong to a different company. Each follows its own protocols and device standards. Interfacing such units is challenging.
6. Change in the user's context: The user interfaces for ubiquitous computing are very different from traditional interfaces. Frequent changes in user requirements involve periodic shifts in interface dynamics that may be difficult to execute. Again, this user's context change is a challenge for interface designers.

1.4 FUNDAMENTALS OF THE INTERNET OF THINGS

The phrase "Internet of Things" was coined in 1999 [6]. It refers to the increasing connection of the physical world to the internet. The IoT is defined as a vast interconnection of devices connected through the internet [7]. These devices use sensors or technologies based on embedding, to sense and gather information from their environment and surroundings. These captured data are shared among the devices connected to the network. To leverage the unknown knowledge collected from the environment, these shared data are analyzed and correlated before being used by the devices to make more intelligent decisions. Undoubtedly, it can be said that the IoT has taken the internet to the next level of processing and extended it to physical devices. The IoT in the present scenario is not only a scientific or technical branch but extends beyond that to become a social phenomenon, a cultural product, and an industrial specialty.

The supportive components and the environment around the IoT form a system. The major components of the IoT system are the things, the data, the process, and the people. It is expected that all these four components work in unison to achieve the desired connected world. Figure 1.2 represents the IoT system, the parts of which can be described as:

- Things: Things are devices connected to the network of things. These devices are capable of communicating with each other by using Bluetooth or Zigbee communication protocols. Apart from communicating, these devices can also collect data and perform operations. Data collection can be carried out using the sensing capability of the things involved, due to sensors embedded within them, or, in some cases, using images captured by the camera. It can also perform tasks related to data transfer and processing.

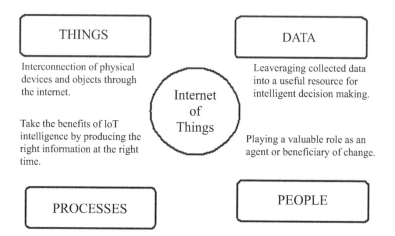

FIGURE 1.2 Components of the IoT system.

Overview of the Internet of Things and Ubiquitous Computing

- Data: Data in the IoT system refers to the content communicated among the things. It also refers to the commands that are passed on to the things. Since a multitude of things participate in the network, the data collection is enormous. These collected data have to be cleaned and checked for errors before being utilized by any process. These tasks can be performed at the edge of the network or on the central server, i.e., the cloud.
- Processes: The power of the IoT can be utilized to make the techniques and methods used by industries more efficient by correctly processing data to obtain the right information at the right time. At this stage, one can realize the benefits of the IoT and its intelligence.
- People: In the IoT ecosystem, people are the beneficiaries of and agents who work for the IoT. People create the network of things, and whatever the data collected and communicated may be, it is for people.

1.4.1 Enabling Technologies of the IoT

As Daniele Miorandi et al. [8] discuss in their work, anything uniquely identifiable that can communicate and perform necessary computations can be used for the IoT network. According to them, the sensing or actuation capability is optional. Over the years, dependency on IoT systems has increased, and with technological advancements and reduced cost, they have become affordable. The enabling technology behind the IoT is indispensable [9]. Some of the enabling technologies are outlined here.

1. Unique identification technique: A large number of nodes interconnect to form the IoT.
2. The capability of each node to generate data: There should be unique identification attached to each thing to resolve any ambiguity. The universally unique identifier (UUID) developed by the Open Source Foundation (OSF) is one of the standards used to provide uniqueness.
3. Sensing technology: Sensing is pivotal in accessing data. In the past few years, with significant progress in technology, sensors have become cheaper and are easily installed in huge numbers. Sensors have enabled the acquisition of data at a swift pace. Processes can further utilize these data to make intelligent decisions.
4. Communication technology: Some communication technology is needed to support the interconnection of nodes and communication among smart devices. These communication technologies can be long-range or short-range. Long-range communication usually supports mobile calls, etc. Bluetooth, Wi-Fi, and Zigbee are short-range communication technologies, which usually support node-to-node data transmission.
5. Cloud computing: The IoT network collects an enormous amount of data. These data are heterogeneous and differ in format, size, layout, etc. Aggregation and processing are required to further utilize this data for decision making. Handling this massive quantity of information is a challenge. Cloud computing provides a centralized storage and processing aspect to the IoT, where data can be aggregated, processed, and stored for further access.

8 Blockchain Technology for Data Privacy Management

6. Service-oriented architecture (SOA): The number of tiny units in the IoT network is vast. This networked system requires the interoperability of each node. SOA considers each device as an individual device, with clearly defined functionality, accessible through interfaces. These interfaces can easily be reconfigured, depending upon other neighboring units and their functionality.

1.5 CHALLENGES AND ISSUES OF THE IOT

The IoT has evolved to an extent where every domain looks for a solution through it. Whether for a business model, education, the environment, or health care, the IoT has marked its presence on a large scale. Despite this success, challenges have to be met for successful IoT implementation. The first challenge is the *lack of interoperability* of smart objects. The problems associated with interoperability can be either technical operability, concerning the standards and protocols of individual IoT devices, or semantic and pragmatic issues, concerning the handling and processing of data received during communication. Another possible challenge is the behavior of IoT devices in the network. IoT usage has great possibilities. But these cannot be utilized, unless objects interact appropriately with each other.

Another challenge is associated with *scalability*, dealing with the massive number of nodes, with differences in their communication behavior. It is challenging to handle the demand of rapidly growing nodes in the network. The need is to accelerate the scalability management protocols, to cater to the demand for resources. *Security* is one of the crucial challenges in the IoT domain. This challenge is due to heterogeneous nodes, each working in its own fashion and interacting with humans and other entities. It is a challenge to secure all the interactions, while preserving optimal system performance. This security concern influences the trust issue associated with the IoT. There should not be stringent government regulations to deal with this issue, as strict regulations would present a barrier to innovation. Instead, defining access control rules, implementing firewalls, and providing end-user authentication could be small steps in this direction. Apart from this, security paradigms need to be followed, including the Low Range Wide-Area Network (LoRaWAN), Bluetooth Low Energy security (BLE), and the Constrained Application Protocol (CoAP) [10].

Another major challenge associated with the IoT, and last in this discussion, is *value for users*. This significant technological shift has generated great opportunities, but it remains difficult to find an effective business model and an excellent value proposition to support it [11].

1.5.1 SECURITY ISSUES OF THE IOT

The drastic increase in the number of smart homes and smart cities has influenced the popularity and acceptance of the IoT in society. Its future importance is evident from the inclusion of smart things in daily life. Due to the growth in hardware technologies pertaining to the communication sector, the IoT is also proliferating [12,13]. The horizon of the IoT has broadened, and technologies such as the wireless sensor network (WSN) and machine-to-machine (M2M) have become an integral part of it.

Overview of the Internet of Things and Ubiquitous Computing 9

Along with that, the IoT has inherited the security issues of WSN and M2M. Generally, the things used in the IoT are small, heterogeneous devices with limited battery power and memory. These constraints impose additional challenges for IoT security and require the adaptation of security solutions to these constrained environments [14]. Before delving into security solutions, it is essential to identify the actual security requirements of the IoT.

1.5.2 SECURITY REQUIREMENTS OF THE IoT

The very essence of the effort to establish a secure network lies in the identification of parameters that are the most vulnerable and need to be taken into consideration by security protocols.

As the IoT network is composed of a diverse integration of devices, a strong encryption technique is essential, along with the protection of data from malicious efforts at data manipulation. In turn, the confidentiality, integrity, and privacy of data should be maintained.

It is always desirable to have secure communication among the devices in the network. Due to the diverse heterogeneous architecture of the network, it is a challenge to define a global protocol that can be standardized, to create a trustworthy environment with proper authorization and authentication of users. Also, resource usage accounting is needed to provide proper network management.

IoT components can easily become victims of attacks, such as sinkhole attacks, denial of service attacks, and replay attacks [15]. These affect the network at different layers and degrade the network's quality of service.

Some attacks can directly impact IoT architectures and may increase energy consumption in the network, which will deplete the network's resources.

1.5.3 CATEGORIES OF SECURITY ISSUES

The IoT network encompasses a variety of devices and components, which range from small sensors to large high-end servers. Due to this variability, a single mechanism cannot be devised to handle all the issues. Based on the IoT deployment architecture [16], security threats are categorized as low-level security issues, intermediate-level security issues, and high-level security issues.

1.5.3.1 Low-Level Security Issues

The lowest level of security is concerned with the physical and data link layers in the communication network. These issues include:

Jamming adversaries: The jamming attack affects the sending and receiving of data in the network, due to radio interference. This is caused by emitting radio frequency signals without following any protocol [17].

Low-level Sybil and spoofing attack: At the physical layer, Sybil nodes use a fake identity to degrade IoT functionality. These nodes use forged MAC values to pose as a different device. Along with the depletion of resources, these cause a denial of service to legitimate nodes [18].

Denial of sleep attack: The devices in the IoT are forced to keep their radios on by increasing their duty cycle, which exhausts the batteries [19].

1.5.3.2 Intermediate-Level Security Issues

Intermediate-level security takes control of the network and transport layers of the IoT and is mainly concerned with routing and session management [20]. Some of the attacks at these layers include:

Buffer reservation attacks: The packets transmitted in the network require reassembly at the receiving node, and hence the receiving node has to reserve space for the purpose. The attacker may send incomplete packets that occupy buffer space. This results in the denial of service to other packets, due to a space crunch [21].

Sinkhole attacks: The attacker node responds to a routing request and directs the node to pass through a malicious route, which can then be exploited for security risks [22,23].

Sybil attacks at the intermediate level: At the communication and network layers, Sybil nodes can be deployed to degrade network performance. These nodes can violate data privacy, which could result in phishing attacks and spamming [24].

Transport level attacks: The aim of the IoT network is to provide secure end-to-end transmission, and, for this, the transport layer end-to-end security mechanism uses a comprehensive security mechanism that includes authentication, privacy, and the integrity of data [25].

Session establishment and resumption: The attacker node can hijack a session with the help of forged messages, which result in a denial of service [26,27]. Sometimes, the attacker can force the victim node to continue the session for an extended period, keeping the network busy.

1.5.3.3 High-Level Security Issues

CoAP security with the internet: The application layer in the network is also vulnerable to attacks. The Constrained Application Protocol (CoAP) is a web transfer protocol of constrained devices [28]. CoAP messages need to be encrypted for secure transmission.

Insecure software and firmware: The codes are written in languages such as JSON, XML, XSS, etc. They should be tested properly, and updates need to be carried out in a secure manner [29]. Insecure software or firmware is equally responsible for vulnerabilities.

Researchers have provided solutions to deal with most of the security issues at the three different levels, in one way or another. The deployment of IoT networks is able to provide secure communication for IoT entities.

1.6 THE IOT'S IMPACT ON UBIQUITOUS COMPUTING

Ubiquitous computing refers to the existence of computers everywhere, i.e., the availability of computational technology throughout the physical environment. This has to be obtained while keeping the computation behind the physical object and making it invisible to the user. During the last two decades, technology has seen

Overview of the Internet of Things and Ubiquitous Computing

tremendous growth and witnessed the evolution of low-cost, powerful processing units, storage, and memory devices. This development has helped achieve the interconnection of physical entities in a cost-effective manner. Smart wearable devices, smart homes, smart cities, and intelligent industries all have become possible, due to the availability of low-cost computational units. But missing interoperability and collaboration, owing to a lack of standardization, forbids the seamless integration of all components. This current direction of the IoT still does not fulfill the characteristics of ubiquitous computing to the fullest. To achieve this, the IoT needs to adopt a standardized middleware protocol that enables peer-to-peer networking. The role of this middleware protocol should be to manage the large-scale IoT with heterogeneous devices and provide an API for an application protocol to develop autonomous applications that meet the user's natural way of interaction with the physical environment. It should make the navigation of things in the system easy. We discuss some of the application domains of the IoT, which are efforts toward ubiquitous computing.

1.7 APPLICATION DOMAINS OF UBIQUITOUS COMPUTING

These days, ubiquitous computing aims to interconnect every field of life. Already, it has brought great changes to people's lifestyles. Ubiquitous computing is being used in various application domains such as health care, education, and smart cities that include smart homes, malls, hospitals, municipalities, etc. This section discusses some of the application domains.

1.7.1 UBIQUITOUS COMPUTING IN HEALTH CARE

One of the main application domains of ubiquitous computing is in the health care sector, which requires smart solutions to deal with societal needs. These days, universal health care solutions are being implemented that provide health care services at any moment, anywhere. This universally available health care is gaining in popularity, due to the growth in health care expenses, aging populations, and rises in the rate of diseases such as heart failure, hypertension, obesity, etc. [30]. The ubiquitous computing approach empowers the monitoring of patient health and patient-related services, which intelligently respond to user needs, such as individual medical care assistance [31,32]. Ubiquitous health care applications may comprise systems for diagnosis, monitoring, and patient report details. In the past, the main focus for health care solutions was on the diagnosis of diseases and diagnosis-based treatment; a person used to contact a doctor only if they felt unwell [31]. These days, due to health care awareness, and the availability of ubiquitous health care devices, a regular health review has become more favorable in our day-to-day lives. Regular monitoring, early diagnosis, and chronic disease management are, first and foremost, essential for avoiding health complications and risks [33]. Recently, an internet-based framework was proposed for skin cancer classification [34].

Ubiquitous computing in health care uses various environmental and patient-based sensors to monitor the physical and mental condition of the patient. Sensors collect data, to help diagnose the disease and monitor a patient. Sensors gather information about an individual, such as temperature, heart rate, blood pressure, the chemical levels of blood and urine, breathing rate, and other activities that are needed to diagnose health problems. These sensors may be installed in a home or the patient's workplace, implanted in the body, or used as a wearable [31].

Ubiquitous computing is being used for various purposes in the health care sector, as follows:

1.7.1.1 Ubiquitous Computing for Cardiac Patient Monitoring

The ubiquitous computing system for cardiac patient monitoring consists of mobile computing, sensors, and communications technologies [35]. Quality of life could be improved by caring for cardiac patients, regular monitoring, and personalizing patient care [35] Doctors primarily use ECGs for diagnostic and medication purposes. An ECG records heartbeats in a waveform graph. This form of electronic monitoring can help detect heart disease, determine a patient's heart rate, monitor the regularity of heartbeats, and determine the size and position of the heart's chambers, all critical areas of focus. Along with these features, an ECG also helps to assess the effects of drugs or specialized devices used for regulating the heart.

As a ubiquitous health care solution for individuals and patients, everyday ECG monitoring is provided on Android mobile devices. It has shown a significant impact on managing heart rates and has enhanced diagnostic capabilities [31]. Wireless technology and wearable sensors provide patients the freedom to be mobile, while being regularly monitored. The impact and usefulness of wearable monitoring devices are as follows [11,35]. In critical situations, they can identify early signs of an individual's health deterioration and send alerts through SMS or another system to the health care professionals concerned. They can also identify correlations between a patient's lifestyle and health problems.

1.7.1.2 Ubiquitous Computing for Cognitive Training and Assessment

Ubiquitous computing has entered into the area of mental health care, as well. Tools have been developed to track cognitive behavior, remotely execute cognitive assessment tests, and detect cognitive decline. Ubiquitous computing can also provide cognitive training and assess cognitive levels through the use of video games [32].

1.7.1.3 Ubiquitous Computing for Rehabilitation Assessment

Ubiquitous health care assessment for rehabilitation gathers data with the help of body sensor networks (BSNs). These wearable computing devices regularly record the motion data for a patient's specific body part [36]. Ubiquitous health care measurements for rehabilitation help patients in self-management, who need regular rehabilitation at home [37]. The physiotherapist can fit a program of specific motions to each patient's requirements and track the patients' daily records [38].

1.7.1.4 Behavior and Lifestyle Analysis

Ubiquitous computing using wearable technology has come up for measuring and analyzing human physical activity, by collecting data from a wearable sensor [32]. In many cases, a person's lifestyle analysis helps identify the root causes of disease.

Overview of the Internet of Things and Ubiquitous Computing

1.7.2 Ubiquitous Computing in Education

The education industry is undergoing significant changes. Online and lifelong learners, and the need for new teaching and learning paradigms for educational institutions, are increasing the demand for ubiquitous computing technologies [39]. Teachers and students use ubiquitous computing for classroom teaching and learning, by utilizing technology and ubiquitous devices. Ubiquitous devices have become an integral part of education. "Ubiquitous" means that computational devices are distributed into the physical world, giving us boundless access to communication and information channels [40]. Network connections such as cellular, Wi-Fi, Bluetooth, and NFC provide wireless communication for different devices [30]. With the development of ubiquitous technology, new approaches to education are becoming popular in teaching and learning procedures. Students are getting the opportunity to gain access to information resources in a timeless and limitless way. In this regard, ubiquitous computing has significant advantages.

1.7.3 Ubiquitous Computing in Houses

Ubiquitous computing is integrated into smart houses to facilitate communications and functionality through the use of sensors [41]. The term "smart home" means a house surrounded by interconnected technologies that respond to the presence of individuals and perform actions. This technology provides an environment by integrating information, communication, and sensing technologies into objects used in our day-to-day lives. Ubiquitous computing for a smart home uses networked sensors, devices, and appliances to build an intelligent environment to automate various activities and support domestic tasks [42].

1.7.4 Ubiquitous Computing for Transportation

Ubiquitous computing is being used for mobility and transportation, where it offers various services to traffic management and users. It provides advantages to users, guiding them in navigation and informing them about transport flow. It has increased user safety, through accident prevention, and user comfort, by providing entertainment, location-dependent information, etc. It offers better networking among transport sectors [43].

1.7.5 Ubiquitous Computing in E-Commerce

Ubiquitous computing using smart objects allows for new business models to provide various digital services. Some examples are location-based services, and renting products rather than selling them. The software agents give a command to ubiquitous computing components for initiation and carry out services and business transactions independently [44]. Ubiquitous computing is producing a new way of doing business by automating the business world, such as the production process and e-commerce. This advancement speeds up the business process and results in manufacturer, supplier, and customer satisfaction [45]. The various business models using the Internet of Things could bring about positive changes for the economy [46].

1.7.6 UBIQUITOUS COMPUTING IN ENVIRONMENTAL CONTROL

Ubiquitous computing has been used for the measurement and control of the environment for a long time, where the measurement and control units of a greenhouse, for example, are divided into nodes that are connected to each other. More recently, it is being used for smart environmental control in smart cities for various purposes, such as smoke detection, air pollution detection, and displaying alert messages on screens [47].

1.8 BLOCKCHAIN FOR IOT SECURITY ENHANCEMENT

The interconnection of massive physical devices has made individual life very convenient. However, this gift by the IoT has also brought about potential risks. An enormous amount of data is generated by applications such as home automation, mobile health care, intelligent vehicles, etc. All these require a robust security mechanism for proper protection, in order to provide ubiquitous computing. In recent years, blockchain has been sought as a descriptive technology that can play a major role in the control, management, and security of IoT devices [48]. In general, these centralized systems face numerous challenges, owing to some of the intrinsic characteristics of the IoT. The vulnerability lies with the centralized server, whose failure could paralyze the entire network [49]. Also, centralized servers are susceptible to revealing sensitive information, due to the need for management authority, and due to the hacking of unencrypted data [50]. Apart from achieving an acceptable level of quality of service (QoS) and energy efficiency, several parameters should be focused upon for the secure deployment of ubiquitous IoT [15].

- Privacy, confidentiality, and data integrity: The reason lies with the multiple hops of data in the network.
- Authenticity, accountability, and traceability: Due to the need for secure communication in the shared network and a record of the usage of shared resources.

1.8.1 BLOCKCHAIN BACKGROUND

Blockchain is a decentralized publicly available immutable transaction database that is replicated in all the nodes [51]. It has revolutionized the way financial transactions are carried out, without the need for a central authority. Also, blockchain-based systems incur minimal security monitoring costs, while providing security against adversaries [49]. Blockchain is a sequence of blocks that store the information related to all transactions. Each block of the chain is linked via the hash of the previous block. Generally, a block consists of a body and a header. The block's body stores a transaction and a counter value for the transaction. The header stores metadata, such as a timestamp, parent block hash, Merkle tree root hash, etc. In order to validate the transaction's authenticity, blockchain employs an asymmetric cryptography mechanism. The Bitcoin and Ethereum blockchains are the most popular application domains. The Bitcoin blockchain is

Overview of the Internet of Things and Ubiquitous Computing   **15**

a cryptocurrency-based application, whereas the Ethereum blockchain implements smart contracts, a computerized transaction protocol that executes the terms of the contract. These smart-contract-based blockchains have the potential to manage, control, and secure IoT devices.

1.8.2 POTENTIAL BLOCKCHAIN SOLUTIONS

This section presents some of the intrinsic features of blockchain that can be utilized for IoT security.

Blockchain Address Space: In contrast to the 128-bit IPV6 address space, the blockchain has a 160-bit address space [52]. This allows around 4.3 billion unique identifications for IoT devices and in turn decreases the collision probability.

Data authentication, integrity and traceability: Due to the design constraints of blockchain, the data transmitted by IoT devices will always be cryptographically proven and signed by the legitimate sender with a unique key. This provides authentication and integrity. Also, the transactions of IoT devices will be recorded on a blockchain ledger that can provide traceability.

Authenticity, authorization, and privacy: Blockchain's smart contract has the ability to set access rules and conditions, to allow users on machines to control, or have access to, data in transit or at rest. It can also assign different rights to IoT devices and change the rules. This way, it provides authorization and privacy. Also, since all devices involved in a transaction possess a dedicated blockchain address, the blockchain-based solution avoids false authentication.

Secure communication: IoT application communication protocols, such as HTTP, and XMPP, and routing protocols such as RPL, and 6LoWPAN, are not secure. To provide secure messaging and communication, these have to be wrapped within security protocols such as TLS or IPsec, for safe routing. All these protocols are complex in terms of storage and computation and are based on centralized management. Blockchain eliminates key management and distribution. This also leads to the evolution of simpler security protocols and hence, reduces the requirements for computation and storage.

Improved availability: There is no vulnerable central point in the blockchain-based solution and hence, it possesses built-in availability features. The decentralized connected device enables the network to remain alive, in case of multiple machine faults [53]. Blockchain is a decentralized system, where the entire transaction database is replicated in all the nodes. Blockchain was initially used for cryptocurrency transactions, but these days, it is coming up with many applications, including in education, health care, business enterprises, and smart computing [50].

1.9 CONCLUSION

With the help of various devices and sensors, ubiquitous computing, where computing is made to appear anytime and everywhere, is entering every part of our life, to make it efficient and smart. It is being used across multiple application domains and provides users with new solutions for efficient and smart living. Ubiquitous computing merges physical and computational infrastructures into an integrated environment.

Substantially, it is producing new ways for convenient and smart living and is also automating the processes of the business world, such as the production process and e-commerce. The Internet of Things is coming up with various opportunities for digitalization. However, it is also being targeted by cyberattacks, and blockchain may work as a security solution for the Internet of Things in various ways.

REFERENCES

1. Arora, A., Kaur, A., Bhushan, B., and Saini, H., 2019, July. *Security concerns and future trends of Internet of Things*. In *2019 2nd International Conference on Intelligent Computing, Instrumentation and Control Technologies (ICICICT)* (Vol. 1, pp. 891–896). Kannur, Kerala, India: IEEE.
2. Bhushan, B., Sahoo, C., Sinha, P., and Khamparia, A., 2020. Unification of Blockchain and Internet of Things (BIoT): Requirements, working model, challenges and future directions. *Wireless Networks*, pp. 1–36. https://doi.org/10.1007/s11276-020-02445-6.
3. Weiser, M., 1991. The computer for the 21st century. *Scientific American*, 265(3), pp. 94–105.
4. Krumm, J., ed., 2018. *Ubiquitous Computing Fundamentals*. Boca Raton: CRC Press
5. Moore, G., 1965. Moore's law. *Electronics Magazine*, 38(8), p. 114.
6. Khodadadi, F., Dastjerdi, A. V., and Buyya, R., 2016. Internet of Things: An overview. In Buyya, R., and Dastjerdi, A. V. (eds.) *Internet of Things* (pp. 3–27). Cambridge, MA: Elsevier.
7. Firouzi, F., Chakrabarty, K., and Nassif, S., (eds), 2020. *Intelligent Internet of Things: From Device to Fog and Cloud*. Cham, Switzerland: Springer Nature.
8. Miorandi, D., Sicari, S., De Pellegrini, F., and Chlamtac, I., 2012. Internet of Things: Vision, applications and research challenges. *Ad hoc Networks*, 10(7), pp. 1497–1516.
9. Sharma, T., Satija, S., and Bhushan, B., 2019, October. *Unifying blockchain and IoT: Security requirements, challenges, applications and guture trends*. In *International Conference on Computing, Communication, and Intelligent Systems (ICCCIS)* (pp. 341–346). Greater Noida, India: IEEE.
10. Goel, A. K., Rose, A., Gaur, J., and Bhushan, B., 2019. July. *Attacks, countermeasures and security paradigms in IoT*. In *2nd International Conference on Intelligent Computing, Instrumentation and Control Technologies (ICICICT)* (Vol. 1, pp. 875–880). Kannur, India: IEEE.
11. Islam, S. R., Kwak, D., Kabir, M. H., Hossain, M., and Kwak, K. S., 2015. The Internet of Things for health care: A comprehensive survey. *IEEE Access*, 3, pp. 678–708.
12. Khan, A. A., Rehmani, M. H., and Rachedi, A., 2017. Cognitive-radio-based Internet of Things: Applications, architectures, spectrum related functionalities, and future research directions. *IEEE Wireless Communications*, 24(3), pp. 17–25.
13. Akhtar, F., Rehmani, M. H., and Reisslein, M., 2016. White space: Definitional perspectives and their role in exploiting spectrum opportunities. *Telecommunications Policy*, 40(4), pp. 319–331.
14. Alaba, F. A., Othman, M., Hashem, I. A. T., and Alotaibi, F., 2017. Internet of Things security: A survey. *Journal of Network and Computer Applications*, 88, pp. 10–28.
15. Khan, M. A., and Salah, K., 2018. IoT security: Review, blockchain solutions, and open challenges. *Future Generation Computer Systems*, 82, pp. 395–411.
16. Conti, M., Dehghantanha, A., Franke, K., and Watson, S., 2018. Internet of Things security and forensics: Challenges and opportunities. *Future Generation Computer System*, 78, pp. 544–546.

17. Xu, W., Trappe, W., Zhang, Y., and Wood, T., 2005, May. *The feasibility of launching and detecting jamming attacks in wireless networks*. In *Proceedings of the 6th ACM International Symposium on Mobile Ad Hoc Networking and Computing* (pp. 46–57). Urbana-Champaign, IL: ACM.

18. Xiao, L., Greenstein, L. J., Mandayam, N. B., and Trappe, W., 2009. Channel-based detection of Sybil attacks in wireless networks. *IEEE Transactions on Information Forensics and Security*, 4(3), pp. 492–503.

19. Bhattasali, T., and Chaki, R., 2011, July. *A survey of recent intrusion detection systems for wireless sensor network*. In *International Conference on Network Security and Applications* (pp. 268–280). Berlin, Heidelberg: Springer.

20. Pajouh, H. H., Javidan, R., Khayami, R., Ali, D., and Choo, K. K. R., 2016. A two-layer dimension reduction and two-tier classification model for anomaly-based intrusion detection in IoT backbone networks. *IEEE Transactions on Emerging Topics in Computing*, 7(2), pp. 314–323.

21. Hummen, R., Hiller, J., Wirtz, H., Henze, M., Shafagh, H., and Wehrle, K., 2013, April. *6LoWPAN fragmentation attacks and mitigation mechanisms*. In *Proceedings of the Sixth ACM Conference on Security and Privacy in Wireless and Mobile Networks* (pp. 55–66). Budapest: Association for Computing Machinery.

22. Weekly, K., and Pister, K., 2012, October. *Evaluating sinkhole defense techniques in RPL networks*. In *2012 20th IEEE International Conference on Network Protocols (ICNP)* (pp. 1–6). Austin, TX: IEEE Computer Society.

23. Pirzada, A. A., and McDonald, C., 2005. *Circumventing sinkholes and wormholes in wireless sensor networks*. In *International Workshop on Wireless Ad-Hoc Networks*. London: Curran Associates, Inc.

24. Zhang, K., Liang, X., Lu, R., and Shen, X., 2014. Sybil attacks and their defenses in the Internet of Things. *IEEE Internet of Things Journal*, 1(5), pp. 372–383.

25. Granjal, J., Monteiro, E., and Silva, J. S., 2013, May. *End-to-end transport-layer security for Internet-integrated sensing applications with mutual and delegated ECC public-key authentication*. In *IFIP Networking Conference* (pp. 1–9). Brooklyn: IEEE.

26. Park, N., and Kang, N., 2016. Mutual authentication scheme in secure Internet of Things technology for comfortable lifestyle. *Sensors*, 16(1), p. 20.

27. Ibrahim, M. H., 2016. Octopus: An edge-fog mutual authentication scheme. *IJ Network Security*, 18(6), pp. 1089–1101.

28. Brachmann, M., Keoh, S. L., Morchon, O. G., and Kumar, S. S., 2012, July. *End-to-end transport security in the IP-based internet of things*. In *21st International Conference on Computer Communications and Networks (ICCCN)* (pp. 1–5). Munich: IEEE.

29. Open Web Application Security Project (OWASP), 2016. Mobile security testing guide. https://www.owasp.org/index.php. Accessed on August 14, 2020.

30. Marinagi, C., Skourlas, C., and Belsis, P., 2013. Employing ubiquitous computing devices and technologies in the higher education classroom of the future. *Procedia-Social and Behavioral Sciences*, 73, pp. 487–494.

31. Omary, Z., Mtenzi, F., Wu, B., and O'Driscoll, C., 2011. Ubiquitous healthcare information system: Assessment of its impacts to patient's information. *International Journal for Information Security Research*, 1(2), pp. 71–77.

32. Brown, I., and Adams, A. A., 2007. The ethical challenges of ubiquitous healthcare. *The International Review of Information Ethics*, 8, pp. 53–60.

33. Sivaraman, K., 2017. Umbitious healthware condition on Android mobile device. *International Journal of Pure and Applied Mathematics*, 116(8), pp. 255–259.

34. Khamparia, A., Singh, P. K., Rani, P., Samanta, D., Khanna, A., and Bhushan, B., 2020. An internet of health things-driven deep learning framework for detection and classification of skin cancer using transfer learning. *Transactions on Emerging Telecommunications Technologies*, p. e3963 https://doi.org/10.1002/ett.3963. Accessed on August 12, 2020 .

35. Hii, P. C., and Chung, W. Y., 2011. A comprehensive ubiquitous healthcare solution on an Android™ mobile device. *Sensors*, 11(7), pp. 6799–6815.

36. Banos, O., and Hervás, R., 2019. Ubiquitous computing for health applications. *Journal of Ambient Intelligence and Human Computing* 10, pp. 2091–2093. https://doi.org/10.1007/s12652-018-0875-3

37. Lymberis, A., 2003, September. *Smart wearable systems for personalised health management: Current R&D and future challenges.* In *Proceedings of the 25th Annual International Conference of the IEEE Engineering in Medicine and Biology Society (IEEE Cat. No. 03CH37439)* (Vol. 4, pp. 3716–3719). Cancun, Mexico: IEEE.

38. Kan, Y. C., Kuo, Y. C., and Lin, H. C., 2019. Personalized rehabilitation recognition for ubiquitous healthcare measurements. *Sensors*, 19(7), p. 1679.

39. Kumar, S., Kambhatla, K., Hu, F., Lifson, M., and Xiao, Y., 2008. Ubiquitous computing for remote cardiac patient monitoring: A survey. *International Journal of Telemedicine and Applications*, 2008. https://doi.org/10.1155/2008/459185

40. Kolomvatsos, K., 2007. Ubiquitous computing applications in education. In Lytras, D., and Naeve, A. (eds.) *Ubiquitous and Pervasive Knowledge and Learning Management: Semantics, Social Networking and New Media to Their Full Potential.* (pp. 94–117). Hershey, PA: IGI Global.

41. Rath, M., 2018. A methodical analysis of application of emerging ubiquitous computing technology with fog computing and IoT in diversified fields and challenges of cloud computing. *International Journal of Information Communication Technologies and Human Development (IJICTHD)*, 10(2), pp. 15–27.

42. Darwish, A., and Hassanien, A. E., 2011. Wearable and implantable wireless sensor network solutions for healthcare monitoring. *Sensors*, 11(6), pp. 5561–5595.

43. Edwards, W. K., and Grinter, R. E., 2001, September. *At home with ubiquitous computing: Seven challenges.* In *International Conference on Ubiquitous Computing* (pp. 256–272). Berlin, Heidelberg: Springer.

44. Sen, J., 2012. Ubiquitous computing: Applications, challenges and future trends. In Santos, R. A., and Block A. E. (eds.) *Embedded Systems and Wireless Technology: Theory and Practical Application* (pp. 1–40). Boca Raton: CRC Press, Taylor & Francis Group.

45. Milbredt, O., Castro, A., Ayazkhani, A., and Christ, T., 2017. Passenger-centric airport management via new terminal interior design concepts. *Transportation Research Procedia*, 27, pp. 1235–1241.

46. Tiwari, R., Sharma, N., Kaushik, I., Tiwari, A., and Bhushan, B., 2019, October. *Evolution of IoT & data analytics using deep learning.* In *International Conference on Computing, Communication, and Intelligent Systems (ICCCIS)* (pp. 418–423). Greater Noida, India: IEEE.

47. Varshney, T., Sharma, N., Kaushik, I., and Bhushan, B., 2019, October. *Architectural model of security threats & their countermeasures in IoT.* In *2019 International Conference on Computing, Communication, and Intelligent Systems (ICCCIS)* (pp. 424–429). Greater, Noida, India: IEEE.

48. Minoli, D., and Occhiogrosso, B., 2018. Blockchain mechanisms for IoT security. *Internet of Things*, 1, pp. 1–13.

Overview of the Internet of Things and Ubiquitous Computing 19

49. Bhushan, B., Khamparia, A., Sagayam, K. M., Sharma, S. K., Ahad, M. A., and Debnath, N. C., 2020. Blockchain for smart cities: A review of architectures, integration trends and future research directions. *Sustainable Cities and Society*, 61, p. 102360.
50. Wang, T., Zheng, Z., Rehmani, M. H., Yao, S., and Huo, Z., 2018. Privacy preservation in big data from the communication perspective: A survey. *IEEE Communications Surveys & Tutorials*, 21(1), pp. 753–778.
51. Ismail, L., and Materwala, H., 2019. A review of blockchain architecture and consensus protocols: Use cases, challenges and solutions. *Symmetry*, 11(10), p. 1198.
52. Antonopoulos, A.M., 2014. *Mastering Bitcoin: Unlocking Digital Cryptocurrencies.* Sebastopol, CA: O'Reilly Media Inc.
53. Goyal, S., Sharma, N., Kaushik, I., Bhushan, B., and Kumar, A., 2020, April. *Precedence & issues of IoT based on edge computing.* In *IEEE 9th International Conference on Communication Systems and Network Technologies (CSNT)* (pp. 72–77). Gwalior, India: IEEE.

2 Case Study on Future Generation IoT

Secure Model for IoT-Based Systems — Health Care Example

Mais Tawalbeh, Muhannad Quwaider
Jordan University of Science and Technology, Jordan

Lo'ai A. Tawalbeh
Texas A&M University, USA

CONTENTS

2.1 Introduction 21
2.2 IoT Challenges 23
    2.2.1 Security 23
    2.2.2 Privacy..24
2.3 Review of Related Work...25
2.4 Existing IoT Security Models and Current Issues.....................................27
2.5 Secure Model for IoT-Based Systems: Health Care Example 29
    2.5.1 IoT Case Study: Wearable Devices for Health Care 29
    2.5.2 IoT Layers and Health Care System Model....................................30
    2.5.3 AWS Platform 32
    2.5.4 Enhanced AWSACIoT Model..32
2.6 Conclusion and Future Work...34
References..34

2.1 INTRODUCTION

Over the last decade, physical devices in business, industrial, mobile, and home markets have been transformed from individual systems to interlaced networks. These networks can connect everything to the internet, communicating through different wired, wireless, and virtual environment technologies (cloud-based virtual connections). The process of associating and connecting devices with each other via the internet is known as the Internet of Things (IoT). The IoT represents growing trends in the ways that corporate entities and individuals interact in their daily lives and operations, including many aspects of life: social interactions, education, finance, industry, transportation, and health. These trends display the shifting means by which people use the internet and how much impact it has on our daily lifestyles. This technological advancement in common utilities is becoming more noticeable and is an integral part of daily human life, where new smart devices are being introduced for individual and corporate usages.

Smart devices are used in our daily life, ranging from home appliances to complicated disaster monitoring sensors. More examples of these devices include: smart refrigerators, smart washers, smart monitors, smart medical equipment, smartphones, and even smart buildings [1]. And new applications are continuously being developed to enhance or adjust these devices' features, in order to make life smarter and easier. A prime example of these advanced features is the content generated and data downloaded onto a user's devices by applications that run in the background. The devices automatically send, receive, and store user data, with little to no input from those operating the device. Often, users are oblivious to the automatic functions of their smart devices and take for granted the simplicity and convenience of the autonomous operation of physical devices and completion of mundane tasks, such as ordering items based on user behavior [2].

The Internet of things (IoT) has been a significant research topic in recent years. It is estimated that there are more than 20 billion IoT devices connected to the internet today [3]. Moreover, the growth of connected IoT devices is expected to generate more than 79 ZB (Zeta Bytes) of data in 2025, according to the recent International Data Corporation (IDC) forecast [4]. With this rapid technological revolution comes the potential to change the way that daily life is carried out, as people can live and work within an interconnected world, where the IoT provides unlimited opportunities in almost every aspect of life. According to Intel's IoT Framework Information Policy [2], the IoT will change global industries, creating jobs in transportation, utilities, manufacturing, health care, etc. This change will also require support from: high-speed networking, huge storage, and efficient computing resources to be able to transmit, store, and analyze the amounts of data generated. The utilization of IoT technology in different sectors is expected to result in great economic growth. CISCO economic research predicts that by 2022, the United States could take a 32 percent stake in the worldwide economy, just because of the use of the IoT [3].

However, despite the benefits of network-based smart devices, it is important to consider all aspects surrounding the IoT. As the IoT is adopted by more parties, the potential benefits, including economic and social growth, as well as the challenges that have arisen by the increased public and private sector usage of IoT-based technology, become more apparent [5].

There is an increase in security threats that face IoT devices, brought about by new hacking techniques, which exploit vulnerabilities associated with the wide spectrum of technologies used by IoT devices. Moreover, cybersecurity threats related to the IoT impact the applications running on it and the services they provide. These security threats pressure and pose a tremendous challenge for information technology specialists, application developers, network administrators, and anyone else working in this domain [6]. Accordingly, IT security experts must be aware of new attacks and implement the protective systems necessary to ensure secure IoT implementation and operational environments. On the other hand, there is the argument that applying more security measures will impose a trade-off between security and performance, which should be taken in consideration. In other words,

Case Study on Future Generation IoT

if more security algorithms are applied, the required speed and the requested performance might not be met by the IoT devices operating in certain environments. So, it is important to make sure that the IoT is not utilized in the wrong way and that it is enhanced in ways that allow the new technology to grow naturally. In order to ensure that the IoT is not abused, an in-depth information policy must be drafted, consisting of guidelines and standards surrounding the network-based devices, data gathered, data use, security standards, privacy, interoperability, legal regulations, economic growth, and ethical standards [7].

In this chapter, we investigate the recent work related to IoT security, challenges, solutions, and existing security models. Moreover, we propose a secure authorization model for IoT architectures. The model can be used to manage the user's authorization securely on many IoT components in different applications. The proposed model is implemented in the Amazon Web Service (AWS) cloud platform. Among the services provided by the AWS platform is an IoT environment that allows the user to set up and test real and virtual IoT devices. As a case study to demonstrate the applicability of the proposed model, we address the health care domain.

The rest of this chapter is organized as follows: Section 2.2 presents some of the main IoT challenges, including security and privacy issues. In Section 2.3, we review related work. Section 2.4 describes the main existing security models for the IoT. In Section 2.5, we propose a secure model for IoT-based systems, taking health care as an example. Section 2.6 concludes the chapter.

2.2 IOT CHALLENGES

With the use of the Internet of Things rapidly increasing among consumers, businesses, and governments, the resulting security complications are a growing concern The exposure of private activity or personal information is a particularly urgent security concern. More and more devices implement some form of IoT technology. However, on some of these new devices, security features might not be up to date. With competition driving short times to market and a cheaper cost of products, for example, many manufacturers of these products are allocating less time and resources to security. Also, consumer knowledge of IoT security is limited, which limits security as a priority in purchasing IoT devices [5]. As a further issue, as the IoT grows, so will more of the data originating from outside devices and traveling over a network. Many of these outside devices are in insecure locations, thus making them potential targets for tampering. Identifying the device that sent the traffic on a network, the location of the device, and the authenticity of the traffic from the device are critical for network security.

Many challenges need to be met before a full and comprehensive understanding of the IoT's benefits and defects can be achieved.

2.2.1 Security

Among information technology concepts, the topic of security is nothing new. However, some new security attributes presented by the implementation of the IoT may have not been previously considered or addressed. Companies and service providers must continually deal with economic and technical barriers and challenges when developing, maintaining, and updating the security features of IoT devices. Often, weak or outdated security measures effectively grant malicious entities free access to any user data. As more IoT devices become commercially available, more security vulnerabilities arise, leading to the need for almost constant software and firmware updates to smart devices. Lifetime security must be a priority for any network-connected device. If a product/service provider cannot guarantee that their customer's devices are secure, the company will lose money, as well as the trust of the general consumer. The consumer must be confident that their data is secure, as smart devices become more passively implemented in the user's daily life [8].

To ensure this, most product/service providers that deal with smart devices are legally required by FTC regulations to secure user data. These regulations are in addition to the slight variations in the different company "Terms of Privacy" statements [6],[7]. Companies should be legally obligated to not cause harm to users when handling their user data. This "Hippocratic" practice ensures that companies will not disclose sensitive user data to third parties without consent from the customer. However, even though the practice of giving out third-party information is illegal in the United States, legal loopholes are used to gather data by funneling it to international sites, effectively allowing service providers to gather any consumer data they want.

2.2.2 PRIVACY

Being able to collect, analyze, and alter gathered user data is an asset of network-connected IoT devices and services. While this may seem beneficial, the mass monitoring of user data can be and is used to create invasive consumer profiles on customers. Examples of this include smart appliances gathering information on what is stored in the device and the levels of what is stored, the tracking of user location, banking log information from mobile applications, browser history, shopping history, etc. The mass storage of user information sparks concerns surrounding the increased surveillance, storage, and use of private/sensitive information from the user's daily routine. Everything from home appliances to workstation PCs is monitored, and in the wrong hands, this information has the potential to destroy a person's life through doxing, the release of information, breaches of integrity, etc. Moreover, sometimes information is gathered without the knowledge or consent of individual users. Such examples include the so-called super-cookies used by Facebook, which were impervious to most commercial virus scans, or the fact that Google Chrome not only tracks what people search for but monitors the rate and flow of individual user keystrokes. While the data gathered undoubtedly helps a smart device owner, the information is more useful to the product manufacturer and/or service provider [8]. The observation and analyzing of individuals are serious privacy concerns and, considering the different privacy policies and standards that each company holds, quite concerning.

Smart devices that gather and store user data can utilize international loopholes to bypass jurisdiction privacy laws to sell information gathered. It would be pragmatic to enable cross-border data flow to be protected by the user via their local

Case Study on Future Generation IoT

jurisdiction's law. This would limit the international privacy breaches that have become so common with network-connected smart devices. Users should be guaranteed that unknown, untrusted third parties cannot see their data. Strategies promoting company transparency surrounding user data collection, storage, and transmission should be implemented to protect an individual's privacy rights, as well as to build peace of mind. The guarantee that user data has not been compromised is a benefit to both the service/product provider and the consumer [9].

It is important to consider policies surrounding the usage, data processing/development, security and infrastructure when considering the further evolution of IoT-enabled devices. As the IoT becomes more common and somewhat all-encompassing in daily life, an information policy plan covering the challenges facing the topic must be drafted to determine new strategies, goals, laws, and initiatives on the topic [10].

Policies should take into consideration the information that gets collected and make sure to provide consent agreements, so that users will not be misled into giving up information about themselves that they do not want to give out. In terms of the government's role in this, they intend to collect some of the user's data, but not all of it. The main problem that big companies and the government are concerned about is the usage and regulations of what is being collected, how it is being monitored, how it affects the user, and whether the user consents to the information they give out.

2.3   REVIEW OF RELATED WORK

The IoT is undergoing rapid development, in different domains and areas. Based on the information in Section 2.2.2, we can say it is a double-edged sword for all groups that have adapted to it to keep abreast of, for example, individuals, organizations, and governments. On the positive side, the IoT can make life easier and transactions faster, which saves time and effort. However, it poses a threat to consumer privacy. It is very important to have legitimate measures to deal with secure purchaser rights without hampering the developments in any field. Unfortunately, the adjustment of legal rules is significantly slower than the enhancement of IoT technology. Therefore, the new features of the IoT consistently outpace current legal standards [11].

As a result, researchers are intensifying their efforts to identify the risks and volatilities that shake the public's trust in IoT technology and find ways to mitigate them. In this section, we discuss the most common IoT security problems and their solutions.

In [12], the authors divided the IoT into four main domains: transportation, health care, personal, and smart environments, such as homes, offices, etc. They discussed in detail the applications that could be included in each domain. At the time, applications were classified in two types: directly applicable applications and advanced applications for the future. In addition, they focused on the most commonly predicted IoT challenges, with security and privacy at the top of the list. They indicated that as long as suspicion of the Internet of Things remains in terms of privacy, opposition from some consumers will remain. For example, an Italian clothing brand faced opposition from customers for using smart devices to track products, because of a lack of privacy. They summarized the causes of the security and privacy issues of the IoT. First, physical assaults on system components, due to the limited resources of IoT devices, which makes them unable to support complex security schemes. Also, the wireless communication in IoT systems provides a greater opportunity for eavesdropping and attacks.

Because of the predicted number of IoT devices and the uncontrolled growth of IoT applications in our daily life, including in business, health care, education, etc. [13], it is difficult to find one related work in the literature that covers all of the different IoT domains and challenges. The researchers in [12] categorized IoT domains as:

- **The IoT in transportation**, where bicycles, cars, vans, buses, trains, planes, and even roads themselves become smarter by including devices such as sensors, actuators, etc. This provides the ability for devices to communicate with each other, or communicate with specific control sites by sending data, controlling traffic, providing a better route, monitoring status, etc. The attacker can take over and change vital information, such as destination city or arrival/departure date. Moreover, he/she could create a completely new trip that does not exist, causing undesired consequences that affect the data's confidentiality, integrity, and availability.

- **The IoT in health care**, where smart IoT devices could be utilized to manage and save people's lives. For example, patients can utilize the IoT for automatic data collection and sensing, and send data periodically to a monitoring system, to identify specific risk signals in the human body. For example, vital risk signals could predict heart attacks. The most common attack in the health care domain is the "ransomware" attack. This is a type of malicious software that denies access to a computer system or database [14]. For example, we can imagine the damage that could be caused if hospital data were exposed to this attack, including staff and patient information. Another attack focuses on damaging the functionality of specific devices, such as tomogram apparatuses, x-ray machines, and so on. All of these attacks are risks to data confidentiality, integrity, and availability.

- **The IoT in smart environments**, one of the most appealing fields of the IoT framework, be it an office, a home, etc. This is where individuals spend most of their time, and they find it more comfortable, and more efficient, to utilize IoT technology [11]. But, like other IoT systems, smart environments are under threat of attacks that can cause great damage. For example, the Ukrainian energy grid suffered a cyberattack on December 23, 2015. The aggressors took control of the grid's Supervisory Control and Data Acquisition (SCADA) framework and caused a power outage for a few hours in Ukraine's enormous Ivano-Frankivsk region, which is populated by 1.4 million inhabitants. This incident demonstrated that the outcomes from assaults on the IoT can be more destructive than data burglary or economic losses [15].

- **The IoT in personal use**, which is related to all applications and systems that provide individuals the ability to communicate and exchange information, to build and protect social relationships. Sometimes, data is accumulated without the information or consent of individual users. Such models incorporate the supposed "super cookies" utilized by Facebook, which were impenetrable to most commercial virus scans, or the fact that Google Chrome tracks not only what

Case Study on Future Generation IoT

individuals search for but monitors the rate and flow of individual user keystrokes. While the information accumulated undoubtedly helps smart devices owners, it is more useful to the product manufacturer and/or service provider [16].

2.4   EXISTING IOT SECURITY MODELS AND CURRENT ISSUES

A breadth of research has been conducted on IoT-related issues, including security and privacy [17]. There are numerous high-quality papers that can be found in the literature, and most of them are available as open access. The literature can be reviewed and categorized by the objective methods used as well as the tools used to validate results [18]. There are various simulation tools, as well as modelers. Moreover, there are many platforms available for producing novel protocols for IoT security. Thus, there is have been advances in IoT security research, supported by many different implementation and simulation environments.

One major concern for the IoT is how to ensure confidentiality and privacy, mainly by using cryptographic functions [19,20]. Also, infrastructures must be configured and implemented so as to maintain the IoT system's service availability. Current research regarding IoT cybersecurity shows that several issues are present that can compromise privacy and cybersecurity. These issues include:

Irregular update: There must be regular updates for the operating systems and applications used in an IoT system. The firmware of the IoT device itself should also be updated, which makes keeping an eye on recent patches and downloading them crucial to maintaining the IoT system's security against possible vulnerabilities, which might be explored by hackers seeking to launch attacks.

Automation: Most individuals and organizations using IoT services select the automation feature to collect, send, and analyze data, or execute a variety of other tasks. Nowadays, artificial intelligence techniques can be used to gain unauthorized access to these automated systems and launch automated attacks, due to the lack of human interaction with the IoT devices [21].

Weak Authentication: Among other factors, this is due to the fact that an IoT system uses applications from different vendors, with different authentication and security settings, which might not be as secure as they should be. Moreover, users access IoT devices from different platforms (PCs, laptops, and mobile devices), using different communications protocols. These factors contribute to distributing the authentication process, which can make it as weak as its weakest link [22].

Remote connectivity: High-speed network connectivity with powerful computing ability is enabling systems to communicate efficiently with other systems on the same network or even on other networks. But at the same time, this means that if any of the devices/systems are connected to the internet, it might be vulnerable to cyberattacks.

The recent rapid adoption of IoT devices in almost every aspect of our lives has helped increase the efficiency of executing tasks in different sectors, from homes to energy grids. But at the same time, it has also increased vulnerabilities and possible attacks that could compromise the security of these connected IoT devices, affecting our privacy as a result. Research shows that more than 90 percent of IoT device users are unsure of their devices' security levels [23]. This research also indicates that sophisticated security measures are necessary to protect data security and privacy.

In another context, researchers proposed using recent advanced technologies for security in IoT systems, among which are cloud and fog computing and blockchain. The authors in [24] proposed that two mechanisms be used in cloud-enabled fog computing in IoT systems: incentive and feedback mechanisms. These techniques allow a cloud provider to eliminate fake edge servers. In other words, cloud servers utilize these two mechanisms to determine which are illegitimate edge nodes and eliminate them. However, the authentication process for IoT devices is not proposed, and the authors assume that all connected devices are trustworthy. In order to reduce latency and increase performance for the two models discussed, the research in [24] used a Proof of Authority algorithm (PoA) for authentication purposes in blockchain, validating every single block in the chain, and thus avoiding unauthorized access to system resources.

As a different attempt to propose security solutions for IoT architectures, the researchers in [25] found that machine learning will be prominently featured in IoT data analytics. The authors in [26] stated that the software program could learn and enhance itself from experience, examples, and analogies. In addition, some common tasks provided by the machine learning, such as features extraction and pattern recognition, are helpful in proposing secure IoT solutions [27]. In [28], the researchers indicated that network architecture consists of three layers to provide intelligent data processing inside the network. These are the application layer, the transport layer, and the sensing layer. Similar parts of the architecture are discussed in [24]. The authors in [28] define the artificial neural network algorithm (part of machine learning) in the transport layer. This algorithm is proposed to build a system that can detect fake data. The researchers go into detail about their approach using mini architecture. This consists of edge devices with low power and low resources, such as sensors, actuators, etc., and gateway devices with greater abilities and resources. Temperature sensors are added for collecting data, while gateway devices aggregate data to enter it into the artificial neural network for processing. The first version of the model uses Device ID and Sensor value as inputs to the ANN, and the output determines whether the record is valid or not. The training and testing processes are completed by four thousand records of valid and invalid data. The second version is done in the same way, except one value is added to the ANN input, which is a delayed time. The proposed neural network was designed to simulate and detect a man-in-the-middle attack.

Authorization and authentication issues are starting points for security issues in the IoT framework [29]. IoT smart devices have the ability to sense, collect and send data, and communicate with each other. So, it is very important to control who can access the device and what they can do with it. In order to conserve data confidentiality, integrity, and availability, authorization and authentication methods must be excellent. In addition, these methods should be suitable to work with devices from different vendors. The authorization and authentication processes can be determined by many techniques. One of these techniques is implementing an access control mechanism, to provide authorization in IoT systems. There are many models that provide access control in IoT systems. One model is Role-Based Access Control (RBAC), in which certain rules are predefined for each network object. Another scheme is Capability-Based Access Control (CapBAC), in which the privilege of the object will determine if it will be enabled or not in a particular network [30]. Moreover, there is a model that uses predefined attributes to grant access for each network object. This model is called Attribute-Based Access Control (ABAC) [31]. Some models consist of a combination of other models. For example, the ARBHAC model [32] combines both the ABAC and RBAC models. The researchers in [33] propose and analyze the suitability of different models of access control to IoT systems.

2.5 SECURE MODEL FOR IOT-BASED SYSTEMS: HEALTH CARE EXAMPLE

2.5.1 IoT CASE STUDY: WEARABLE DEVICES FOR HEALTH CARE

As time passes, devices and technologies continue to advance, one such being "wearable devices." These have become a part of many people's everyday life, from making calls to tracking vital signs. Wearable technologies include glasses, watches, sleep trackers, fitness trackers, and many more, including "smart contact lenses," which are currently in development. The concept of wearable devices is to provide people with all the functions that they get from their computer or smartphone, without the inconvenience of the devices being too big or bulky. Instead, a person can just look at their watch to check the time and also see whether they have received missed calls or messages. They could go so far as to check how many steps they have taken, or whether their sleeping pattern the previous night showed any irregular activity. Smart glasses are used in the same way, except that they can turn a view of the world into augmented reality, showing a screen in front of the user's face, to display a "to-do" list for the day or to allow the user to browse the internet. Considering there are so many types of wearable devices, multiple policies have been put into effect to account for the dangers of these devices [34].

Since the average person has access to such devices, companies, as well as governments, these devices can be seen as a way to regulate the general public. While such public regulation in the future seems likely, information security policies should be put into place. Information security policies should be developed to protect the data being collected from the users, especially in the case of health care applications. For example, wearable devices that track vital signs could be important for doctors, giving them the ability to check on their patients. On the other hand it would not be necessary for doctors to collect information being tracked from someone's jog around the block. Other policies have been put into effect with third party vendors. When companies work with third parties, they need to communicate with those vendors to control how the data will be used and whether or not it will be distributed.

Although not perfect, most of these policies provide guidelines as to what a company should seek to do, and what they could improve on. One advantage these policies have provided to users has been the ability to trust their health care provider to allow them a safeguard, so that they know they are being looked after. Health care companies invest in wearables to improve workforce productivity, cut absenteeism, and reduce health care costs [35]. Wearables give users the freedom to post, or update, what they want to social media without being limited in what they can and cannot share.

While health care providers are working to keep customers and patients satisfied with their services, some issues still need to be addressed. One concerns the storage of users' data and the procedures to be followed when data is exposed to a third party for legitimate or illegitimate reasons. This task entirely depends on the health care provider, which should develop and impose data privacy policies to identify the responsibilities of third-party vendors. Similar polices are required to clearly indicate a patient's rights and how to protect their confidentiality. In many cases, third parties, for example, insurance companies, can receive a user's information if that has been "consented" to. This might include exposing confidential patient information that is not required to carry out the operations involved by the third-party employee, making this private data vulnerable to unauthorized access.

2.5.2 IoT Layers and Health Care System Model

Figure 2.1 below shows the top-level design of the IoT health care system model. As seen in the figure, the model is composed of three layers: a cloud layer, a device layer, and an end-user layer. The cloud layer hosts the data gathered by sensors, in order to process it. The processing step includes feature extraction and noise removal.

This data will be fed at a later stage into a decision-support system that applies data mining and machine intelligence techniques to extract appropriate decisions for health care providers, in terms of the patient's health.

On the other hand, the device layer is composed of a pool of sensors. These sensing devices are connected to the internet though wireless technologies (4G/WiFi). This layer also has circuitry for data acquisition and communication protocols that enable sending the data to storage to be processed. These sensing devices enable users to collect data at different acquisition frequencies. in real time.

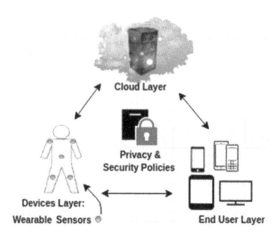

FIGURE 2.1 The generic architecture of the IoT health care model.

The end-user layer consists of the receiving user and can take different forms. One of these forms, that of smart devices, poses great challenges for security and privacy. Within the boundaries of these three layers, a list of sub-layers or modules can be added to ensure the robustness of the health care decision support system. For example, this step ensures that data is sent and processed in a timely manner for critical health decisions that cannot wait until data has been sent to the cloud. In this section, we propose the capability of edge computing. This feature (through a completely new edge layer) performs more than one task on the collected data, at the same time: it sends a copy of the cloud layer for processing and long-term storage, and conducts decision making bases on the allied data. Sometimes, we need to send commands or instructions to wearable devices to update their acquisition rate, or to perform a certain functionality, which requires another protocol and security procedure.

As can be seen from Figure 2.2, the top-level design of the proposed IoT model contains the new additional edge layer. This layer serves two purposes: first, it overcomes the extra delay due to complete dependency on the services provided by the cloud layer. Second, it enables faster decision making, especially in time-constrained environments. The edge computing layer is helpful in IoT systems where devices have sensors attached or where sensors are very close to the IoT device. The proposed edge layer can manage the sources collecting data and provide decisions to users in real time. Moreover, the edge layer can transfer sensed data to all other layers, for purposes such as storage, fusion, and analysis. Edge computing tasks are more powerful when physically more distant from the sources and sensors collecting data; but at the same time, the edge is connected to local area networks. These advantages impose additional information privacy challenges.

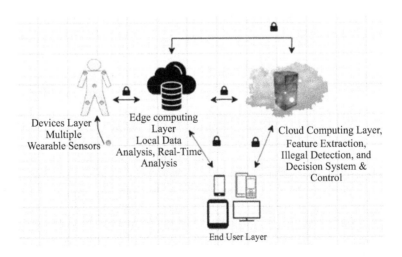

FIGURE 2.2 Top-level design of the proposed IoT model: Wearable devices case study.

2.5.3 AWS PLATFORM

The Amazon company has provided a platform for cloud and related technologies implementation, such as edge and IoT implementation environments. The service is called the Amazon Web Service (AWS). This service can either be free (providing limited options) or paid, through subscriptions determined by company or personal needs and requirements. Users can benefit from AWS services such as storage, computing, networking, and tools to support developing simulations of IoT and edge computing environments. One well-known service is the EC2, which is the main cloud platform provided by Amazon [36]. By using the EC2, users can create virtual machines with different computing, networking, and memory characteristics, as required by the tasks to be executed.

Besides EC2, another useful service by Amazon for IoT platforms is AWS-IoT. This platform has many options for users to choose from, including a broker for the messages, gateways, and engine to set up certain rules as needed by the design constraints. Also, AWS-IoT provides a secure development platform for different IoT devices, including actuators, sensors, and even home smart devices. These devices are allowed to interact with the AWS Cloud and with other components in the network using secure protocols such as HTTPs and MQTT. Moreover, and as an extra measure of security, authentication must be completed for each device before connecting it to the system. For authentication purposes, certificates are used (X.509).

There are also other powerful features that can be utilized to increase efficiency and facilitate the management of the implemented IoT system. For example, the registry feature allows the user to manage devices easily though grouping related devices with their certificates and resources. In other words, this feature can be used to gather many devices and manage them at once, with minimal effort and overhead. Also, users can create mobile applications to manage their IoT devices and monitor their activities (data acquisition for example) from anywhere, at any time. For details on the characteristics and features of this platform, please see the company's documentation [36].

2.5.4 ENHANCED AWSACIoT MODEL

In this section, the enhanced AWSACIoT layers and the configuration of a simple health care case study are shown, using the AWS Platform. In order to enhance security and make more real-time decisions, we add the edge computing concept to the proposed models. To utilize the AWS cloud provider platform services successfully, we need to create an account on it. Then, we use the AWS management console, which is shown in Figure 2.3.

In this proposed model, as shown in Figure 2.4, the cloud level is proposed using the AWS platform like a virtual machine with a specific Amazon Machine Image (AMI) and specific characteristics, such as CPU and memory storage. Each physical device, such as sensors, is proposed as an IoT virtual machine. We have three sensors, as an example, collecting specific patients' data. Each IoT virtual machine has specific characteristics, such as device type and device attributes. This information about each virtual machine is set during the IoT machine launching and can be changed. Also, each device has its own X.509 certificate, which is used to authenticate by AWS cloud services.

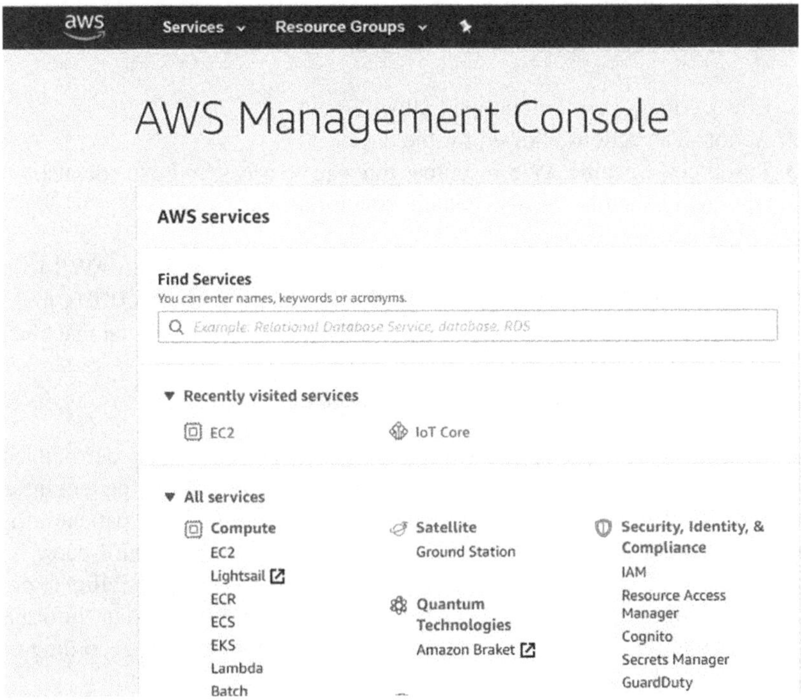

Case Study on Future Generation IoT

FIGURE 2.3 AWS management Console.

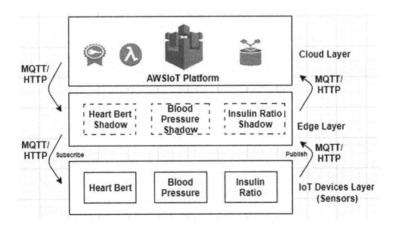

FIGURE 2.4 Simple health care case study using the AWS Platform: Enhanced AWSACIoT.

We used a policy-based access control mechanism, JavaScript Object Notation file (JSON). The JSON file contains three main components:

1. Effect: Means permission type (Allow, Deny).
2. Action: The actions allowed for the device.
3. Resources: Specific AWS resources that can be accessed by the device.
4. Optionally, the file can also contain conditions.

However, all sensors use the MQTT protocol to communicate with AWS IoT services, and this should be simulated as MQTT Clients using the MQTT.fx tool. The edge concept can be proposed as a device shadow or as another virtual machine.

2.6 CONCLUSION AND FUTURE WORK

With the tremendous increase in using the IoT in many aspects of our daily lives, including the health care sector, we need a policy to identify which patient information can be shared and which cannot, to provide better security over patient information. On the other hand, while focusing on the legitimacy of a third party is one important element, health care service providers have a great responsibility to protect the privacy of their patients' medical records. For example, important information can be encrypted, to maintain the required level of confidentiality, according to the classification of the data.

While health care organizations generally have existing information security policies, a very limited number of these policies specifically address networked wearable devices. The increasing number of cyberattacks on medical devices, IoT wearable or implanted devices, and patients' records, impose the necessity for clear policies that identify patient rights, classify medical data, impose the use of encryption functions, and clarify other important parameters. Moreover, there should be an increasing effort to standardize the operation of IoT devices, not only in the health care domain, but also in other fields where these devices are used to gather information and share it over inter-networks or even over the internet. Such standardization will help unify and improve information security and privacy requirements. Also, taking into consideration the health care perspective, a suggested enhancement for the AWSACIoT model is proposed. This enhanced model can be used as a basis to propose and implement real-life cases that involve real data fusion and analysis as a future area of work.

REFERENCES

1. Ray, P. P., 2018. A survey on Internet of Things architectures. *Journal of King Saud University-Computer and Information Sciences*, 30(3), pp. 291–319.
2. Intel Corp., n.d. Internet of Things policy framework. https://www.intel.com/content/www/us/en/policy/policy-iot-framework.html. Accessed on March 3, 2020.
3. Cisco IoT., n.d. What do you need to succeed with IoT? https://www.cisco.com/c/en/us/solutions/internet-of-things/overview.html. Accessed on February 25, 2020.
4. International Data Corp., 2019, June 18. The growth in connected IoT devices is expected to generate 79.4ZB of data in 2025. https://www.idc.com/getdoc.jsp?containerId=prUS45213219. Accessed on February 20, 2020.

Case Study on Future Generation IoT 35

5. Tawalbeh, L. A., and Somani, T. F., 2016. *More secure Internet of Things using robust encryption algorithms against side channel attacks.* In *2016 IEEE/ACS 13th International Conference of Computer Systems and Applications (AICCSA)* (pp. 1—6). Agadir, Morocco: IEEE.

6. Alsmadi, I., Easttom, C., Tawalbeh, L. A., 2020, April. *The NICE Cyber Security Framework, Cyber Security Management.* Switzerland: Springer. ISBN 978-3-030-41987-5. https://www.springer.com/gp/book/9783030419868#otherversion= 9783030419875. Accessed on April 4, 2020.

7. Estrada, D., Tawalbeh, L. A., and Vinaja, R., 2020, March. How secure having IoT devices in our home. *Journal of Information Security,* 11(2), pp. 81–91. https://doi. org/10.4236/jis.2020.112005. Accessed on May 6, 2020.

8. Maleh, Y., Shojafar, M., Alazab, M., and Romdhani, I., 2020. *Blockchain for cybersecurity and privacy: Architectures, challenges, and applications.* Taylor & Francis Group. https://doi.org/10.1201/9780429324932. Accessed on May 5, 2020.

9. Zheng, S., Apthorpe, N., Chetty, M., and Feamster, N., 2018. User perceptions of smart home IoT privacy. *Proceedings of the ACM on Human-Computer Interaction,* 2 (CSCW), pp. 1–20.

10. Barrera, D., Molloy, I., and Huang, H., 2018. *Standardizing IoT network security policy enforcement.* In *Workshop on Decentralized IoT Security and Standards (DISS),* (pp.1–6). San Diego, CA: NDSS.

11. Tiwari, R., Sharma, N., Kaushik, I., Tiwari, A., and Bhushan, B., 2019. *Evolution of IoT & data analytics using deep learning.* In *2019 International Conference on Computing, Communication, and Intelligent Systems (ICCCIS)* (pp. 418–423). Greater Noida, India: IEEE.

12. Atzori, L., Iera, A., and Morabito, G., 2010. The internet of things: A survey. *Computer Networks,* 54(15), pp. 2787–2805.

13. Tawalbeh, L. A., Tawalbeh, M.A., and Aldwairi, M., 2020. Improving the impact of power efficiency in mobile cloud applications using cloudlet model. *Concurrency and Computation: Practice and Experience,* 32(21), e5709. https://doi.org/10.1002/cpe.5709.

14. Tawalbeh, L. A., Somani, T. F., and Houssain, H., 2016. *Towards secure communications: Review of side channel attacks and countermeasures on ECC.* In *2016 11th International Conference for Internet Technology and Secured Transactions (ICITST)* (pp. 87–91). Barcelona: IEEE.

15. Electronic Sharing and Information Analysis Center, 2016, March 18. Analysis of the Cyber Attack on the Ukrainian Power Grid: Defense Use Case. Washington, DC: E-ISAC. https://ics.sans.org/media/E-ISAC_SANS_Ukraine_DUC_5.pdf. 2016. Accessed on May 5, 2020.

16. Pagliery, J., 2015, January 9. Super cookies track you, even in privacy mode. *CNNMoney.* https://money.cnn.com/2015/01/09/technology/security/super-cookies/index.html. Accessed on March 13, 2020.

17. Hassija, V., Chamola, V., Saxena, V., Jain, D., Goyal, P., and Sikdar, B., 2019. A survey on IoT security: Application areas, security threats, and solution architectures. *IEEE Access,* 7, pp. 82721–82743.

18. Hassan, W. H., 2019. Current research on Internet of Things (IoT) security: A survey. *Computer Networks,* 148, pp. 283–294.

19. Tenca, A. F., and Tawalbeh, L. A., 2004. Algorithm for unified modular division in GF (p) and GF (2n) suitable for cryptographic hardware. *Electronics Letters,* 40(5), pp. 304–306.

20. Al-Haija, Q. S. A., 2009. Efficient algorithms for elliptic curve cryptography using new coordinates system. Master's Thesis, Computer Engineering Department, Jordan University of Science and Technology.

21. Dalipi, F., and Yayilgan, S. Y., 2016, August. *Security and privacy considerations for IoT application on smart grids: Survey and research challenges.* In *Future Internet of Things and Cloud Workshops (FiCloudW), IEEE International Conference* (pp. 63–68). Vienna: IEEE.
22. Jararweh, Y., Al-Ayyoub, M., and Song, H., 2017. Software-defined systems support for secure cloud computing based on data classification. *Annals of Telecommunications*, 72(5–6), pp. 335–345.
23. Tawalbeh, L., AMuheaidat, F., Tawalbeh, M., and Quwaider, M., 2020. IoT Privacy and security: Challenges and solutions. *Applied Sciences*, 10(12), pp. 4102–4114.
24. Mahmoud, R., Yousuf, T., Aloul, F., and Zualkernan, I., 2015. *Internet of Things (IoT) security: Current status, challenges and prospective measures.* In *2015 10th International Conference for Internet Technology and Secured Transactions (ICITST)* (pp. 336–341), London: IEEE.
25. Jindal, M., Gupta, J., and Bhushan, B., 2005. *Machine learning methods for IoT and their Future applications.* In *2019 International Conference on Computing, Communication, and Intelligent Systems (ICCCIS)* (pp. 430–434). Greater Noida, India: IEEE.
26. Negnevitsky, M., 2005. *Artificial Intelligence : A guide to intelligent systems*, 2nd ed. New York: Pearson.
27. Moh, M., and Raju, R., 2018. *Machine learning techniques for security of Internet of Things (IoT) and fog computing systems.* In *2018 International Conference on High Performance Computing & Simulation (HPCS)* (pp. 709–715). Orléans, France: IEEE.
28. Canedo, J., and Skjellum, A., 2016. *Using machine learning to secure IoT systems.* In *2016 14th Annual Conference on Privacy, Security and Trust (PST)*, (pp. 219–222). Auckland, New Zealand: IEEE.
29. Varshney, T. , Sharma, N. , Aushik, I., and Bhushan, B., 2019 *Architectural model of security threats & their countermeasures in IoT.* In *2019 International Conference on Computing, Communication, and Intelligent Systems (ICCCIS)* (pp. 424–429). Greater Noida, India: IEEE.
30. Hernández-Ramos, J. L., Jara, A. J., Marin, L., and Skarmeta, A. F., 2013. Distributed capability-based access control for the internet of things. *Journal of Internet Services and Information Security*, 3(3/4), pp. 1–16.
31. Bhatt, S., Tawalbeh, L. A., Chhetri, P., and Bhatt, P., 2019. *Authorizations in cloud-based internet of things: Current trends and use cases.* In *2019 Fourth International Conference on Fog and Mobile Edge Computing (FMEC)* (pp. 241–246). Rome, Italy: IEEE.
32. Tawalbeh, M., Quwaider, M., and Tawalbeh L., 2020. *Authorization Model for IoT Healthcare Systems: Case Study.* In *2020 11th International Conference on Information and Communication Technology (ICICS)* (pp. 337–442). Irbid, Jordan: IEEE.
33. Atlam, H. F., Alenezi, A., Walters, R. J., Wills, G. B., and Daniel, J., 2017. *Developing an adaptive risk-based access control model for the Internet of Things.* In *IEEE International Conference on Internet of Things (iThings)* (pp. 655–661). Exeter, UK: IEEE.
34. Singh, J., Pasquier, T., Bacon, J., Ko, H., and Eyers, D., 2015. Twenty security considerations for cloud-supported Internet of Things. *IEEE Internet Things Journal*, 3(3), pp. 269–284.
35. Tawalbeh, L. A., and Saldamli, G., 2019. Reconsidering big data security and privacy in cloud and mobile cloud systems. *Journal of King Saud University-Computer and Information Sciences* (in press).
36. Singh, J., Pasquier, T. F. J.-M., Bacon, J., Ko, H., and Eyers, D. M., 2016. Twenty security considerations for cloud-supported Internet of Things. *IEEE Internet of Things Journal*, 3(3), pp. 269–284.

3 Spy-Bot
Controlling and Monitoring a Wi-Fi-Controlled Surveillance Robotic Car using Windows 10 IoT and Cloud Computing

A. Diana Andrushia, J. John Paul, K. Martin Sagayam, S. Rebecca Princy Grace
Karunya Institute of Technology and Sciences, India

Lalit Garg
L-Universita, Malta, Malta

CONTENTS

3.1 Introduction 38
3.2 Background 39
    3.2.1 Smart Parking Systems 39
    3.2.2 Surveillance Systems 43
3.3 Spy-Bot: Wi-Fi Controlled Robotic Car.................................45
    3.3.1 Block Diagram of Wi-Fi Controlled Robotic Car............45
    3.3.2 Major Components Used 47
    3.3.3 Raspberry Pi 2..47
    3.3.4 Digital Web Camera..48
    3.3.5 Geared DC Motor 48
    3.3.6 Dual Full-Bridge L298N Driver 48
    3.3.7 Ultrasonic Sensors 49
    3.3.8 Microsoft Azure (Windows Cloud Platform)...................49
3.4 Flow Diagram...50
3.5 Experimental Conditions...50
3.6 Results and Discussion...53
    3.6.1 Hardware Implementation 54
    3.6.2 Results...54
3.7 Conclusion..55
3.8 Future Development..56
References..56

3.1 INTRODUCTION

The Internet of Things (IoT), a booming technology, will change the way we look at the world today. IoT-based solutions facilitate business by improving and simplifying tasks. The IoT consists of a huge number of sensors, each with a unique address, grouped to form a network entity that exchanges data internally or with other inter-networks and intranetworks, using different internet protocols. The IoT paves ways to do business and services in almost every domain, such as health, logistics, agriculture, etc., by connecting these objects/things to microcontrollers, such as a Raspberry Pi, or Node MCU, which are Wi-Fi or Bluetooth compliant.

The IoT helps to broadcast the information from a single source to multiple clients. Hence, IoT implementation is incomplete without cloud computing, which can be considered the backbone of the IoT. In the early days, the buzzword "cloud" was just used as a symbolic representation of the internet. Later, the term cloud computing was coined, to signify data storage, data processing, or data analysis carried out remotely from various sensor nodes (i.e., a doctor remotely accessing patients' health records, or remote monitoring of the physical condition of a smart car). Nowadays, with huge functionalities such as the mammoth storage of data for data analytics, managing a wide arena of networks, operating systems (OS), and virtual machines (VMs), a bunch of key market players are providing cloud computing services, such as Google Drive, Microsoft OneDrive, IBM Bluemix, Amazon Web Services, etc. IoT systems with various sensor networks will surely take advantage of these cloud-based services, because the market for IoT devices is expected to grow to $3.7 billion, comprised of 30.7 billion devices. Cloud services provide much more flexibility and speed for devices to operate, and we pay only for the service provided by the cloud, not for the infrastructure.

In 2017, the Gartner research firm predicted that by 2020, there would be an excessive amount of data due to the genesis of billions of IoT devices. The disruptive technology of the IoT, and swift improvisations in existing technologies, are creating an extensive aggregation of things, which leads to enormous collections of data to be handled and stored in the cloud. For instance, if an employee has to submit a report to a manager, it is now easy, with no need worry that memory space will run out. Instead, if his device is connected to the cloud, he can push most of his data/reports to the cloud platform.

Cloud robotics [24] is a field of robotics, wherein technologies such as cloud computing, cloud storage, etc., are used as beneficial services. Through these services, robots can be made to share information with other machines wirelessly, so that they can be controlled and monitored from anywhere, anytime, around the globe. Also, humans can assign tasks to robots remotely. Thus, simplification and ease of access will be top priorities. The security issues of robotic cars are now leveraged in terms of deep learning strategies [30,31,32,33]. Gesture-features are also added in most robotic cars, to control the overall operation [34,35,36,37]. The IoT's numerous applications range from smart appliances, smart health care, smart farming, smart wearables, smart buildings, smart cars, and smart cities to smart robots, paving the way to a fully networked ecosystem [25].

In the proposed system, a prototype of a surveillance assistive robot (spy-bot) is developed, which helps to monitor or identify any change in the usual behavior and records it in order to manage or protect one's asset or position. The spy-bot is an intelligent system, with Raspberry Pi as its controller (brain), four cameras, two DC motors, and a secured Windows 10 Azure cloud platform. The Raspberry Pi is an intelligent controller, with Raspbian OS installed with Noobs installer. The four cameras help provide a 360° view of the ambience. The DC motors attached to the Raspberry Pi controller help the spy-bot move forward, backward, right and left .

The Azure cloud platform stores images or livestreams videos, with built-in security services. A webpage is developed using the HTML scripting language, which gives access only to authenticated users. These authenticated users can control the spy-Bot using control commands/button widgets available in the webpage; the spy-bot's movements can be controlled remotely through Wi-Fi, by means of a secured Windows 10 cloud platform [26]. The motivation behind this is to capture images and videos remotely and store or retrieve them from a secured Azure cloud platform.

The organization of this chapter is as follows: Section 3.2 discusses related works by other researchers, on using wireless robotic vehicles for different applications. Section 3.3 presents the proposed work on a WiFi controlled robotic car, with block diagrams, and operational details. Section 3.4 presents a flow diagram. Section 3.5 shows the hardware implementation of the proposed work, with experimental conditions. Section 3.6 presents the results of the work, with a discussion. Section 3.7 concludes the proposed research, and Section 3.8 describes the future scope and possibilities for enhancing this robotic technology.

3.2 BACKGROUND

This section discusses works by various authors using wireless robotic vehicles in different application areas. It concentrates mainly on two major applications: smart parking systems and surveillance systems.

3.2.1 SMART PARKING SYSTEMS

In today's world, vehicle parking is one of the key challenges faced when people visit shopping malls, movie theaters, restaurants, etc. As a result, researchers have come out with smart car parking prototypes. These still face difficulties, however, in manipulating the data received from sensors, including the fusion and filtering of data. "Data fusion" refers to the accumulation and consolidation of data from many sensors. "Data filtering" refers to a process that removes unused data transmitted bv the sensors, to reduce the delay time in data transmission. Both data fusion and data filtering remain open challenges for today's researchers.

Another major challenge lies in choosing appropriate algorithms, to process the information gathered. In addition, all of the node sensors require exceptional functionality, to prevent errors as much as possible.

In 2018, Wael Alsafery et al. [1] proposed a system wherein initially, the sensed data is obtained from various sensors distributed inside the parking lot and also outside, for on-street parking. The data obtained from the sensors is then manipulated and handled locally, with the assistance of IoT edge gadgets, thus making the system

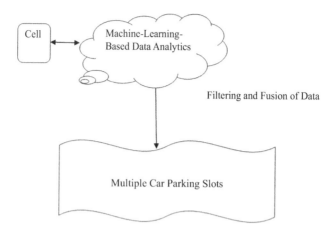

FIGURE 3.1 General IoT parking system.

work in real time, because it continuously receives data from the sensors. This cumulative sensor data is analyzed using artificial intelligence (AI) algorithms, as shown in Figure 3.1, through pre-characterized conditions defined in the algorithms. This framework is easy to understand and consists of a versatile application (app), which encourages clients to effectively discover the closest empty vehicle leaving zone by declining conceivable traffic check through Google API, which gives a constant perusal of traffic status.

Muftah Fraifer and Mikael Fernstrom [2] have proposed a cloud-based intelligent vehicle parking system for developed cities. The entire arrangement consists of three main layers: the sensor, communication, and application layers. The server searches for the finest vacant zone, based on the user's preference. After searching, the server sends the user the route that he or she has to drive, to arrive at that vacant parking zone. The prime idea of the paper is to use the CCTV camera systems already installed. Initially, the authors aimed to make the existing CCTV cameras smarter, but later, they proposed a low-cost and reliable model of an intelligent parking control system, using an embedded micro-web server, with IP connectivity for remote evaluation and monitoring. This system has improved reliability in finding the preferred parking zone without wasting time.

Vaibhav Hans et al. [3] have proposed a smart parking system that is very convenient for the users. Their main objective is to find nearby vacant parking zones and allocate them at the entry point itself, identifying the vehicle through a mobile app. They have designed their system in such a way that payment for the parking can be processed through an online transaction. Also, they give first preference to aged/physically challenged people, so that a parking zone can be reserved at the earliest point in time and near the elevator. They use image processing techniques to recognize the license plate, to identify the vehicle. IBM Blue Mix applications have used for cloud storage [4].

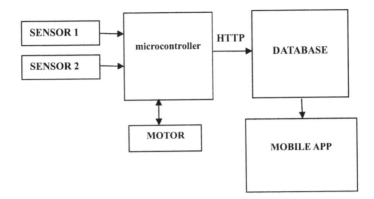

FIGURE 3.2 IoT-based sensor-empowered intelligent parking system.

Abhirup Khanna and Rishi Anand [5] have proposed an IoT-based intelligent vehicle parking system, which uses a mobile app to extract information through sensors about available parking spaces and send the information to a cloud platform in real time, as shown in Figure 3.2. Ultrasonic sensors were used to determine the availability of parking spaces. This information was sent to the cloud through a Wi-Fi chip attached to the Raspberry Pi, which acted as a gateway to the cloud, using the Message Queuing Telemetry Transport Protocol (MQTT) protocol. This is a lightweight data-centric protocol. It is 93 percent faster than the Hypertext Transfer (HTTP) protocol. Thus, by using the mobile app, a user can see vacant parking locations. But the execution charge of this framework is very high, because each sensor is connected to the Wi-Fi chip.

B. M. Mahendra et al. [6] have proposed an IoT-based sensor-empowered smart car, as shown in Figure 3.3. The proposed system is customer authenticated. It also excludes the false charges that happen when a customer parks a car in a zone other than the one he or she reserved. The overall implementation cost is less than for some other systems, since the application uses low-cost IR sensors, which proves to be advantageous.

Yanxu Zheng et al. [7] have proposed a parking vacancy forecast system, wherein each parking slot is sensor enabled, to forecast the standby time based on such parameters as temperature, humidity, weather, day and time, etc. Using these parameters, the standby time is forecast by algorithms such as a regression tree, support vector regression, and neural network [8,9,10].

Juan Rico et al. [11] have proposed a simple parking system, using the framework information of the city. The proposed structure allocates four parking conditions: vacant parking zone, booked parking zone, in-use parking zone, and load/unload parking zone. Here, the amount due can be paid wirelessly through near-field communication technology. The four parking conditions can be done by using geomagnetic sensors, which help to detect the presence of a vehicle. The main drawback of geomagnetic sensor-based vehicle occupancy detection is that the sensors' reactions are prone to magnetic intervention.

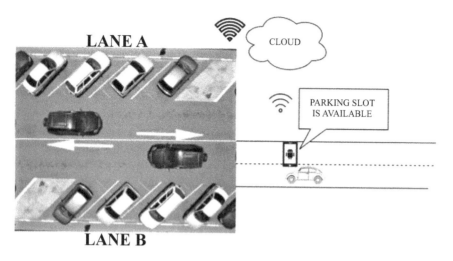

FIGURE 3.3 Overview of a smart parking system.

F. Zhou al. [12] have proposed an intelligent parking assortment. In this method, magnetic and ultrasonic sensors are used wirelessly, to detect the presence or absence of vehicles in the parking zones accurately. They have also described a customized version of the Min-Max algorithm, to discover vehicles using magnetometers.

R. Shyam et al. [13] have proposed a technique that utilizes the current structure of a cloud server, such that a boundless amount of space might be incorporated, with no change in the code. The developed mobile application (app) can run on Windows, Android, and iOS. In addition, the code can be utilized for multiple boards, making the proposed framework cost efficient, reliable, and resourceful.

P. S. Saarika et al. [14] have proposed a system wherein the sensor's data is collected at the fog controller (decentralized computing/edge computing), as shown in Figure 3.4. The collective data is sent to the edge devices that are close to the clients' area, to begin examining and taking care of information. This refined/processed information is then sent to the corresponding clients, to call attention to the adjacent vacant space least obstructed by traffic. The client also receives a response from the cloud showing the route with the least traffic, heading toward the parking space in the client's locality.

Burak Kizilkaya et al. [15] have proposed a system using the progressive placement algorithm with a binary search tree (BST) to formulate and identify the nearest possible locations. Initially, they locate parking zones nearby. Once the nearest parking space is discovered, it can be used in the hunt for empty parking spaces. This progressive technique with BST extensively improves the inquiry system, in terms of the required hunt time and energy effectiveness. The main aim of their proposed system is to reduce time in finding vacant parking zones. By utilizing a progression in placement algorithm, less time is expended in finding the nearby

Spy-Bot

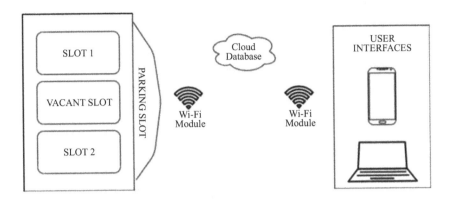

FIGURE 3.4 Overview of a smart transportation system.

vacant zone in a car park. A decrease in time while parking a vehicle involves less power and fuel usage and even creates less CO_2.

Felix Jesus Villanueva et al. [16] have proposed a framework using a magnetometer sensor, which is implanted in a large portion of the present cell phones. They proposed an idea where they designed an app that empowers individuals to share their sensor data, in exchange for receiving data on empty stopping zones, whenever needed.

3.2.2 Surveillance Systems

Surveillance cameras are widely accepted and acknowledged today and comprehensively utilized to detect unlawful activity and conduct inspections, such as in shopping malls, airports, bus and railway stations, individual assets, etc. The surveillance camera is a proficient development whose purpose can be properly utilized in all locations.

Nowadays, government agencies recommend and emphasize the importance of using these cameras to make the transportation system easy, as traffic details are reported on radio stations all the time [17]. Traffic's random conditions are well known to regular drivers, who would abstain from routes that frequently have more traffic obstructions. The major advantage in this regard is to convey safety measures to the public. Surveillance cameras also should to be correctly placed. This alleviates the problems of misbehavior and the burglary of vehicles in houses, strip malls, and cafés. [17]. The next sections present some of the methodologies previously adopted for smart vehicles used in surveillance.

Poonsak Sirichai et al. [17] have proposed a system wherein, during the early development phase of the car, the camera and radio receiver are deployed to ensure safety. Along with the main purpose of driving, this also serves the purpose of monitoring vehicle disasters, investigating offenders, and direction-finding. Figure 3.5 presents a block diagram of the proposed system. Using radio receivers, drivers can

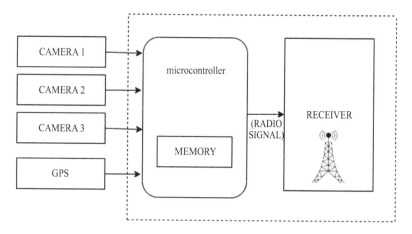

FIGURE 3.5 Block diagram of a smart car with a surveillance camera for native land protection.

also evade commuter traffic jams. Besides these, the information collected can be kept undisclosed until the examination is introduced by the administration authority required for giving reasonable judgment between two gatherings. In addition, alarms indicating suspected wrongdoing can be added to the framework, which is helpful for safety networks, homeland security, and the insurance business.

Chein-Hung Chen et al. [18] have proposed a smart camera that is a split mobile cloud model for deep learning, capable of shooting and processing videos. It then refines the vital information from the video clips and sends it to the cloud platform. The authors also showcased the usage of the Git protocol in portable cloud design.

In their work, Sirichai et al. [19] propose a technique known as a vehicular ad hoc network (VANET), as shown in Figure 3.6. VANET is an intelligent network technology, where each car acts as a mobile node to share information, such as road status and traffic congestion, without any central access point as a safety measure. Thus, mobile vehicles can assist transportation safety by avoiding congestion in traffic systems through VANET technology. In the future, VANET technology can help capture pictures that can be recorded, analyzed, and utilized for transportation security and broadcasting purposes. Furthermore, for security purposes, the transmitted pictures can be encrypted through algorithms, which can be accessible only by the authorities.

As the literature shows, the data or images communicated over a wireless environment can be made available remotely, over a cloud platform. Then, these datas or images are processed using various algorithms, either locally or in the cloud environment. However, the problem or gap in these treatments is how far the data are secure. That is the billion-dollar question to be answered.

This research gap applies to how securely our data/images are available in cloud platforms. Our focus is to have secure data communication in a wireless environment. Data security is the need of the hour. Protecting information from malicious

Spy-Bot

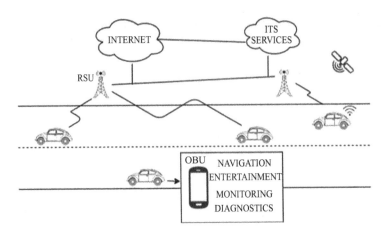

FIGURE 3.6 Structure of VANET.

attacks in networks or by data hackers, both on the server side and the client side, is the prime concern.

Our main focus is to develop a surveillance robot (spy-bot), whose movement can be controlled remotely and which captures images that are stored and wirelessly communicated over a secured Windows 10 Azure cloud platform. This platform provides platform-as-a-service (PaaS) cloud computing services, which allows customers to develop, run, and manage applications without the complexity of developing and maintaining the infrastructure. Azure cloud has built-in security services that include unparalleled security intelligence to help in identifying rapidly evolving threats at an early stage [29].

3.3 SPY-BOT: WI-FI CONTROLLED ROBOTIC CAR

The development of a robotic car plays a vital role in many areas. The main aim of the proposed system is to design a Wi-Fi controlled robotic car (spy-bot) that can be used for surveillance applications. A complete description of the block diagram, the flow chart, and the working of the proposed system are discussed in the following sections.

3.3.1 BLOCK DIAGRAM OF WI-FI CONTROLLED ROBOTIC CAR

A 12v lithium-ion battery supplies power to the spy-bot. The spy-bot has three main sections:

1. Webpage development (HTML script)
2. Windows 10 Azure cloud platform
3. Moving the robotic car

The Raspberry Pi acts as the brain behind the spy-bot. After the Raspberry Pi (Raspbian OS boots) is initialized, it waits for input commands from the webpage. The authenticated user controls the various movements of the spy-bot by issuing various commands through a webpage to the Raspberry Pi. All the commands (Left/Right/Forward/Backward/Stop) are issued wirelessly through Wi-Fi, which is based on the IEEE 802.11 family.

Geared DC motors are used to direct the movement of the spy-bot. Four DC motors are used in the proposed system to control the different directions, such as left, right, forward, and stop. DC motors cannot be interfaced directly with the Raspberry PI, due to a mismatch in the current and voltage ratings. So, we require motor driver ICs to drive the DC motors. The L298N motor driver IC [28] is a motor driver IC that drives the DC motor with respect to the commands issued by the Raspberry Pi. In our prototype, we require two motor driver ICs. With the help of two driver ICs, we can control four DC motors, as shown in Figure 3.7. The Raspberry Pi directs the user commands from the webpage to control the required set of motors, which directs the movement of the surveillance bot.

The spy–bot has four digital cameras, with built-in Complementary Metal Oxide Semiconductor (CMOS) sensors. CMOS technology has the great advantages of blooming and smear performance. Choosing the right camera for our application is also a challenging task, because cameras capture a lot of real-time pictures and also stream live videos for surveillance systems. In the proposed system four digital

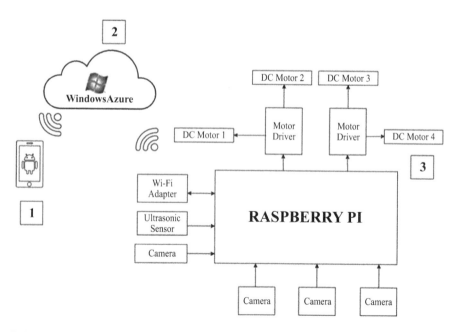

FIGURE 3.7 Overall block diagram of the proposed system.

Spy-Bot 47

cameras are used, to capture images in all directions, a most efficient way to capture images with a 360° view. The captured videos or images are stored in a secure cloud platform, for retrieval whenever needed for an investigation.

3.3.2 MAJOR COMPONENTS USED

The following are the various hardware and software components that have been used in the proposed system:

1. Raspberry Pi 2
2. Digital web camera with CMOS sensors
3. Geared DC motors
4. Dual full-bridge L298N motor driver ICs
5. Ultrasonic sensor
6. Microsoft Azure cloud

3.3.3 RASPBERRY PI 2

The proposed system uses Raspberry Pi 2. Raspberry Pi 2 has attractive features, such as a 900 MHz clock speed, four USB ports, 1 GB RAM, a camera interface connector (CSI), etc.

One of the most promising features for system reliability is the clock speed. Table 3.1 compares various versions of the Raspberry Pi, based on their clock speed, RAM memory, the processor used, and the USB ports. Raspberry Pi 2 does better when compared to the other versions. Raspberry Pi 2 promises 900 MHz processing speed, which is faster than the other Pis and makes the system work more reliably in real time [27]. Furthermore, the Pi 2 also has three additional USB ports when compared with the other Pis. This is very important, because we have to connect four digital webcams to the Raspberry Pi, which requires four USB ports. Hence, most interestingly, among the various versions, the Raspberry Pi 2 is preferred for our wireless surveillance system.

TABLE 3.1
Comparison of Raspberry Pi Versions

	Raspberry Pi 2	Raspberry Pi A+	Raspberry Pi B
Processor	Broadcom BCM2836-32bit ARM v7 Processor	Broadcom BCM2835-32 bit ARM1176JZF-S Processor	Broadcom BCM2835- 32 bit ARM1176JZF-S Processor
Processor Speed	900 MHz	700 MHz	700 MHz
RAM	1 GB	256 MB	512 MB
USB Ports	4	1	2 × USB 2.0

3.3.4 Digital Web Camera

A driverless PC webcam with 5.0 mega pixels, USB 2.0, and microphone is used. Four digital web cams that stream videos or images in real time are interfaced with the Pi through the USB slots. The camera can be used in such a way that whenever an image or video is captured, it can be either viewed or stored in the computer. It can also be sent to the other users, via internet, in the form of mail. These captured videos or images can also be stored in the cloud for future reference. This also has a microphone that can be used to capture videos with sound. It has a special feature for face detection that is used to detect human faces in order to offer sharp focus automatically. These computerized web cameras can be utilized for security reconnaissance, PC vision, video broadcasting, and recording social videos.

Thus, these digital web cameras are very compatible for the proposed system, in terms of picture or video quality, memory, etc. In the proposed system, four digital web cameras are used to capture the surrounding images.

3.3.5 Geared DC Motor

A high-quality, low-cost 100 RPM Center Shaft Economy Series DC geared motor is used in the proposed system. This geared DC motor has a long lifetime, because it has metal gears and pinions. Despite the fact that the motor gives 100 RPM at 12 V, the motor turns smoothly from 4 V to 12 V and gives a wide scope of RPM and torque. For the proposed system, four DC motors are used in order to control four wheels in the robotic car. The course of the DC motor turn can be controlled by the user, by giving either a high or a low voltage to the GPIO pins of the Raspberry Pi, where the driver module L298N is connected. This L298N driver module is used to interface the DC motors with Raspberry Pi. The experimental conditions for the DC motor are described in Table 3.2.

3.3.6 Dual Full-Bridge L298N Driver

The L298N utilized is a dual full-bridge motor driver that permits controlling the speed and bearing of two DC motors at the same time. It has a high voltage and a high ebb and flows double full-connect drivers intended to acknowledge standard TTL logic levels and drive inductive loads, for example, relays, solenoids, DC, and stepping motors. These driver modules are utilized to interface the DC motor with the

TABLE 3.2
Control Pattern of DC motor

Input A	Input B	Motor Condition
0	0	Break or Stop
0	1	Anti-Clockwise Movement
1	0	Clockwise Movement
1	1	Break or Stop

Spy-Bot

Raspberry Pi, in light of the fact that the yield flow from the GPIO pins of the Raspberry Pi is insufficient to drive the motor. Thus, these driver modules are used, which can give an output current that is sufficient to drive a motor. For the proposed system, it is therefore enough to use two dual full-bridge L298N driver modules to drive the required four DC motors.

3.3.7 Ultrasonic Sensors

Ultrasonic sensors [20] are used to find the presence of any obstacle in the pathway that the robotic car explores. In the proposed system, only one ultrasonic sensor is used. This sensor finds the presence of the obstacle using ultrasonic signals. The ultrasonic sensors emit ultrasonic waves. When these waves hit any obstacle, it is reflected back. In this manner, the separation between the sensor and the obstruction is estimated. The estimation time is calculated between the discharge and the gathering of the ultrasonic waves.

3.3.8 Microsoft Azure (Windows Cloud Platform)

Gartner [21] predicts a 38 percent growth in the usage of cloud platforms in the coming years. Forty Clouds [21] reviewed the security abilities of the top five IaaS suppliers—Amazon Web Services (AWS), Google Cloud Platform (GCP), IBM Cloud, Rackspace, and Azure—with a particular spotlight on information and system security, as well as identity and access management (IAM). A comparison report is shown in Figure 3.8.

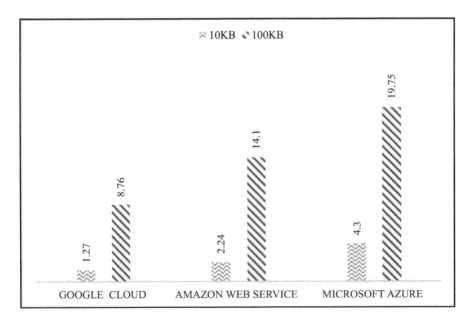

FIGURE 3.8 Comparison of various cloud platforms (AWS, Azure, Google Cloud, and IBM Cloud).

Microsoft's Azure Blob Storage has topped Amazon's S3 and Google Cloud Storage in a progression of benchmark tests performed by NAS and capacity organization Nasuni [22]. In their 2015 white paper [22], *The State of Cloud Storage*, Nasuni revealed that for the second year straight, Microsoft topped tests that measure usefulness, cost, and execution, as shown in Figure 3.9. Amazon took the second spot, and Google finished a distant third. Microsoft Azure is an open cloud administration verified stage. It bolsters a wide assortment of working frameworks, programming dialects, and gadgets. It can go about as work back-closes for Windows, iOS, and Android-bolstered gadgets. One of the top motivations to utilize Azure for the proposed framework is to exploit its broad range of security instruments and abilities. Additionally, it offers the capacity to control them to modify security to meet what is needed to design an application. These devices and capacities help make it conceivable to make secure arrangements on the protected Azure stage. Microsoft Azure gives secrecy, trustworthiness, and accessibility to client information, while empowering the responsibility of the client.

For this very reason, Microsoft Azure is used in the proposed system, because it is very important that the images and videos captured be stored in a secure environment, since they are often used for surveillance purposes. Thus, Microsoft Azure acts as the best cloud platform, in terms of system security.

3.4 FLOW DIAGRAM

The overall flow diagram is shown in Figure 3.10. Once the Raspberry Pi boots, the digital camera interfaced with the Pi is initialized and starts to capture the required images. The movement of the spy-bot awaits commands from the user, made through the webpage, which guides the bot to remotely capture images or livestream videos and stores them in the cloud platform for future investigation.

3.5 EXPERIMENTAL CONDITIONS

Once the Raspbian Pi is initialized, it waits for input commands through the webpage. But the webpage allows only authenticated users to log in and take control of the spy-bot. This restricts unauthorized users from accessing the Spy-Bot. As soon as a user receives authentication, the user continues to the webpage, with button icons representing the commands that control the spy-bot and screen-space to view images or stream live videos, as shown in Figure 3.11.When the user directs commands through the webpage, they are received by the PHP client (spy-bot) through a byte array over a socket connection and accordingly control the direction in which the spy-bot moves.

If the ultrasonic sensor detects any obstacle in its path, image capturing sends the information to the Raspberry Pi. The Raspberry Pi then directs the DC motors accordingly, to avoid any collision locally, on the client end. The Python programming in the Raspberry Pi provides control signals to the DC motor whenever it receives a command from the webpage, and also autonomously avoids collision locally, in the client end. Each DC motor has two wires for its control. The control pattern of the DC motor is shown in Table 3.2.

Spy-Bot

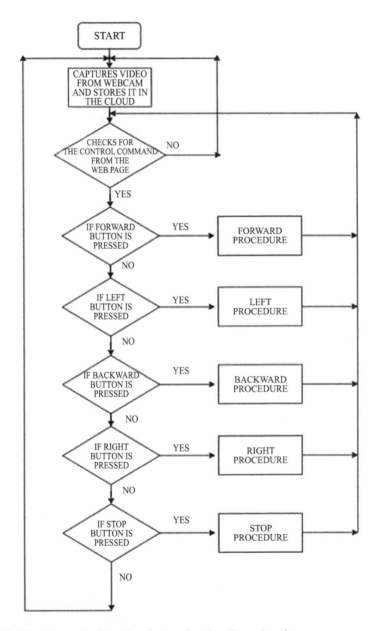

FIGURE 3.9 Azure cloud tops the chart on functionality and performance.

The proposed system uses five control commands to control the direction of the spy-bot. They are forward, backward, right, left, and stop. Table 3.3 gives the values that have to be given to DC motors with the corresponding control commands, in order to move as directed by the user.

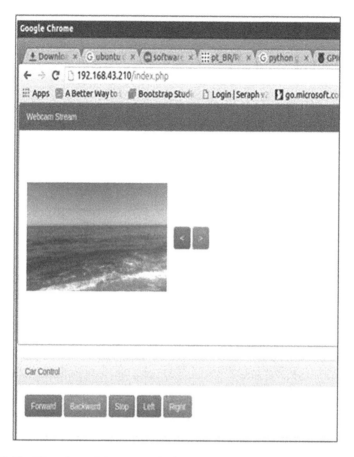

FIGURE 3.10 Flow chart of the proposed Wi-Fi-controlled robotic car.

Lagrange equations [38] for motion are expressed in following equations:

$$n_2 y_2 = a(y_1 - y_2) + b(y_1 - y_2) \tag{3.1}$$

$$n_1 y_1 = b(y_2 - y_1) + a(y_2 - y_1) - a_t(y_1 - q) \tag{3.2}$$

where y_1 and y_2 are the displacements of the wheel and motor, n_2 is the weight of the robot, a is the stiffness coefficient, b is the damping ratio, n_1 is the mass of the wheel and a_t is the equivalent stiffness coefficients. q is road roughness. The output can be rewritten with respect to Euler's equation,

$$y_1 = y_{10} e^{iwt + \varphi} \tag{3.3}$$

$$y_2 = y_{20} e^{iwt + \varphi} \tag{3.4}$$

Spy-Bot 53

FIGURE 3.11 Proposed webpage to control the robotic car.

TABLE 3.3
Experimental Conditions for DC motor

CONDITION	DC MOTOR 1	DC MOTOR 2	DC MOTOR 3	DC MOTOR 4
FORWARD	HIGH	LOW	HIGH	LOW
BACKWARD	LOW	HIGH	LOW	HIGH
RIGHT	HIGH	LOW	LOW	LOW
LEFT	LOW	LOW	HIGH	LOW
STOP	LOW	LOW	LOW	LOW

The webpage design is one of the prime elements in the development of our prototype. The HTML scripting language is used to design our webpage, a platform through which we can access the spy-bot. To receive the images taken by the Spy-Bot in the byte array format, a webcam client is used at the client slide. This enables images to be displayed in the webpage at a rate of 10 to 15 images per second, so that they can be seen as a continuous video. The viewing angle of the camera can also be altered by using the predefined arrow buttons in the webpage, i.e., changing the direction in which the video is streamed. These videos are streamed and recorded simultaneously. The recorded videos are uploaded to the cloud for future reference.

3.6 RESULTS AND DISCUSSION

The proposed spy-bot system was successfully implemented and tested in real time. The overall set up and hardware implementation of the spy-bot are explained in the following sections. The important building blocks of the wireless robotic car are the

54 Blockchain Technology for Data Privacy Management

Raspberry Pi, the camera, DC motors, and the Windows Azure platform. If the movement functionalities are increased, then the design complexity of the car also increases. So, much effort has been taken to design a prototype of a spy-bot that allows end-users to instruct the vehicle. Two major contributions are involved in this work: the design of intelligent controllers and the interconnection of cloud platforms to access live video inputs.

3.6.1 Hardware Implementation

The overall hardware setup of the wireless robotic car (spy-bot) is shown in Figure 3.12. The figure shows that there are four digital web cameras, with four geared DC motors, used in the proposed system. The Raspberry Pi act as a brain for the whole setup. 360^0 of rotation are possible through the DC motors. Forward and backward actions are controlled by the Raspberry Pi. As soon as the spy-bot is powered, it is connected to the webpage client through the TCP protocol to livestream videos in the webpage by providing the static IP address of the Raspberry Pi. It also initiates to capture the surrounding images, and records and stores them in the secure Azure cloud. The spy-bot travels, according to commands given by the user in the webpage. Authenticated users can control the spy-bot remotely through the webpage. The live images and videos are accessed via an encrypted cloud platform.

3.6.2 Results

The Raspbian OS gets booted up, and all the required drivers are installed, before the Python program is dumped into the Raspberry Pi. The Python script runs in the

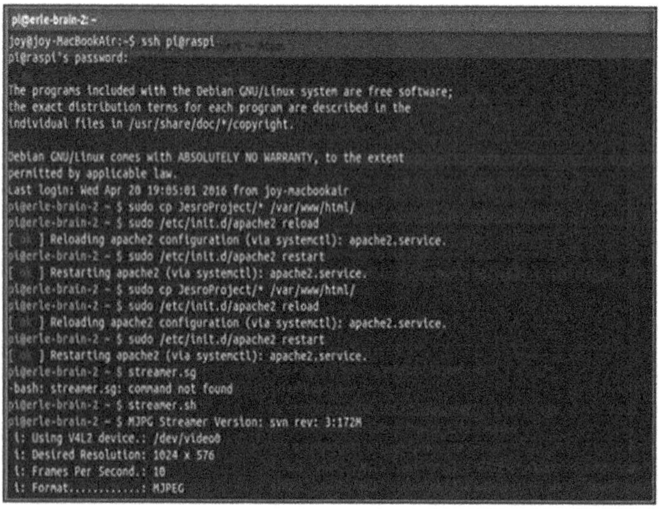

FIGURE 3.12 Hardware setup of the proposed system.

Spy-Bot 55

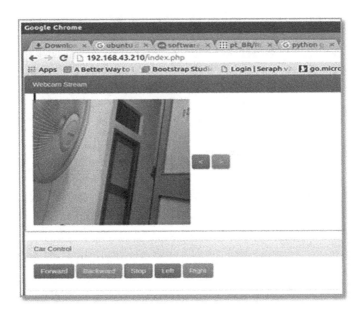

FIGURE 3.13 Raspberry Pi console window.

console window, as shown in Figure 3.13. The Apache server service is reloaded. It reboots when it is done. The streamer.sh file is uploaded in the MPEG format, with a resolution of 1024x576, 10 frames per second.

Figure 3.14 shows the webpage where the control commands are given by the user. It acts as input for the Raspberry Pi, to control the direction of the robotic car. The webpage has five buttons to control the direction of the robotic car and two buttons to control the camera's viewing angle. The five buttons that are used to control the robotic car are forward, backward, left, right, and stop. Thus, the forward button is used to start the vehicle, which moves, followed by the other directions. The buttons with a left and right arrow mark, seen in the left side of the display, are used to control the camera directions to livestream video.

3.7 CONCLUSION

This chapter presents the prototype of a spy-bot, which is Wi-Fi controlled for surveillance applications. The prototype was implemented and tested successfully with real-time images streaming over an Azure cloud storage platform. The spy-bot can be compliant for military purposes in remotely monitoring intrusions along the border. Compared to previous work, our system is more reliable and secure, with the advantage of four digital web cameras that can gather information in all four directions. Also, our proposal has the chance to avert major catastrophes, by tracking and monitoring enemies' plans and allowing authorities to act swiftly. Other major applications include health care monitoring, where it assists doctors in diagnosing patients remotely.

FIGURE 3.14 Webpage design with direction controls for the proposed spy-bot.

3.8 FUTURE DEVELOPMENT

As a future development, more features could be added to the wheel movement, in order to increase the total efficiency of the spy-bot. The device can be augmented with machine learning features, to build next-generation applications. By adding a few more applications, these spy-bots could also be used to map catastrophe areas, for example, after they have been affected by an earthquake. These robotic cars provide the best research platform, as they are powered with LINUX and the Robot Operating System (ROS), with infinite options. As a whole, the spy-bot will be the best system to perform remote surveillance applications with improved interconnections and cost-effective components.

REFERENCES

1. Alsafery, W., Alturki, B., Reiff-Marganiec, S., Jambi, K., 2018. *Smart car parking system solution for the Internet of Things in smart cities*. In *International Conference on Computer Applications & Information Security (ICCAIS)*. Riyadh, Saudi Arabia: IEEE. doi: 10.1109/CAIS.2018.8442004
2. Fraifer, M., and Fernstrom, M., 2016. *Smart car parking system prototype utilizing CCTV nodes: A proof of concept prototype of a novel approach yowards IoT-concept based smart parking*. In *IEEE 3rd World Forum of Internet of Things* (pp. 649–654). Reston, VA: IEEE. doi: 10.1109/WF-IoT.2016.7845458.
3. Hans, V., Sethi, A. S., and Kinra, J., 2015. *An approach to IoT based car parking and reservation system on cloud*. In *International Conference on Green Computing and Internet of Things* (pp. 352–354). Noida, India: IEEE. doi: 10.1109/ICGCIoT.2015.7380487.

4. Husni, E., Hertantyo, G. B., Wicaksono, D. W., Candrasyah Hasibuan, F., Rahayu, A. U., and Triawan, M. A., 2016. *Applied Internet of Things: Car monitoring system using IBM Blue Mix.* In *International Seminar on Intelligent Technology and Its Application* (pp. 417–422). Lombok, Indonesia: IEEE doi: 10.1109/ISITIA.2016.7828696.

5. Khanna, A., and Anand, R., 2016. IoT based smart parking system. In *International Conference on Internet of Things and Applications* (pp. 266–270). Pune, India: IEEE. doi: 10.1109/IOTA.2016.7562735.

6. Mahendra, B. M.,Sonoli, S., Bhat, N., and Raghu, R., 2017. *IoT based sensor enabled smart car parking for advanced driver assistance system.* In *IEEE International Conference On Recent Trends in Electronics Information & Communication Technology* (pp. 2188–2193). Bangalore, India: IEEE. doi: 10.1109/RTEICT.2017.8256988.

7. Zheng, Y., Rajasegarar, S., and Leckie, C., 2015. *Parking availability prediction for sensor enabled car parks in smart cities.* In *IEEE International Conference on Sensor Networks and Information Processing (ISSNIP)* (pp. 1–6). Singapore: IEEE. doi: 10.1109/ISSNIP.2015.7106902 .

8. Galicia, A., Talavera Llames, R., Troncoso, A., Koprinska, I., and Martínez-Álvarez, F., 2018. Multi-step forecasting for big data time series based on ensemble learning. *Knowledge-Based Systems.* 163, pp. 830–841. https://doi.org/10.1016/j.knosys.2018. 10.009.

9. Andrushia, A. D., and Thangarajan, R., 2017. An efficient visual saliency detection model based on ripplet transform. *Sadhana-Academy Proceedings in Engineering Sciences*, 42(5), pp. 671–685.

10. Andrushia, A. D., and Thangarajan, R., 2020. RTS-ELM: An approach for saliency-directed image segmentation with ripplet transform. *Pattern Analysis and Applications*, 23, pp. 385–397.

11. Rico, J., Sancho, J., Cendon B., and Camus, M., 2013. *Parking easier by using context information of a smart city: Enabling fast search and management of parking resources.* In *IEEE International Conference on Advanced Information Networking and Applications Workshops (WAINA)*, (pp. 1380–1385). Barcelona, Spain: IEEE. doi: 10.1109/WAINA.2013.150.

12. Zhou, F., and, Li, Q., 2014. *Parking guidance system based on ZigBee and geomagnetic sensor technology.* In *IEEE International Symposium on Distributed Computing and Applications to Business, Engineering and Science* (pp. 268–271). Xian Ning, Hubei, China: IEEE.

13. Shyam, R., and Nrithya, T., 2017. *Cloud connected smart car park. IEEE International Conference on I-SMAC (IoT in Social, Mobile, Analytics and Cloud)* (pp. 71–74). Palladam, India: IEEE. doi: 10.1109/I-SMAC.2017.8058269.

14. Saarika, P. S., Sandhya, K., and Sudha, T., 2017. *Smart transportation system using IoT. In IEEE International Conference on Smart Technologies for Smart Nation.* (pp. 1104–1107). Bangalore, India: IEEE. doi: 10.1109/SmartTechCon.2017.8358540.

15. Kizilkaya, B., Caglar, M., Al-Turjman, F., and Ever, E., 2019. Binary search tree based hierarchical placement algorithm for IoT based smart parking applications. *Internet of Things*, 5, pp. 71–83 .

16. Villanueva, F. J., Villa, D., Santofimia, M. J., Barba, J., and Lopez, J. C., 2015. *Crowd sensing smart city parking monitoring.* In *IEEE Second World Forum on Internet of Things.* Milan, Italy: IEEE. doi: 10.1109/WF-IoT.2015.7389148

17. Sirichai, P., Kaviyaa, S., and Yupapin, P. P., 2010. Smart car with security camera for homeland security. *Procedia-Social and Behavioral Sciences*, 2(1) pp. 58–61.

18. Chen, C.-H., Lee, C.-R., and Lu, C.-H., 2017. Smart in-car system using mobile cloud computing framework for deep learning. *Vehicular Communications*, 10, pp. 84–90.

19. Sirichai, P., Kaviya, S., Fujii, Y., and Yupapin, P. P., 2011. Smart car with security camera for road accidence monitoring. *Procedia Engineering*, 8, pp. 308–331.
20. Panda, K.-G., Agarwal, D., Nshimiyimana, A., and Hossain, A., 2016. Effects of environment on accuracy of ultrasonic sensor operates in millimetre range. *Perspectives in Science*, 8, pp. 574–576.
21. Edwards, J., 2015, May 27. How secure is your IAAS? Compare the top 5 CSP's security. https://solutionsreview.com/cloud-platforms/how-secure-is-your-iaas-compare-the-top-5-csps-security. Accessed on May 16, 2020.
22. Edwards, J., 2015, May 26. Microsoft beats AWS Google on cloud storage benchmark test. https://solutionsreview.com/cloud-platforms/microsoft-beats-aws-google-on-cloud-storage-benchmark-test/. Accessed on May 16, 2020.
23. Gartner Inc., 2017, February 7. Gartner says 8.4 billion connected "Things" will be in use in 2017, up 31 percent from 2016. https://www.gartner.com/en/newsroom/press-releases/2017-02-07-gartner-says-8-billion-connected-things-will-be-in-use-in-2017-up-31-percent-from-2016. Accessed on May 17, 2020.
24. Kehoe, B., Patil, S., Abbeel, P., and Goldberg, K., 2015. A survey of research on cloud robotics and automation. *IEEE Transactions on Automation Science and Engineering*, 12(2), pp. 398–409.
25. Analytics Vidyha, 2016, August 26. 10 real world application of Internet of Things (IoT): Explained in videos. https://www.analyticsvidhya.com/blog/2016/08/10-youtube-videos-explaining-the-real-world-applications-of-internet-of-things-iot/. Accessed on May 17, 2020.
26. Calder, B., 2011, November 21. SOSP paper–Windows Azure storage: A highly available cloud storage service with strong consistency. https://azure.microsoft.com/en-in/blog/sosp-paper-windows-azure-storage-a-highly-available-cloud-storage-service-with-strong-consistency/. Accessed on May 17, 2020.
27. Elementzonline, 2016, March 1. Raspberry Pi 3 model B: Wireless Pi released! https://elementztechblog.wordpress.com/2016/03/01/raspberry-pi-3-model-b-wireless-pi-relaesed/. Accessed on May 17, 2020.
28. Saifur Rahman Faisal, S. M., Ahmed, I. U., Rashid, H., Das, R., Karim, M., and Taslim Reza, S. M., 2017. *Design and development of an autonomous floodgate using arduino uno and motor driver controller.* In *2017 4th International Conference on Advances in Electrical Engineering (ICAEE)* (pp. 276–280). Dhaka, Bangladesh: IEEE. doi: 10.1109/ICAEE.2017.8255366
29. Microsoft Azure, n.d. Security: Strengthen the security of your cloud workloads with built-in services. https://azure.microsoft.com/en-in/product-categories/security/. Accessed on May 17, 2020.
30. Varshney, T., Sharma, N., Kaushik, I., Bhushan, B., 2019. *Architectural model of security threats & their countermeasures in IoT.* In *International Conference on Computing, Communication, and Intelligent Systems (ICCCIS)* (pp. 424–429). Noida, India: IEEE. https://doi.org/10.1109/ICCCIS48478.2019.8974544.
31. Tiwari, R., Sharma, N., Kaushik, I., Tiwari, A., and Bhushan, B., 2019. *Evolution of IoT & Data Analytics Using Deep Learning. International Conference on Computing, Communication, and Intelligent Systems (ICCCIS)* (pp. 418–423). Noida, India: IEEE. https://doi.org/10.1109/ICCCIS48478.2019.8974481.
32. Arora, A., Kaur, A., Bhushan, B., Saini, H., 2019. *Security concerns and future trends of Internet of Things.* In *International Conference on Intelligent Computing, Instrumentation and Control Technologies (ICICICT)* (pp. 891–896). Kannur, Kerala, India: IEEE. https://doi.org/10.1109/ICICICT46008.2019.8993222.

33. Lin, J., Yu, W., Zhang, N., Yang, X., Zhang, H., and Zhao, W., 2017. A survey on Internet of Things: Architecture enabling technologies security and privacy and applications. *IEEE Internet of Things Journal*, 4(5), pp. 1125–1142.
34. Ullah, S., Mumtaz, Z., Liu, S., Abubaqr, M., and Madni, A. M. H. A., 2019. Single-equipment with multiple-application for an automated robot-car control system. *Sensors*, 19, p. 662. https://doi.org/10.3390/s19030662.
35. Budheliya, C. S., Solanki, R. K., Acharya, H. D., Thanki, P. P., and Ravia, J. K., 2017. Accelerometer based gesture-controlled robot with robotic arm. *International Journal for Innovative Research in Science & Technology*, 3, pp. 92–97.
36. Ankit, V., Jigar, P., and Savan, V., 2016. Obstacle avoidance robotic vehicle using ultrasonic sensor, Android and Bluetooth for obstacle detection. *International Research Journal of Engineering and Technology*, 3, pp. 339–348.
37. Zha, X., Ni, W., Zheng, K., Liu, R. P., and Niu, X., 2017. Collaborative authentication in decentralized dense mobile networks with key pre distribution. *IEEE Transactions on Information Forensics and Security Information*, 99, p. 1.
38. Ning, M., Xue, B., Ma, Z., Zhu, C., Liu, Z., Zhang, C., Wang, Y., and Zhang, Q., 2017. Design, analysis, and experiment for rescue robot with wheel-legged structure. *Mathematical Problems in Engineering*, 2017. https://doi.org/10.1155/2017/5719381.

4 Multi-Antenna Communication Security with Deep Learning Network and Internet of Things

Garima Jain
Swami Vivekanand Subharti University, India

Rajeev Ranjan Prasad
Ericsson Global Services Pvt. Ltd., India

CONTENTS

4.1 Introduction: Background 61
4.2 Literature Survey 63
4.3 Internet of Things (IoT) with Multi-Antenna Beam 65
    4.3.1 Radio Frequency Identification System (RFID) 66
    4.3.2 Sensor Technology 66
    4.3.3 Smart Technology 67
    4.3.4 Nanotechnology 68
4.4 Deep Learning with Multi-Antenna Beam 68
4.5 Multi-Antenna Beam with the IoT and Deep Learning 69
4.6 Mathematical Background 74
4.7 Security Assessment 75
    4.7.1 TCP Scan 76
    4.7.2 SYN Scan 76
    4.7.3 FIN Scan 76
4.8 Conclusion and Future Scope 77
References 78

4.1 INTRODUCTION: BACKGROUND

Multiple-Input Multiple-Output (MIMO) technology increases the limit of a radio connection utilizing various transmissions and arranging reception antennas to feed multipath propagation. MIMO has gotten a fundamental component of remote correspondence principles, including IEEE 802.11n (Wi-Fi), IEEE 802.11ac (Wi-Fi), HSPA+ (3G), WiMAX (4G), and Long-Term Evolution (4G LTE). In current usage, MIMO explicitly refers to a main method for sending and receiving more than one information signal all the while over similar radio channels by exploiting multipath broadcasts. MIMO is, at a very basic level, not the same as the smart antenna strategies created to improve the performance of a signal data signal, for example, beam formation, and diversity. In the area of wireless communication, the MIMO framework plays an important role. The development of multi-antenna selection was proposed to reduce the MIMO system and equipment costs, while keeping the benefits of MIMO correspondence frameworks [1]. As of late, research has been initiated to focus on applying AI (ML) developments in a wide range of communications, because of the way that ML is capable of transferring a conventional network into information-driven networks to accomplish lower constant online computational complexity [2,3,4]. The Convolutional Neural Network (CNN) strategy for multi-antenna frameworks [5] is the newest benchmark of the telecommunications industry. In telecommunications, the base station is characterized as a radio transmitter, or a receiver, or an antenna that can be cast off in a mobile telecommunications network. The base station preserves the communication between the network and mobile users through a radio link. A radio link is a wireless connection between two nodes, or radio units, in a data network. Each radio unit consists of a transceiver and a highly directive antenna, typically operating at microwave frequencies in the range of 6–23 GHz. Depending on the frequency, the maximum communication range for a radio link varies from a few meters to hundreds of kilometers. With the rapid rate of 5G development activities, wireless multiple-input, multiple-output applications are taking advantage of the high data rate and bandwidth; the power saved compared to a single antenna element, which requires a great deal; link improvement, and multipath propagation [6,7]. The QoS prerequisites in 5G and its normalization issues, and the effect of the reconciliation of 5G with the IoT and AI [8], are still under discussion. It has already been proven that beams can be reconfigured to include extensive beams, with respect to various antennas and beam designs. The half-a-wavelength inter-element spacing is used by MIMO. The antenna system framework has been generally utilized in numerous application fields [9].

For multi-antenna telecommunications systems, the base station uses limited protocols on the sender side for encoding, while the receiving end uses protocols to decode the message. MIMO antenna systems are among the most promising solutions to fulfill the demands of upcoming fifth-generation (5G) wireless mobile communication systems [10]. On the sender side, to start any kind of sending operation, the sender opens the host and port. If they are not available, then the process is terminated on both the host and the port sides. Once the host and port are available, then the string from the database is read by the reader. There are several receiving processes that occupy the value read from the database. Once this task is completed, the host and the port are closed. At the receiver end, the receiver's local port is opened and starts the message receiving operation, using several routine processes. Once these routine processes receive them, the messages are stored in the database. An acknowledgement is sent to the sending operation that the messages have been received.

Multi-Antenna Communication Security with Deep Learning Network   63

A GOB command originates with a data structure that consists of a couple of fields, a slice, a map, and even a pointer directed toward itself. When it is being executed, the gob package drives through the network without a problem.

It contains a rather ad hoc protocol, where the client and the server agree that a command is a string, followed by a new line, followed by data. For each command, the server should be aware of the exact data format and how to process that data. An optical message framework utilizes a transmitter, which encodes a correspondence into an optical sign channel, which conveys the sign to its goal, and a receiver, which repeats the message from the optical sign. When electronic hardware is not utilized, the receiver is an individual outwardly observing and deciphering a sign, which might either be straightforward (e.g., the nearness of a reference point fire), or complex (e.g., lights utilizing shading codes or flashed in a Morse code grouping). The GOB function uses a fixed beam that forms the table to generate responses to the UE's request. But due to the complexity of the architecture and the process, the GOB is not optimal. It cannot handle the user's request on time, and it goes through a great deal of computation followed during the control process to get the desired performance, carrier distribution, power consumption, downlink and uplink throughput, etc.

This chapter emphasizes the use of DL and the IoT with GOB to have better performance, the best carrier distribution, low power consumption, and better downlink and uplink throughput, etc. The DL model will directly control the base station to get the anti-optical beam (phase, amplitude, width) values. This concept will cooperate with the GOB function to choose most optical types, to achieve a faster response (cache the module suitable for different situations and quickly use them), more accurate data control (decrease or increase the data layers and beam with different sites and periods), and epically high power efficiency and low data consumption. This concept emphasizes having one gNodeB or eNodeB that can be a master node station, and other neighbor stations working as slave stations. They connect with each other with IoT standards. The master station connects to all slave station locations and the user's requests and then sends it to the DL system, which compares the GOB and DL optimized models and chooses the best one to control that location's beams as one's intelligent network. It will be lowered down the network complexity level and optimize the whole network.

The chapter is organized as follows. Section 4.2 provides a survey of the literature. Section 4.3 presents theory and concepts of the IoT with a multi-antenna beam and is used to analyze the algorithm on which our future implementation will be based. Section 4.4 discusses DL in a multi-antenna beam. Section 4.5 provides a detailed analysis of a multi-antenna beam with the IoT and DL and compares proposed models. Section 4.6 provides mathematical background. Section 4.7 provides a security assessment. Section 4.8 provides a conclusion and Section 4.9 explains the suggestions for further research work.

4.2 LITERATURE SURVEY

According to Kaisa Zhang et al. [11], with the improvement of remote gadgets and the expansion of portable clients, the administrator's center has moved from the development of the corresponding system to the activity and support of the system.

Administrators are anxious to know about the conduct of portable systems and the constant experience of clients, which requires utilizing authentic information to precisely anticipate future system conditions. Examining extensive information and figuring out which is generally received can be utilized as an answer. Nonetheless, difficulties remain in information examination and forecasting for portable system enhancement, for example, the practicality and precision of the expectations. The chapter recommends a traffic investigation and expectation framework that is reasonable for urban remote correspondence, organized by consolidating real call detail record (CDR) information examinations and multivariate forecast calculations. From that point forward, a causality examination is applied to corresponding information investigations. In view of causal examination, multivariate long momentary memory models are utilized to foresee future CDR information. Finally, the forecast calculation is utilized to process genuine information about various scenes in the city to check the display of the whole framework.

According to Alberto Mozo et al. [12], 90 percent of system traffic will experience a delay in telecommunications in the coming years In this setting, they propose the utilization of CNNs to gauge momentary changes in the measure of traffic crossing a server farm. This value is a marker of virtual machine action and can be used to shape the server farm framework likewise. The conduct of system traffic at the seconds scale is exceptionally tumultuous and thus, customary time-arrangement examination approaches, for example, the autoregressive incorporated moving normal (ARIMA) model, fail to acquire exact gauges. They show that our CNN approach can abuse the non-straight regularities of system traffic, furnishing huge enhancements for the mean total and standard deviation of the information, and beating ARIMA by a noteworthy edge, as the gauging granularity is over the 16-second goals. To expand the precision of the determining model, they approve their methodology with a broad arrangement of trials utilizing an informational index gathered at the central system of an internet service provider over a period of five months, totaling 70 days of traffic at the one-second goals.

According to Guiyang Yu et al. [13], switching powerful direct models is a generally utilized technique to depict change in an advancing time arrangement, where the exchanging model is an extraordinary case. The momentary determination of traffic streams is a fundamental piece of canny traffic frameworks (ITS). These authors apply the exchanging ARIMA model to a traffic stream arrangement. They have seen that the customary exchanging model is unseemly to portray the example that is evolving. Therefore, a variable of length is presented, and they utilize the sigmoid capacity to portray the impact of span to the likely progress of the examples. In view of the exchanging ARIMA model, an estimating calculation is introduced. They apply the proposed model to genuine information from UTC/SCOOT frameworks in Beijing's rush hour gridlock. The trials show that our proposed model is material and successful.

According to Yuri Hua et al. [14], a time-arrangement expectation can be summed up as a procedure that concentrates helpful data from verifiable records and afterward decides future qualities. Adapting long-go conditions, which are installed in time arrangements, is frequently an obstruction for most calculations. However, long-short-term memory (LSTM) arrangements, as a particular sort of plan in profound

Multi-Antenna Communication Security with Deep Learning Network

learning, are guaranteed to adequately overcome the issue. In the field of media transmission organization, the forecasting of traffic and client portability could straightforwardly profit by this development, as they influence a practical dataset to demonstrate that for RCLSTM, an expectation execution similar to LSTM is accessible, while extensively less calculating time is required.

According to Khadija Mkocha et al. [15], building traffic is at the core of media communications designing. History shows a cozy connection between the progressions in the two broadcast communications systems and their related design strategies. This review utilized a subjective report investigation to sequentially investigate the advancement of, and relationship between, traffic building and moving cell systems, from the 1990s to the present.

According to Mourad Nasri et al. [16], the basic essentials of an LTE broadband remote system are to develop internet applications to clients. This technique permits advancement designers to have precise ideas regarding the present nature of the administration given to clients and take proper actions to deal with system assets. Present mainstream versatile internet applications, such as gaming, voice administration, gushing, and long-range interpersonal communication applications, have various traffic models and, therefore, extraordinary QoS prerequisites. Finally, these authors examine some contextual investigations to show the capability of the ideal straight capacity in depicting the system QoS development design and the tuning of existing highlights.

In a 2019 article, Asad Arfeen et al. [17] present an investigation based on extensive information about two-route congestion in different connectivity and spine center connections and diagnostically display the reason internet traffic unites with Weibull appropriation, as traffic moves from access to center connections. Likewise, they show the adaptability of Weibull dispersion in catching the stochastic properties of internet traffic at the parcel, stream, and session levels in different access and spine center connections. The outcomes were supported by utilizing genuine traffic information wellness tests.

4.3 INTERNET OF THINGS (IOT) WITH MULTI-ANTENNA BEAM

With the help of the IOT, objects can make intelligent decisions and collaborate without human intervention [18]. IoT networks normally depend on antennas with fixed radiation qualities, which on a very basic level restrains the working distances and information rates of individual radio connections and limits the general network [19,20]. Multi-antennas transform IoT organization through cutting-edge computerized signal handling and radio wire beam-forming procedures. These advantages can be acknowledged without unsatisfactory capital use or progressions to IoT gadgets or radio conventions. Only IoT entryway terminals should be overhauled. Some smart antennas adjust their radiation qualities to changing sign situations. They make wireless beams that are consequently shown in a multi-antenna beam [21].

A multi-antenna is formed to remove obstacles that hinder the presentation of each radio connection. When conveyed at an IoT network entryway, a computerized smart antenna can make autonomous pillars that serve various IoT devices consistently and give helpful data on the signals received, for example, exact data about the

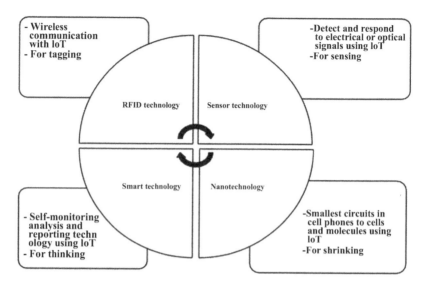

FIGURE 4.1 IoT enablers.

headings of the devices. Simultaneously, smart radio wires can improve the exhibition of remote IoT devices. The radio handset inside an IoT device is regularly the prevailing customer of energy. The capability utilized by the handset is straightforwardly identified with the installation of the radio wires on both the gadget and the door. The capacity of a smart antenna at the IoT entryway terminal to improve the radio connects to all gadgets, therefore, permits the transmission power and thus the energy used by each IoT device especially to be decreased. This could increase the battery life of an IoT gadget, which commonly works from batteries or with collected energy; however, it can likewise reduce the expense by using small batteries and a less innovative radio transceiver. Figure 4.1 shows a description of the IoT enablers for multi-antenna beams.

4.3.1 Radio Frequency Identification System (RFID)

A radio frequency identification system (RFID) is a remote procedure used with the IoT to label user data. An RFID framework is comprised of labels (transmitters/responders) and readers (transmitters/collectors). Figure 4.2 shows a typical RFID system involved of labels and readers. An RFID is often seen as a prerequisite for the IOT.

4.3.2 Sensor Technology

Sensors detect electrical or optical signals used by IoT devices. Sensors form the structure of the front end in the IoT. These devices detect and respond to changes in an environment. Figure 4.3 shows the typical working of sensor technology with the IOT. This figure depicts a sensor that can measure an occurrence and convert it into a signal.

Multi-Antenna Communication Security with Deep Learning Network 67

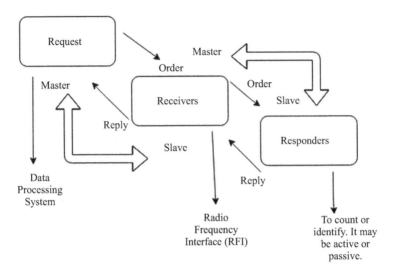

FIGURE 4.2 RFID Technology.

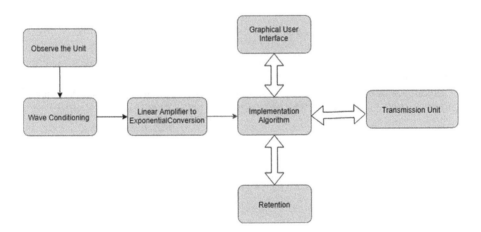

FIGURE 4.3 Sensor Technology Figure.

4.3.3 SMART TECHNOLOGY

Smart technology is a self-checking protocol that shows the details of the utilization of IoT devices. Smart technology methods are utilized prominently in signal handling and tracking and radio telescopes, and generally in cell frameworks, such as W-CDMA, UMTS, and LTE. Smart technology has numerous capacities: beamforming, impedance nulling, and consistent modulus preservation. Two of the principal kinds of smart antenna incorporate switched-beam smart antennas and versatile antennas. A choice is made with respect to which bar to access, at some random point in time, in view of the prerequisites of the framework.

4.3.4 NANOTECHNOLOGY

Nanotechnology is the building of useful frameworks on the subatomic scale of the IoT. It meets with IoT frameworks in different manners, from the assembling of solid sensors, and processes the information gathered by IoT sensors. Nanotechnology (or "nanotech") is the control of the issue on a nuclear, subatomic, and supramolecular scale. A more general explanation of nanotechnology was settled by the National Nanotechnology Initiative, which characterized nanotechnology as the control of issues within any event that one measurement estimated from 1 to 100 nanometers. This definition mirrors the way that quantum mechanical impacts are significant at this quantum-domain scale. Thus, the definition moved from a specific innovative objective of an exploration class comprehensive of a wide range of examinations and advances that manage the unique properties of the issue that happen underneath the given size's edge. It is, in this manner, normal to see the plural structure "nanotechnologies," as well as "nano-scale innovations," to indicate the wide scope of the examination and applications whose regular attribute is size.

4.4 DEEP LEARNING WITH MULTI-ANTENNA BEAM

Deep learning is a powerful machine learning approach that provides function approximation, classification, and prediction capabilities [22].

Figure 4.4 depicts that in a DL model, the first step is to train the user data that is coming to the master station. The data may contain the following information:

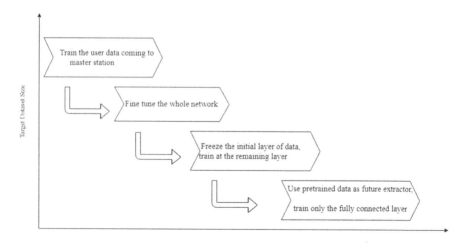

FIGURE 4.4 Deep learning model.

Multi-Antenna Communication Security with Deep Learning Network

1. User-request location
2. Data rates
3. Status
4. Base-station capacity
5. Throughput
6. Response time

In next step, the user data gets fine-tuned by the whole telecom network. There is a certain area for network tuning, whether it is working well or there are issues and some parameters need to be tweaked.

In the next step, we freeze the initial layer of data. In this case, 20 percent of the data is frozen and 80 percent of the data is trained. In the following stage, we use pretrained information as a future extractor, and train just the completely associated layer.

4.5 MULTI-ANTENNA BEAM WITH THE IOT AND DEEP LEARNING

Among the many machine learning approaches, deep learning (DL) has been actively utilized in many IoT applications in recent years [23]. An advanced antenna system (AAS) is a blend of an AAS radio and a set of AAS highlights. An AAS radio consists of receiving wire displays firmly incorporated into the equipment and programming required to transmit and gather radio signals, and signal handling calculations to help with the execution of AAS highlights. Sensors should be able to deliver diverse services. To make that possible, machine learning techniques have been introduced, which can help in efficiency [24].

Figure 4.5 depicts the basic communication between the base station and the slave station communicating with each other through the IoT, which means that every signal is notified to and from in between the base station and slave station communicating devices. The DL model shown which will be added at the base station handles users' requests, data rates, status, and the capacity of the base station and uses the GOB function to send users' requests to the base station. This process is designed to avoid overhead on communications and reduce the time needed to get a response from the server to a user.

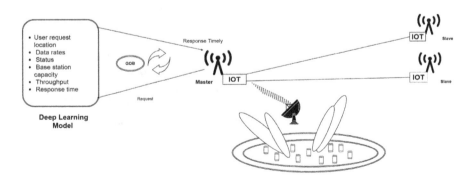

FIGURE 4.5 Communication between the stations.

Contrasted with regular frameworks, this arrangement gives a lot more noteworthy adaptivity and steerability, as far as adjusting the reception apparatus radiation examples to quickly time-shifting traffic and multi-way radio proliferation conditions. Also, various signals might at the same time be received or transmitted by various radiation designs.

The multi-radio wire procedures here alluded to as AAS highlights incorporate beam-forming and MIMO. Such highlights are now utilized with customary frameworks in present LTE systems. Applying AAS highlights to an AAS radio outcome gains in critical execution, due to the higher degrees of opportunity given by a larger number of radio chains, likewise alluded to as Massive MIMO.

Figure 4.6 depicts the shrouded layers of a proposed solution that commonly involves a development of convolutional layers that convolve with replication or another spot thing. The authorization work is commonly an RELU layer and is, in like manner, followed by additional convolutions, for instance, pooling layers, totally related layers, and normalization layers, suggested as covered layers, because their wellsprings of information and yields are hidden by the incitation limit and last convolution. The last convolution, therefore, frequently includes back propagation, to even more precisely weigh the conclusive outcome.

Figure 4.7 shows an overall picture of the proposed idea. Although a lot of prominent DL algorithms were discussed by researchers, CNN is the best solution for the underlying problem. This chapter utilizes the concept of the CNN algorithm to design the solution efficiently.

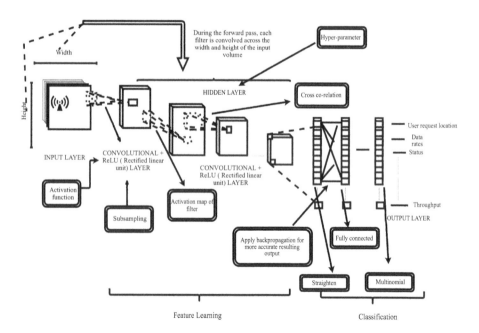

FIGURE 4.6 Shrouded layer of proposed solution.

Multi-Antenna Communication Security with Deep Learning Network 71

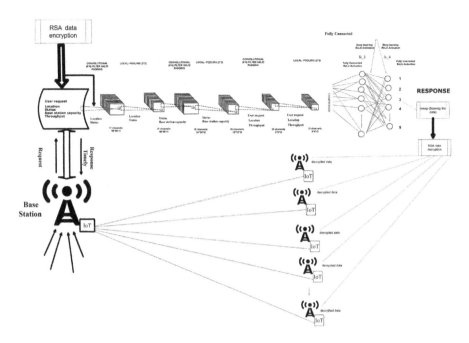

FIGURE 4.7 Multi-antenna with deep learning and IoT communications.

A CNN is a deep learning algorithm that acquires different data from the user, coming from a base station, and has the option to distinguish the forms of data one from another [25,26]. The CNN layer is the central structural square of the algorithm. The layer's parameters include of many learnable channels (or bits), which have a little responsive field coming as the part of a user data request, however stretched out with the full perspicacity of the information capacity. The reason behind the popularity among developers of deep learning is that conventional machine learning techniques were ineffective, to use with IoT applications [27].

During the onward pass, each channel is convolved over the size and importance of the user data, figuring the spot piece between the sections of the network and the information and distributing a two-dimensional actuation monitor of that channel. Accordingly, the system absorbs channels that actuate when it recognizes some specific sort of highlight at some spatial situation in the info. The preparation mandatory in a CNN is minor, when compared to other order calculations.

Let's take an example of the user data where:

- Location is encoded as 1 3 0 4
- Status is encoded as 0 6 7 1
- Throughput is encoded as 9 8 5 2
- Capacity is encoded as 0 1 3 2

In CNN, this information is represented as an image that is a matrix of pixel values, which in turn can be represented as an array of flattened values. Figure 4.8 depicts one such representation of CNN data in a flattening format.

Further, the CNN algorithm decreases the pixels into a format that is much simpler to process. Figure 4.9 depicts an example of how the user data pixel is represented,

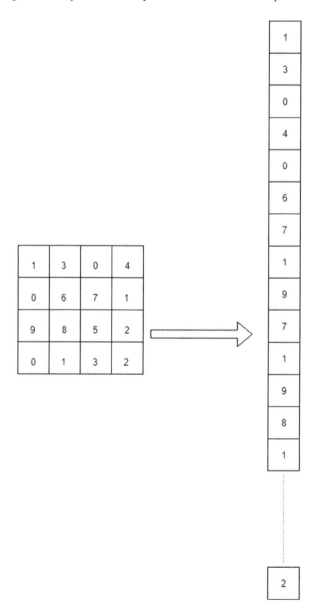

FIGURE 4.8 User data presentation in flattened format.

Multi-Antenna Communication Security with Deep Learning Network 73

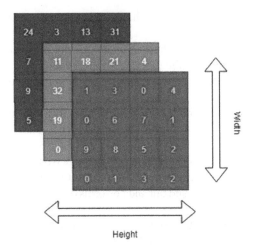

FIGURE 4.9 User data pixel in simpler format.

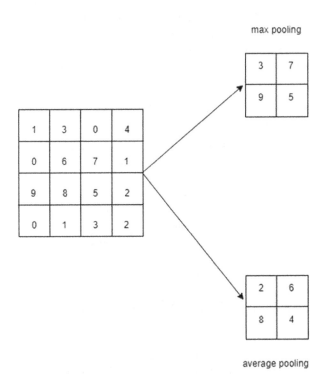

FIGURE 4.10 Max pooling and average pooling of user data.

in a simple format. We have an image that has been described by its three planes of color: blue, green, and pink.

A CNN can effectively detect the spatial and temporal dependencies in an information demand image through the usage of applicable channels. The work achieves a better fit for the image dataset because of the reduction in the number of parameters comprised and the reusability of loads.

The pooling of data from the image to the simpler format is carried out internally and in two ways:

1. Max Pooling: restores the most extreme incentive from the bit of the image held by the kernel. It also likewise continues as an outlier suppressant. It arranges the energetic beginnings out and out and performs de-noising, together with dimensionality reduction.
2. Average Pooling: reestablishes what is typical of the significant number of characteristics from the fragment of the image verified by the piece. Average pooling just implements dimensionality decline as a sound smothering segment.

The pooling layer and the CNN layer, together with configuration user data coming from the base station, to the ith the layer of a CNN. Once the input image goes through a channel, it is converted into the column vector, and the flattened output becomes the input of the next channel of a neural network. After the training model has been successfully prepared with the help of user data (user request location, data rates, status, base station capacity, throughput, and response time), the model can be used with future data sets, to generate the response on time.

4.6 MATHEMATICAL BACKGROUND

The mathematical background of the proposed solution is divided as follows:

Step 1: CNN works with the idea of bit convolution and can be characterized numerically as a procedure where a small matrix of numbers is taken and defined as a channel, transferred over the signal, and changed, according to dependent qualities from the channel. The resulting antenna beam highlight map values are determined by the accompanying equation, where the given input is indicated by l and our bit by p. The values of rows and columns of the outcome grid are set apart with m and n, separately.

$$G[m,n] = (l * p)[m,n] = \sum_j \sum_k p[j,k] l[m-j,n-k] \qquad 4.1$$

Step 2: Through every convolution, images shrivel. So, the procedure of convolution cannot be performed repeatedly, as there is a chance that the picture will totally vanish. The idea of suppression comes into the signal to determine these situations and chooses two distinctive sorts of convolution: valid and same. "Valid" is defined as using the original signal. "Same"

Multi-Antenna Communication Security with Deep Learning Network 75

means we are working on a signal that has borders around it. The goal is that the signals at the incoming and outgoing network beams are similar in size. In the subsequent case, the padding width should meet the accompanying condition, where q is cushioning and l is the channel measurement.

$$q = \frac{l-1}{2}$$ 4.2

Step 3: Sensor Convolution.

The resultant output matrix, which includes both padding and stride, is evaluated using the formula given.

$$n_{out} = \frac{n_{in} + 2q - l}{s} + 1$$ 4.3

Step 4: The transition to the third dimension.

The idea that is proposed is to use multiple filters on the same network antenna. Convolution is carried out separately for each incoming signal beam. The dimensions of the received tensor meet the given equation which contains following attributes given as: n – signal size, l – filter size, nc – number of channels in the signal, q – used padding, s – used stride, and nl – number of filters.

$$s\left[n,n,n_c\right] * \left[l,l,n_c\right] = \left[\left[\frac{n+2q-l}{s}+1\right],\left[\frac{n+2q-l}{s}+1\right],n_l\right]$$ 4.4

4.7 SECURITY ASSESSMENT

The security and protection conservation of AI methodologies are the most significant components for utilizing these techniques in the IoT. IoT security is concerned with protecting networks and devices that are connected in the IoT [28]. IoT security is the area concerned to see that the system is being ensured of networks and signals of antenna beams that are associated in the IoT. It allows antenna signals to set up an association, with some issues in the signal if they are not protected. The IoT systems can promptly communicate conventions; for example, messages are transferred signal by signal, by a CNN layer in matrix format, using average and max pooling, until they arrive at their destination. Solutions have been proposed to deal with objective function attacks, some of which include a black-box attack, which uses DE technique and is based on the statistic that there is no valid statement on the optimization problem of finding effective goal roles giving to the assumption, but that directly works on increasing the chance of labeling values of the destination target. One kind of attack is scalability, which is able to attack different types of CNN algorithms [29]. In this research, various kinds of cybersecurity attacks can be seen in the context of the proposed idea; it also provides the nmap solution that suits the current technical solution. The network topology is determined based on wireless connections between relay nodes, which are the IoT devices [30,31,32].

The proposed security countermeasure, network mapper (nmap), is a system disclosure and security evaluation device. It is known for its straightforward and easily

76 Blockchain Technology for Data Privacy Management

remembered signals, which provide a powerful filtering approach. A scan of outputs can be performed utilizing nmap. These are:

4.7.1 TCP SCAN

A TCP scan is commonly used to check and complete a three-route handshake between two different frameworks. A TCP filter is commonly extremely loud and can be distinguished with practically no effort.

This is "noisy" in light of the fact that administrations can log the sender antenna signal, which may trigger intrusion detection systems.

4.7.2 SYN SCAN

This is another type of TCP examined. The thing that matters is, it is not normal for a typical TCP filter; nmap itself creates a syn packet, which is the principal signal that is sent to build up TCP mapping between incoming and outgoing resultant signals.

4.7.3 FIN SCAN

This is similar to the SYN scan, except that it sends a TCP FIN parcel. Almost all antenna signals send back RST signals, in the event that they get this information. So, the FIN scan shows positives and negatives, yet it might get in under the radar of certain IDS programs and different countermeasures. As seen in Table 4.1, various commands can be proposed using nmap along with their flag and the usage of the antenna signal transmitting, defined as:

Nmap, which is the new cybersecurity scan paradigm, has been introduced in this chapter to provide security with the different aspects of scanning, such as a TCP scan, cipher scan, etc. Figure 4.11 depicts the overall flow of security measures in the idea. When the expectation is completed, the encoded information and ports will experience nmap filtering, and on effective checking of the equivalent, the encrypted information will be decoded, utilizing the same RSA decryption algorithm. All things considered, the information will be in cypher text throughout the channel.

TABLE 4.1

Nmap commands

Command	Determination	Identifier
nmap–sA 192.168.1.1	acknowledgement port	–sA
nmap–sS 192.168.1.1	syn port	–sS
nmap–sT 192.168.1.1	connect port	–sT
nmap–sU 192.168.1.1	port scan	–sU

This is the command to scan for running service. Nmap contains a database of about 2,200 well-known services and associated ports.

Multi-Antenna Communication Security with Deep Learning Network 77

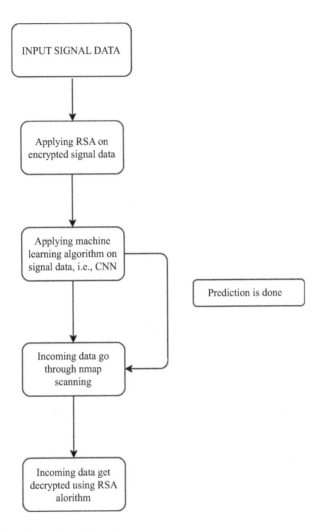

FIGURE 4.11 Flow of security countermeasure.

4.8 CONCLUSION AND FUTURE SCOPE

We have proposed a novel solution to provide better performance, the best carrier distribution, low power consumption, and better downlink and uplink throughputs, etc., for a multi-antenna beam with the help of DL and the IoT. This chapter leverages the use of the CNN algorithm to parse the user data coming from the base station, and prepares a model that generates timely responses, which are distributed to different slave stations through the IoT devices. The basic concept adopted by this chapter is very similar to the client-server communication protocol, to provide the interaction between the base station and the CNN layers. The IoT and DL are current trends in the communications industry. These techniques are being used heavily in the telecom sector.

Multi-antenna technology is at the heart of the communications industry. Therefore, this chapter goes to the heart of the telecom industry, with modern trend technologies. This chapter also deals with the mathematics behind the proposed idea. In this chapter, nmap, which is the new cybersecurity scan paradigm, has been introduced to provide security to the different aspects of scanning, such as a TCP scan, cipher scan, port scan, etc. Future research can be done on providing multi-antenna beam communication with the help of NLP, where the slave stations can be addressed by the master station with the help of the language based on regions, locations, etc. The IoT will play an important role in this communication, as well.

REFERENCES

1. Cai, J.-X., Ranxu, Z., and Yan, L., 2019. Antenna selection for multiple-input multiple-output systems based on deep convolutional neural networks. *PloS One,* 14(5), e0215672. doi:10.1371/journal.pone.0215672.
2. Joung, J., 2016. Machine learning-based antenna selection in wireless communications. *IEEE Communications Letters*, 20(11), pp. 2241–2244.
3. Yao, R., Zhang, Y., Qi, N., Tsiftsis, T.A., and Liu, Y., 2019, June. *Machine learning-based antenna selection in untrusted relay networks.* In *Proceedings of the 2019 2nd International Conference on Artificial Intelligence and Big Data (ICAIBD)* (pp. 323–328). Chengdu, China: IEEE..
4. Leeladhar Malviya, R., Panigrahi, K., and Kartikeyan, M. V., 2017. Four element planar MIMO antenna design for long-term evolution operation. *IETE Journal of Research.* 64(3), pp. 367–373. doi: 10.1080/03772063.2017.1355755.
5. Mahey, R., and Malhotra, J., 2014. *Multi antenna techniques for the enhancement of mobile wireless systems: Challenges and opportunities.* In *IEEE 2014 International Conference on Advances in Engineering & Technology Research (ICAETR – 2014)* (pp. 1–5). Unnao: IEEE. doi: 10.1109/ICAETR.2014.7012900.
6. Patwary, M. N., Nawaz, S. J., Rahman, M., Sharma, S. K., Rashid, M. M., and Barnes, S. J., 2019. The potential short-and long-term disruptions and transformative impacts of 5G and beyond wireless networks: Lessons learnt from the development of a 5G testbed environment. *IEEE Access*, 8, pp. 11352–11379. arXiv Preprint arXiv:1909.10576.
7. Shafique, K., Khawaja, B. A., Sabir, F., Qazi, S., and Mustaqim, M., 2020. Internet of things (IoT) for next-generation smart systems: A review of current challenges, future trends and prospects for emerging 5G-IoT scenarios. *IEEE Access,* 8, pp. 23022–23040. doi: 10.1109/ACCESS.2020.2970118.
8. Ishfaq, M. K., Abd Rahman, T., Himdi, M., Chattha, H. T., Saleem, Y., Khawaja, B. A., and Masud, F., 2019. Compact four-element phased antenna array for 5G applications. *IEEE Access*, 7, pp. 161103–161111.
9. Wang, J., Wang, Y., Li, W., Gui, G., Gacanin, H., and Adachi, F., 2020. Automatic modulation classification method for multiple antenna system based on convolutional neural network. TechRxiv. Preprint. pp. 1–5, https://doi.org/10.36227/techrxiv.12129801.v1.
10. Amiri, A., Carles Navarro, M., and de Carvalho, E., 2020. Deep learning based spatial user mapping on extra large MIMO arrays, pp. 1–6. arXiv preprint arXiv:2002.00474.

11. Zhang, K., Chuai, G., Gao, W., Liu, X., Maimaiti, S., and Si, Z., 2019. A new method for traffic forecasting in urban wireless communication network. *EURASIP Journal on Wireless Communications and Networking*, 2019. doi:10.1186/s13638-019-1392-6.

12. Mozo, A., Ordozgoiti, B., and Gómez-Canaval, S., 2018. Forecasting short-term data center network traffic load with convolutional neural networks. *PLOS ONE*, 13(2), e0191939. doi:10.1371/journal.pone.0191939.

13. Yu, G., and Changshui, Z., 2004. *Switching ARIMA model based forecasting for traffic flow*. In *IEEE International Conference on Acoustics, Speech, and Signal Processing* (Vol. 2), , pp. ii–429). Montreal: IEEE.

14. Hua, Y., Zhao, Z., Li, R., Chen, X., Liu, Z., and Zhang, H., 2019. Deep Learning with long short-term memory for time series predictionn. *IEEE Communications Magazine*, 57(6), pp. 114–119. doi:10.1109/mcom.2019.1800155.

15. Mkocha, K., Kissaka, M. M., and Hamad, O. F., 2019, May. *Trends and opportunities for traffic engineering paradigms across mobile cellular network generations*. In *International Conference on Social Implications of Computers in Developing Countries*, pp. 736–750. Cham: Springer.

16. Nasri, M., and Hamdi, M., 2019. *LTE QoS parameters prediction using multivariate linear regression algorithm*. In *IEEE 22nd Conference on Innovation in Clouds, Internet and Networks and Workshops (ICIN)*. (pp. 145–150). Paris: IEEE. doi:10.1109/icin.2019.8685914.

17. Arfeen, A., Pawlikowski, K., McNickle, D., and Willig, A., 2019. The role of the Weibull distribution in modelling traffic in Internet access and backbone core networks. *Journal of Network and Computer Applications*. 141, pp. 1–22. doi:10.1016/j.jnca.2019.05.002.

18. Alam, N., Vats, P., and Kashyap, N., 2017. *Internet of Things: A literature review*. In *Recent Developments in Control, Automation & Power Engineering (RDCAPE)* (pp. 192–197). Noida, India: IEEE. doi:10.1109/rdcape.2017.8358265.

19. Jindal, M., Gupta, J., and Bhushan, B., 2019. *Machine learning methods for IoT and their future applications*. In *International Conference on Computing, Communication, and Intelligent Systems (ICCCIS)* (pp. 430–434). Greater Noida, India: IEEE. doi:10.1109/icccis48478.2019.8974551\.

20. Farhady, H., Lee, H., and Nakao, A., 2015. Software-defined networking: A survey. *Computer Networks,* 81, pp. 79–95.

21. Wang, T., Wen, C.-K., Wang, H., Gao, F., Jiang, T., and Jin, S., 2017. Deep learning for wireless physical layer: Opportunities and challenges. *China Communications*, 14(11), pp. 92–111. doi:10.1109/cc.2017.8233654.

22. Puri, D., and Bhushan, B., 2019. *Enhancement of security and energy efficiency in WSNs: Machine Learning to the rescue*. In *International Conference on Computing, Communication, and Intelligent Systems (ICCCIS)* (pp. 120–125). Greater Noida, India: IEEE. doi:10.1109/icccis48478.2019.8974465.

23. Tiwari, R., Sharma, N., Kaushik, I., Tiwari, A., and Bhushan, B., 2019. *Evolution of IoT & data analytics using deep learning*. In *2019 IEEE International Conference on Computing, Communication, and Intelligent Systems (ICCCIS)*. (pp. 418–453, 418–423). Greater Noida, India: IEEE. doi:10.1109/icccis48478.2019.8974481

24. Mohammadi, M., Al-Fuqaha, A., Guizani, M., and Oh, J.-S., 2018a. Semisupervised deep reinforcement learning in support of IoT and smart city services. *IEEE Internet of Things Journal*, 5(2), pp. 624–635. doi:10.1109/jiot.2017.2712560.

25. O'Shea, T. J., Erpek, T., and Clancy, T. C., 2017. Deep learning based MIMO communications. arXiv preprint pp. 1–9. arXiv:1707.07980.

26. Uysal, A. K., and Gunal, S., 2012. A novel probabilistic feature selection method for text classification. *Knowledge Based Systems,* 36(6), pp. 226–235.

27. Blot, M., Cord, M., and Thome, N., 2016, September. *Max-min convolutional neural networks for image classification*. In *IEEE International Conference on Image Processing (ICIP)* (pp. 3678–3682). Phoenix: IEEE.
28. Mohammadi, M., Al-Fuqaha, A., Sorour, S., and Guizani, M., 2018b. Deep learning for IoT big data and streaming analytics: A survey. *IEEE Communications Surveys & Tutorials*, 4, pp. 2923–2960. doi:10.1109/comst.2018.2844341.
29. Su, J., Vargas, D. V., and Sakurai, K., 2019. Attacking convolutional neural network using differential evolution. *IPSJ Transactions on Computer Vision and Applications*, 11, pp. 1–16. doi.org/10.1186/s41074-019-0053-3.
30. Choi, Y.-S., Park, J.-S., and Lee, W.-S., 2020. Beam-reconfigurable multi-antenna system with beam-combining technology for UAV-to-everything communications. *Electronics*, 9(6), p. 980. doi:10.3390/electronics9060980.
31. Arora, A., Kaur, A., Bhushan, B., and Saini, H., 2019. *Security concerns and future trends of Internet of Things*. In *2nd International Conference on Intelligent Computing, Instrumentation and Control Technologies (ICICICT)*. (Vol 1, pp. 891–896). Kannur, Kerala, India: IEEE. doi:10.1109/icicict46008.2019.8993222.
32. Kwon, M., Lee, J., and Park, H., 2019. Intelligent IoT connectivity: Deep reinforcement learning approach. *IEEE Sensors Journal*, 20, pp. 2782–2791. doi:10.1109/jsen.2019.2949997.

5 Security Vulnerabilities, Challenges, and Schemes in IoT-Enabled Technologies

Siddhant Banyal, Amartya, and Deepak Kumar Sharma
Netaji Subhas University of Technology, India

CONTENTS

5.1 Introduction 82
5.2 IoT Architecture and Systemic Challenges 85
    5.2.1 Sensing Layer: Introduction and Challenges in End-Nodes 86
    5.2.2 Threat Based on Network Layer 87
    5.2.3 Service-Layer Based Threats 88
    5.2.4 Application Interface Layer 88
    5.2.5 Cross-Layer Challenges 89
5.3 Challenges and Associated Vulnerabilities in IoT-Enabled Technologies 89
    5.3.1 Authentication- and Authorization-Related Challenges 89
    5.3.2 Insecure Access Control...93
        5.3.2.1 Role-Based Access Control Systems 93
        5.3.2.2 Access Control List-Based Systems.....................................93
        5.3.2.3 Capability-Based Access 93
        5.3.2.4 Challenges in Access Control...95
    5.3.3 Physical Layer Security 95
    5.3.4 Encryption Based on Transport of Data..95
        5.3.4.1 TLS: Transport Layer Security...96
        5.3.4.2 HTTPS...96
        5.3.4.3 Transport Trust in IoT 96
    5.3.5 Secure Cloud and Web Interface...97
    5.3.6 Secure Software and Firmware 97
    5.3.7 Estimating the Cost of a Cyber-Breach: Case Study on Intellectual Property Breach for the IoT...98
5.4 Existing Cyber-Attack Detection Software and Security Schemes 100
    5.4.1 Conventional Cyber-Security Schemes...100
        5.4.1.1 Access Control Technologies 100

5.4.1.2	System Integrity	101
5.4.1.3	Cryptography	101
5.4.1.4	Audit and Monitoring	102
5.4.1.5	Configuration Management and Assurance Tools	103

5.4.2 Embedded-Programming-Based 103
5.4.3 Agent-Based Approach 104
5.4.4 Software-Engineering- and Artificial-Intelligence-Based 105
5.5 Conclusion and Open Challenges 105
References 106

5.1 INTRODUCTION

The last two decades have been catalyzed by developments on a myriad of technological fronts, and this development has severely affected social functioning. Technology has been increasingly integrated with our way of living and daily life, ranging from the moment we wake up at home and use smart home appliances, to the usage of integrated technology at the workplace, to health monitoring, and the analytics of our sleep. This development has asymmetrically changed the ways industries perceive and use technology, and they have been trying more and more to integrate the incumbent developments into their operations, for efficiency. Reports suggest that the estimated count of connected IoT devices is set to rise to fifty billion by the end of this decade [1]. The ecosystem involves a myriad of elements, such as IoT devices, sensors, actuators, network elements (servers, routers, etc.), and associated industrial machinery. In this pursuit of connecting the conventional devices across networks, the Internet of Things (IoT) and the Web of Things (WoT) have been pivotal in catalyzing and catering to this need. As an emerging technology, the IoT offers a novel solution and optimizing paradigms to both and unconventional industrial operations. One such example of this innovative behavior is the case of innovative transportation in the field of intelligent transportation systems (ITS), where the IoT and associated technology have enabled smart traffic management and traffic prediction, through monitoring and predicting traffic location.

The IoT's origin can be traced back to the emergence of radio frequency identification technology (RFID), where it was envisioned to refer to identifiable interoperable connected objects. In common parlance, the IoT encompasses a network infrastructure that is global, dynamic, and self-configuring on the basis of communication protocols working in tandem with a myriad of physical and virtual entities [2]. Tracing the history of the IoT, from RFID technology in the 1980s, the technology transformed to Wireless Sensor Networks (WSN) in the 1990s. WSN encompasses intelligent sensor networks, health care monitoring, industrial monitoring, environmental monitoring, and a few other sectors. This is what eventually amalgamated with other emerging trends into what we identify as modern IoT technology. The authors of this chapter conducted a literature review across major academic research databases, including but not limited to IEEE Xplore, the Web of Knowledge, the ACM digital library, and Science Direct, in order to thoroughly identify the breadth of trends. We found a number of IoT publications still growing quickly. It is this

Security Vulnerabilities, Challenges, and Schemes

proliferation that has led to such a boom in the technology. The development is limited not just to the IoT as an industry but has contributed to the development of several other industries, especially blockchain [3], which in turn has led to increasing usage of cryptocurrency across the globe [4]. As mentioned earlier, these fifty billion devices are on a path to exceed the number of humans soon, in 2020, and this proliferation has given birth to the Internet of Things that we know now. With this increased dependency on the technologies comes an increased susceptibility to the threats present in cyberspace that also transcend to the physical world through IoT devices. A news magazine aptly noted how weapons of destruction are being maneuvered by GPS, UAVs are being monitored and guided by pilots remotely, and the conventional soldier is now being upgraded by the use of technologies such as an exoskeleton, etc. [3]. This wiring up or digitization has been a double-edged sword, as it comes with augmenting the threat landscape. Bits and bytes are now more damaging than bombs and bullets. This digital reliance leaves society open to the prospect of a cyberattack, which, as noted, can be severely damaging. The advent of modern technological advancement in IoT-based information-sharing technologies has asymmetrically changed the prospects in the domain of informatics. With the growing scope of networking and information exchange come data security concerns. Data security is a necessity in contemporary times, with the growing advent of information exchange and networking. This has been explained by Mansi Jindal et. al. [5], with special emphasis on the future perspective. Recognition in patterns for information corruption and data breaches has been studied, with attacks being defined into categories: phishing, Dos attack, brute force attack, malware, etc. Each of these employs different methodologies for information exchange hampering or corruption, having different intentions for network encroachment. Figure 5.1 covers various types of prevalent cyberattacks and their categories.

Various examples of cyberattacks are given in Table 5.1. Security against each kind of attack needs to be considered, for security paradigm design. This lays the foundation of modeling techniques for cyber-intrusion detection and countermeasures, tailored to curtail actions of the standard paradigm used for intrusions and information attacks. "Cybersecurity" is defined as technologies and processes developed to safeguard computer networks, hardware, and software from malicious activities and vulnerabilities, such as unauthorized access catalyzed via the cyberattackers. The primary goal of this is to ensure confidentiality, availability, and integrity, which are tenets for all data handling, networks, information, and technology applications. Gitika Babar and Bharat Bhushan [6] detail the relevant frameworks in this regard, with special emphasis on security in Industry 4.0. This involves safeguarding the internet and networks from unauthorized alteration. The internet is not only a forum to communicate, but it impacts our work and life severely, as it is integrated with the physical, social, and financial spectrum of our lives. The fractious challenges we face on this front can be bifurcated into broad systemic challenges and specific security challenges. Ideas have been put forth by several authors regarding the malicious attacks in opportunistic IoT networks, along with various approaches inculcating futuristic technologies, such as machine learning, to better the understanding and prevention of such attacks [7–9]. The factors that lead to security challenges are specialty, diversity, and principles. By drawing relevant

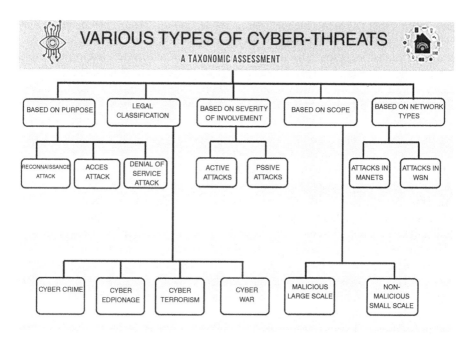

FIGURE 5.1 Various Categories of Cyber Threats.

TABLE 5.1
Examples of Cyberattacks

Name of security threat	Example
Reconnaissance Attacks	Packet sniffers, Port scanning, Ping sweeps, and DNS (Distributed Network Services) queries
Access Attacks	Port trust utilization, Port redirection, Dictionary attacks, Man-in-the middle attacks, and Phishing
Denial of Service	Smurf, SYN flood DNS attacks, DDOS (Distributed Denial of Service)
Cyber Crime	Identity theft, Credit card fraud
Cyber Espionage	Tracking cookie, Rat controllable
Cyber Terrorism	Crashing the power grids by Al-Qaeda via a network, Poisoning of the water supply
Cyberwar	Russia's war on Estonia (2007) and Georgia (2008)
Active Attacks	Masquerade, Reply, Modification of message
Passive Attacks	Traffic analysis, Release of message contents
Malicious Attacks	Sasser attacks
Non-Malicious Attacks	Registry corruption, Accidental erasing of hard disks
Attacks in MANETS	Byzantine attacks, Black hole attacks, Flood rushing attacks, Byzantine wormhole attacks
Attacks in WSN	Application layer attacks, Transport layer attacks, Network layer attacks, Multilayer attacks

Security Vulnerabilities, Challenges, and Schemes

insights from various instances of cyber-breaches, we can chart out the following security practices for IoT devices [10]:

- The development and lifecycle of all IoT devices should be covered under security
- Authentication and authorization are pivotal for IoT devices and associated data management
- Prior to initiating sending or receiving data, an IoT device must perform authentication when it is turned on
- Owing to computation constraints and limited buffering, the requisite firewalls must be present so as to filter data packets and protect networks from cyberattacks

The primary contribution of this chapter includes a delineation of the methodologies used in the assessment and analysis of vulnerabilities and security challenges. The significance and contribution of this piece are manifold, as this chapter not only gives details for IoT-based devices but also introduces the systemic vulnerabilities and identifies the bottlenecks present with regard to cybersecurity. Further, this chapter undertakes a comprehensive mathematical case assessment to estimate the various factors associated in an intellectual property (IP)-based breach, in the event of a cyberattack. Last, this work aims to provide insight into various application domains of cybersecurity. The trends highlighted could enable the research community to develop more robust cybersecurity schemes for IoT-based systems and optimize the existing ones.

The organization of this chapter is as follows: Section 5.2 explores elementary IoT architecture and the systemic threats pertinent to it. Section 5.3 presents the challenges in IoT-enabled devices and networks, whereas Section 5.4 focuses on existing cyberattack detection software and security schemes and various technological trends in this area. The last section summarizes the chapter by drawing relevant conclusions and subsequently, covers open challenges in this domain and the possibility of future work to improve the existing technologies in security schemes.

5.2 IOT ARCHITECTURE AND SYSTEMIC CHALLENGES

The following section elaborates on the multiple-layer structure of the IoT and the numerous factors, parameters, and conditions that form the skeletal framework for the IoT architecture. For its varied uses, the IoT has become an indispensable tool in every industry, ranging from energy conservation in smart buildings [11] to its use in green buildings management and smart automation. In all of these, it is necessary to identify challenges at every step of the way. Tanishq Varshney et al. [12] have described these challenges in detail, in "Architectural Model of Security Threats & their Countermeasures in IoT." The section also presents the threats faced at different strata of the architecture.

5.2.1 SENSING LAYER: INTRODUCTION AND CHALLENGES IN END-NODES

Amidst the vast variety of IoT devices that surround humans, the most common are sensors, actuators, RFID readers, RFID tags, etc. These devices form a set of devices that are collectively termed as the sensing layer of the IoT architecture. The critical contribution of this layer in the IoT can be broadly summed up as the sensing of ambient parameters and the transmission of sensed data for processing in the next layers [13]. A few parameters need to be considered in the sensing layer:

- Cost, resource and energy consumption: The devices are equipped with minimal energy resources and memory, in order to reduce cost.
- Communication: The devices act as the receiving ends of information and are designed to communicate with other devices on the network.
- Networks: WSNs (Wireless Sensor Networks) and WMNs (Wireless Mesh Networks) connect a unique category of things in a complex, wireless and autonomous networks are employed for data acquisition, transmission, and operation.

Figure 5.2 explains the fundamentals of service-oriented architecture in an IoT network and its interaction with the other layers of IoT infrastructure. Coupled with synchronized computing and communication capabilities, the IoT is credited with tapping into the potential offered by these individual sensors, turning them from classic into smart. In this regard, the security of the end nodes of the sensing layer of this network becomes of prime importance, particularly owing to the uncertainty regarding data control. The foremost prerequisite for the security mechanism in the Internet of Things is to have the rationale to make its own decisions, which includes approving a command to accept, execute, or terminate it. However, the confines of "Things" set up for minimal energy consumption and

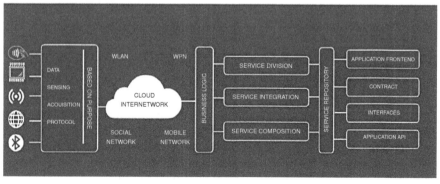

FIGURE 5.2 Service-Oriented Architecture IoT.

Security Vulnerabilities, Challenges, and Schemes

limited memory pose an extended range of security vulnerabilities at the sensing layer and end node. Upon sorting the various insecurities and threats faced by the sensing layer of the IoT, a few security preconditions are essential, including security prerequisites in IoT end nodes: confidentiality, integrity, privacy, access control, authentication, physical security protection, and nonrepudiation. Security Prerequisites in the IoT sensing layer are device authentication, authentication of the information source, availability, integrity, and confidentiality.

In order to achieve the above-mentioned requirements in the sensing layer of the IoT network, the actions suggested include: the creation of a trustworthy data sensing system and reinstatement of the privacy and confidentiality of all devices in a network, the identification of the source of users forensically and further tracing of them, designing the software or firmware of the IoT to secure end-nodes, and administering security standards for all IoT devices.

5.2.2 THREAT BASED ON NETWORK LAYER

For the optimum utilization of data procured in the sensing layer, it is equally important to transmit data among the IoT infrastructure. The network layer, thus, provides the necessary medium to exchange information. For smooth functioning and coordination among IoT devices, the proper arrangement, organization, and management of networks are important, for which the prerequisites include: effective network management, such as wireless networks, fixed networks, or mobile networks; energy efficiency within the network layer; QoS requirements; the maintenance of privacy, confidentiality, and security, and a mechanism for mining and searching.

Among these requirements, the maintenance of privacy, confidentiality, and security lies within the purview of the chapter, and its importance is critical, based on the complexity and mobility of IoT networks. Although existing security protocols and frameworks have provided security against threats and vulnerabilities until now, a multitude of concerns need to be addressed, entailing broad security provisions to ensure confidentiality, integrity, and privacy for group authentication, the protection of keys, and the availability of data. Second, IoT security entails protection against privacy leakage: The location and complexity of certain devices in an IoT network often troubles developers, who fear the susceptibility of attacks upon sensitive data, such as user identity and credentials. Third, it has secure communication: For an IoT system to exist, it must be fortified against the attacks and reinforced with robustness, trustworthiness, and confidentiality. Fourth, it has fake network messages: Creating fake signals propagates miscommunication among the devices from the entire network. And last, it includes MITM attacks: Attacks are carried out independently by attackers over networks, to forge a private connection while the attacker is controlling the entire conversation.

Although the innovations and technology currently available have kept the major threats at bay until recently, the growing influence of attackers has sent shockwaves across the globe. A series of steps in the following directions could help to provide greater security in the future. These include a stringent authentication/authorization process and secure transport encryption.

5.2.3 SERVICE-LAYER BASED THREATS

Upon sensing and transmission, the data procured requires operation while utilizing and integrating the services of hardware and software platforms. The service layer, hence aptly touted to be the middleware technology, is designed based on the application requirements, application programming interface, and service protocols, along with the standards of service providers, vendors, and organizations. This layer is responsible for the integration, analysis, security, and management of UI and event processing services [14]. To render these services, the steps taken include service detection, locating the optimal infrastructure necessary to conduct services efficiently. Second is service combination and integration, where the goal of further broadening the scope of interaction among services and drawing out more reliable ones is achieved by interaction, by scheduling, or by recreating. Third,is authentication management, where the focus is on the verification of trusted devices by other services. Last, service APIs help to improve the interconnections between services.

To tackle the numerous challenges and threats, developers and corporations have contributed relentlessly to offering solutions to augment and improve services within a connected network. The ambitious SOCRADES integration architecture aims to ameliorate the interactions between the application and service layers efficiently [15]. The "things" in the interconnection of devices are often limited to delivering services while exploiting these devices for the discovery of networks, the exchange of metadata, and the asynchronous publication and subscription of events [16]. In Pedro Peris-Lopez et al. (2006) [17], to increase the interoperability of loosely coupled devices and distributed applications, a representational state transfer is set up. In Pedro Peris-Lopez et al. (2013) [17], a service-provisioning process is introduced in the service layer that could strengthen ties and buttress cooperation between applications and services.

In light of the above-mentioned challenges and the solutions offered to counter them, it is imperative to understand that certain security precautions, requirements, and protocols, if undertaken, could shield against attacks in the service layer. These include: dedicated authorization methods for service verification, the authentication of groups, the protection of privacy and integrity for the upkeep and storage of keys, protection against privacy leakage and location tracking, tracking services involving unauthorized use and unsubscribed services, and last, the prevention of potential threats, such as DoS attacks, node identification masquerades, replay attacks, service information manipulation and communication, and service repudiation. This section broadly covers the solutions to major potential security threats.

5.2.4 APPLICATION INTERFACE LAYER

This interface is the most visible and interactive layer of the IoT network and encompasses a myriad of utility-based implementations, ranging from radio-frequency-identification-based tracking to intelligent home management that is enabled via standardized protocols and other technologies [18]. Application maintenance requires certain security preconditions, such as: safety-based isolation, secure methodologies for the acquisition of software and updates, patches for augmenting

Security Vulnerabilities, Challenges, and Schemes

security, verification means for the administrators, and integrated platforms for the enhancement of security. Different layers of IoT architecture require the following, so as to sustain security in communication between layers: maintaining the three tenets of security (privacy, confidentiality, and integrity) for interlayer communication, the verification and approval of administrators cross-layer, and the isolation of critical data.

Regulations could prove helpful in designing security solutions. The safety of these nodes should be attended to carefully, as most of the nodes in question are unsupervised, and the energy efficiency of nodes is of utmost importance while designing security solutions, considering their large numbers.

5.2.5 CROSS-LAYER CHALLENGES

Across all layers of IoT architecture through which data is shared, certain standards are to be maintained to ensure that a network remains secure and fully interoperable. With the growing number of things in a network, it is the prerogative of users to ascertain that their data is guaranteed protection against challenges among the layers of the architecture. The security needs across layers are virtually the amalgamation of the challenges faced across the IoT network: protecting security in terms of design and execution time, ensuring high privacy standards to protect personal data through enhancement technologies, and reinstating trust in IoT architecture.

5.3 CHALLENGES AND ASSOCIATED VULNERABILITIES IN IOT-ENABLED TECHNOLOGIES

5.3.1 AUTHENTICATION- AND AUTHORIZATION-RELATED CHALLENGES

With the growing ubiquitous presence of IoT devices, it has become the need of the hour for each system to possess the secure authentication of the sensitive sensor data being shared. Although the password is one of the most secure authentication methods widely in use in the virtual world, it is considered incompatible with and difficult to implement in IoT systems, owing to the sheer size of the IoT networks at work. One of the industries that heavily employs IoT is the health care industry. Data transmission occurs within the IoT architecture in the health care industry by verifying to the sink node (Medical Services Gateway) before relaying the procured data out. This data is then verified by the cloud upon forwarding the data. Finally, the analysis and operations upon the procured data are performed once the IoT applications and services authenticate to the system. The authentication method in the IoT scheme is based on a security token. The two most frequently used processes for authentication include one-way authentication and mutual authentication.

OAuth-based authentication schemes have been gradually gaining traction in the last few years. This is an open standard authentication, which invokes third-party websites for logging in, without exposing passwords. The key benefit offered in this scheme is that the client is granted a secure delegated access to the server. The OAuth flow is explained in Figure 5.3.

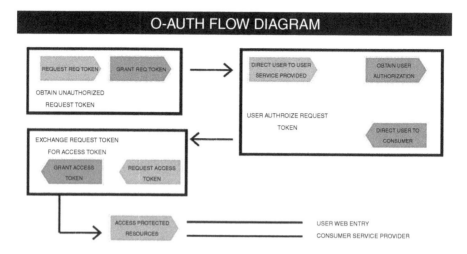

FIGURE 5.3 O-Auth Flow Diagram.

Application program interfaces (APIs) are made available for IoT apps by the OAuth-based scheme, which aids the IoT service and application layers and, by extension, the users, by:

- Permitting actions to be performed by an untrusted application that is in agreement with the IoT user or end node at the API provider;
- Granting the authority to perform actions, by verifying the devices'/user's permission without revealing the user password;
- Giving certain permissions to untrustworthy users.

OAuth 2.0, which is an authentication technique that is remarkably helpful for web applications, mobile phones, and IoT clients for next-generation, comes with its own set of challenges, involving credible APIs and entrusted credentials. Certain applications in the IoT system use digital IDs furnished by state authorities to identify the user/device. There is a lack of central authorization management. The diverse security management interfaces across various platforms make it difficult for IoT end nodes to manage cloud interactions. Considering both OpenID and OAuth, certain parallels can be drawn between them in terms of basic operations: log-in request; requester authentication; redirecting the URL for the identity provider; identity provider authentication of the user; processing request and the response of the provider by sending a redirect; and, sending the URL to the requester and the response of the requester.

Upon closer inspection of the network, we find that the authentication layer forms the bedrock of the IoT architecture, which relies on identity information for providing and verifying an IoT device. These IoT/M2M devices build a trust relationship, which allows them to access the infrastructure upon the identity of information. Although there exists a large difference between storing and offering

Security Vulnerabilities, Challenges, and Schemes

identity information to the user predominantly in an enterprise network, the endpoints are identified using human biometrics or passwords. There have been efforts to reduce or eliminate human interaction by using fingerprinting IoT endpoints. The identifiers that have evolved in this quest include RFID, the MAC address of the endpoint, or X.509 certificates. Among these, X.509 certification is regarded as having a strong authentication system, but the paucity of memory to store a certificate and the CPU power used to compute cryptographic operations for authentication pose a challenge to IoT devices as a whole. Along with X.509, there exist a few other identity footprints, such as 802.1AR, and protocols for authentication, such as IEEE 802.1X, which are beneficial for devices that can handle both the CPU power and memory needed to store strong credentials. These challenges have instigated fresh research, to develop smaller footprint credentials, which is sustainable in less computation-intensive cryptographic frameworks.

The second determining factor of the IoT framework is an authorization that manages a device's access throughout the entire network architecture. The authorization to access data is based on an authentication operation performed, which employs identity information from a device to initiate and augment trust alliances and trust relationships, to relay suitable information. A trust relationship allows the minimal, obligatory transfer of data, while a trust relationship allows a relatively unconstrained flow of data. The existing architecture and policy mechanisms cater to the needs of endpoints efficiently, both in consumer and enterprise networks. The biggest bottleneck is to further enhance the size of IoT/M2M devices with diverse trust relationships in the construct. The focus might shift toward improving policies and controls, to minimize and segregate network traffic and improve end-to-end communication.

The primary difference in security management in an IoT network is that prior to a new device being added in, a network must authenticate themselves prior to sharing data, unlike websites that are authenticated using Secure Socket Layers (SSL) on browsers or users who have to authenticate using a password. Since the driving force to create an IoT network is to reduce human intervention and increase the coordination among things to render useful data to the user, introducing humans to input credentials to access the network would be taking a step back. So, what are the measures that can ensure devices to correctly ensure authorization to access into an IoT network? Based on user authentication, machine authentication uses a similar set of credentials, which are stored securely in memory

Ensuring that the data transmitted, stored and operated upon remains secure in an IoT network is of prime importance to fend off threats from taking advantage of IoT in an unauthorized manner. Vulnerable points that are susceptible to threats from external actors include:

- Vehicles, health care systems and control systems that may even be used to access the human body (WBAN) to manipulate and cause injury or even worse.
- Manipulated health care diagnosis, which could lead to improper treatment or modified health information rendered to a patient
- Homes or commercial businesses may face attacks against electronic, remote-controlled door lock mechanisms and allow physical access

The devices in a network are often at risk at various levels. First is the single-user level, where illegal surveillance through the persistent remote monitoring offered by small-scale IoT devices is possible. The examination of a network, geographic tracking, and IoT metadata leads to inappropriate profiles and categorization. Asset usage and management develop a usage pattern that can lead to the unauthorized tracking of users' locations. User location and tracking personal behavior, choices and activities through location-based sensing information could expose patterns that may be stored without prior notice to the user. The second level is at commercial business, with the unauthorized tracking, analysis, and manipulation of financial transactions via POS and POS access. The inability to provide service can lead to monetary losses. The destruction or theft of IoT assets that are employed in obscure areas and lack security controls can further exacerbate the damage. The third and final level is a degree of access in the IoT, with the potential to acquire unauthorized access to IoT end-nodes to exploit data unscrupulously by exploiting the drawbacks concerning upgrading the software and firmware of embedded devices, such as those present in cars, houses, or health care facilities. The potential exists to acquire unauthorized access to the IoT in enterprises to compromise IoT end-nodes by often manipulating, violating, or exploiting the trust relationships and alliances. There is the potential to compromise IoT end-nodes to form botnets. There is also the potential to join a network by faking IoT devices by acquiring keys and other confidential data stored in these devices. The compromised gadgets in the IoT network are unknowingly fielded upon their security concerns.

The most pivotal operation in an IoT system is the authentication of devices and data transmitted at each layer. Beginning from granting access to IoT devices to relaying procured data or to operating upon this data, to the application layer being authorized when requesting data, the entire IoT is centerd around security and authentication. For this purpose, one of the most compatible and optimal methods available in the IoT is security tokens, which verify one application or user or device to the device or application at another layer, using a token received. It helps to verify the first actor to another operational at another layer, thus granting access and authorization. To keep authorization and authentication under check, it is imperative for the user to be aware of how the data gets transmitted, analyzed, or operated on and the extent to which each actor authorizes another to access data.

OpenID Connect 1.0 and OAuth 2.0 are the two most widely used authentication tools that operate along with the above-mentioned model. Both frameworks use a token for authentication to grant privileges and control. One of the salient features of OpenID Connect 1.0 is that it offers a discovery and registration mechanism, which is apposite in scaling the network. OpenID Connect 1.0 and OAuth face the biggest challenges through HTTP, an IoT protocol. The inadequacy of HTTP to operate among various "Things" has been one of its major setbacks. The development of MQTT (Message Queuing Telemetry Transport) and CoAP (Constrained Application Protocol) are proving to be suitable alternatives to HTTP, yet there remain a number of roadblocks ahead to combine IoT-centered frameworks.

Security Vulnerabilities, Challenges, and Schemes 93

5.3.2 Insecure Access Control

Most online services and computer-based networks use the following algorithm to authorize a user: The user furnishes his or her identity, which upon being established, yields him or her privileges and controls, as per the role specified within an organization. This algorithm is followed by nearly every protocol and framework available (LDAP, SSH, Kerberos, and RADIUS) and is called a role-based authorization process. A similar case is online services that utilize HTTP cookies, which are saved in a browser once their identity is established. There might be differences in the approach to establishing the identity of the user, but the central idea revolves around authenticating the details and granting privileges upon verification.

The next most important task is granting access and applying limiting controls for the user. This is to ensure that the user has access to a minimum number of mandatory resources to perform their operations. In case the system is compromised, access controls curtail privileges and thus mitigate damage. This is in line with the device-based access control mechanism, which in turn forms the basis for network-based access control systems, such as MAD (Microsoft Active Directory). The greatest advantage of these mechanisms is that they restrict the right to obtain data from a network to only those authorized by particular credentials, to minimize any breach of security. This is called the principle of least privilege.

5.3.2.1 Role-Based Access Control Systems

The wide popularity of role-based access control is not replicated in IoT systems, unlike computers, since the identity of an individual device may be unknown in role-based systems. Furthermore, access control focuses on other criteria, such as position, coordinates, architecture, etc. Therefore, a broader attribute-based access control system is required. The only drawback to OAuth in this sphere is that it verifies applications, and not users, using tokens.

5.3.2.2 Access Control List-Based Systems

To further maintain a record in order to determine the rights granted to each user, application, or end node, an access control list (ACL) is maintained. Figure 5.4 portrays the ACL-based system used for accessing or granting the rights of IoT devices/ applications/users. In other words, it is a set of rules that governs rights, privileges, and controls and lists the permitted IoT users/applications.

5.3.2.3 Capability-Based Access

An FTP receives information using a given port, leaving the port highly vulnerable to attacks. To limit the accessibility of the port, usernames and passwords are set up, although this is not feasible for the growing audience of the IoT. This limits its usage, and hence, the approach is being scrapped. Also, the limited complexity of the device and end node raises concerns. Figure 5.5 provides a relevant example.

A secure and optimal alternative to the status quo is to utilize "capabilities," which are cryptographic keys used to access certain activities, such as the communication among devices.

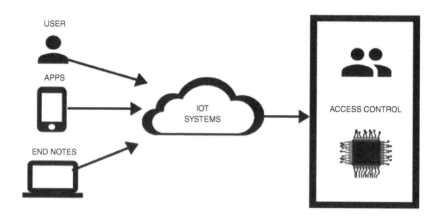

FIGURE 5.4 ACL-Based System.

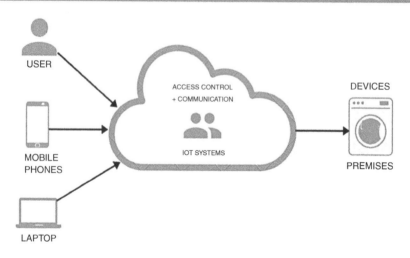

FIGURE 5.5 Capability-Based Access Control System.

Security Vulnerabilities, Challenges, and Schemes

5.3.2.4 Challenges in Access Control

A number of day-to-day challenges pose a serious threat to access control in the IoT, including weak passwords, weaker password encryption software, unreliable protocols, etc. The tested mobile apps that do not use SSL connections to the cloud account for nearly 19 percent of the total number and are at risk from man-in-the-middle attacks or attacks from the connection. Most devices do not have a secure authentication system between user and server. Stronger passwords are generally not supported in many IoT devices. There are no highly secure and encrypted firmware and software updates available. There is a lack of two-factor authentication (2FA) in the IoT cloud interface. Powerful attacks, which can be prevented by lockout or delaying measures, are unavailable in many IoT services. Weaker passwords are a critical security issue.

5.3.3 Physical Layer Security

In the last few years, increasing attention has been paid toward physical layer security As the name suggests, in this technique, secrecy is achieved by deploying physical layer properties, rather than relying on traditional cryptographic methods such as channel security, thermal noise, and interference to ensure security within an IoT system. Physical layer security is operational at the physical layer and here, an adversary's computation power and information are not assumed. Computational power is confirmable and measurable in bits per second per Hertz and could be implemented using programming, signal processing, and communication.

There are several vulnerabilities that the physical layer is prone to, including IoT device capture, where the end-node entities of an IoT system are essentially at risk of being captured, manipulated, and tampered with by attackers. This might in turn cause the leaking of sensitive information, such as communication keys, and further threaten the entire IoT network. A second type of vulnerability entails the addition of fake IoT devices where, by accessing other users' data by entering a fake code to gain access into the system, the addition of a fake device to the system is possible. A third vulnerability is attacks from a side-channel. Devices are often hacked into through side channel leakage while operating the device; this includes radio interference and energy and time consumption. Last, the vulnerabilities include attacks based on timing, where highly precise attacks are performed by assessing the time required by the system to encrypt/decrypt algorithms to access key data. A perfect IoT device in a network would thus be an amalgamation of energy efficiency, performance, flexibility, and cost-effectiveness. The most preferred algorithms used are 3DES, AES, RSA, and ECC.

5.3.4 Encryption Based on Transport of Data

Encryption based on the transport of data has been a key area of interest for safeguarding security and privacy in IoT devices. The prerequisites to design a cryptographic transport protocol are securing, defending, and encrypting data integrity and

Blockchain Technology for Data Privacy Management

key handshaking. Some widely popular transport protocols include TLS and SSL. SSL, TLS, and HTTPS are algorithms that form the basis for commonly used encryption methods.

5.3.4.1 TLS: Transport Layer Security

The user and server form a TLS connection by following the steps discussed below; such a connection is called a "handshake." The user requests access to the server. The client receives a certificate from the server. The server's decryption of a "pre-master key" transmitted by the user ensures the HTTP server's identity. A concluding message for final verification is sent by the user and the server, to ensure that the same session key is possessed by both the parties.

5.3.4.2 HTTPS

The primary function of the HTTPS protocol is to authenticate the websites visited and protect the privacy, integrity, and confidentiality of the data exchanged. The different attacks secured against HTTPS are forging content, manipulating and contorting content, eavesdropping, man-in-the-middle attacks (MITM), etc. HTTPS provides two-way encryption between the sender and receiver.

5.3.4.3 Transport Trust in IoT

One of the foremost challenges in a majority of IoT devices and sensors is to operate within the limited resources for security and transmission. Message Queuing Telemetry Transport (MQTT) and the Constrained Application Protocol (CoAP) are protocols that have proved their worth and encompass multiple features, such as serving as an open standard, being easily implementable, and equipping systems with bandwidth-efficient and energy-efficient communications. Figure 5.6 depicts the transport security in the IoT.

TRANSPORT SECURITY IN IOT

	MQTT + MUTUAL AUTH TLS	AWS AUTH + HTTPS
SERVER AUTH	TLS + CERT	TLS + CERT
CLIENT AUTH	TLS + CERT	AWS API KEYS
CONFIDENTIALITY	TLS	TLS
PROTOCOL	MQTT	HTTPS

FIGURE 5.6 Transport Security in IoT.

Security Vulnerabilities, Challenges, and Schemes

5.3.5 Secure Cloud and Web Interface

An efficient, smart IoT network is one that consists of thousands of intelligent gadgets operating in tandem to improve business management; the cloud is considered to be the jugular vein of an efficient IoT network. The cloud helps provide secure interconnections between devices procuring data and those rendering useful data. This is illustrated in Figure 5.7. An M2M (machine-to-machine) gateway contributes to linking devices and applications.. The IoT M2M gateway is required to obtain telemetry and reconnaissance data. Rules govern how the data would be made available to users, analyzed, and controlled in the IoT cloud.

5.3.6 Secure Software and Firmware

Hardware security has many benefits to offer for the security of an IoT system, such as a strongly built infrastructure. The secure network offered by hardware security offers seamless management for users. Another benefit is multilateral protection, to protect, secure, and defend the overall IoT architecture and Quality of Services (QoS) in the application layer. Also, due to its diverse protections and readily flexible structure, hardware security is getting popular in security-seeking industries such as health care services, smart homes, industrial automation, etc.

The security controls in an IoT infrastructure involve the following basic elements:

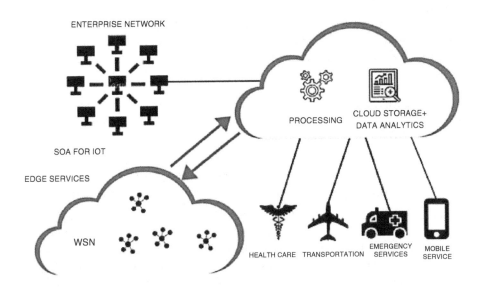

FIGURE 5.7 WSN-Based IoT System.

- Cryptography and Management of Keys
  a. Includes confidentiality, integrity, authentication, authorization, and encryption, which are the cornerstones of cryptography
  b. Crypto-variables, which include entropy source/pool, symmetric keys, and random numbers
  c. Management of keys by storing and maintaining keys and key material accounting
- Protocols for various layers
  d. Device authentication/authorization, ensuring integrity and confidentiality
  e. Network layer authentication, signal communication authentication, and confidentiality
  f. Application layer integrity and signal communication authentication

Multiple software-centric innovations in the past were focused on secure devices. However, the software itself has certain drawbacks and weaknesses.

- To secure software, it is imperative to understand what software is. Software is a program code that is read, analyzed, and executed, and hence, it can be unscrupulously accessed or/and disassembled;
- Therefore, in software-centric protection systems, secret keys are easily recognized by attackers.

Considering the aforementioned nuances of software and their drawbacks, a combination of software security and robust, protected hardware makes a system more reliable and secure. Further, software can be protected by hardware. For a software/firmware update, IoT devices are to be designed in order to be updated with the new version of the installed software/firmware, which could be done using the following steps: Software/firmware decryption ≥ Signature verification ≥ Update the initiation process ≥ Updating the signed software/firmware. These updates of software/firmware offer benefits, in that the trustworthiness of the secured systems and devices is improved. Also, the updated device software that is periodically rolled out helps curb technical problems and fix bugs. Cost efficiency is enhanced by avoiding expensive software update and support calls. Further, there is secure service delivery from service providers to authorized users.

5.3.7 ESTIMATING THE COST OF A CYBER-BREACH: CASE STUDY ON INTELLECTUAL PROPERTY BREACH FOR THE IoT

So far, we have seen exactly why there is a need for cybersecurity; to understand exactly how important it is, we will present a case study. Let us consider the hypothetical case of a company named "Things of Things," in the field of information technology. Based in the United States, the company has roughly 50,000 employees and is evaluated at $40 billion. The primary area of operation for this company is developing software-based management tools for a myriad of IoT technologies. It has a profit margin of roughly 12 percent and has made significant developments in intellectual property through its Research and Development Division. Six months

Security Vulnerabilities, Challenges, and Schemes

before the release of a new product, the company is notified via a federal agency regarding a cyber-breach at one of its facilities that were responsible for the innovation. The product was expected to contribute 25 percent to total revenues over the next half-decade. Although the intention of the attackers is debatable, with this newfound information, they have become a threat to the existing technologies of the company. Further, an investigative media outlet has reported that the attackers are attempting to reverse-engineer the network-based product, hence potentially destroying the market of Things, as the company has invested millions of dollars in its development.

In response, the company hires a high-level public relations firm to coordinate a response and reach out to all those concerned, such as the clients, end-users, and various others, in an attempt to mitigate the damage on the PR front. Things of Things also retains top attorneys and a cyber-forensics team to probe further into the event, so as to identify the cause of the breach and additional vulnerabilities. Further, they hire a cybersecurity firm to help with triage and remediate the breach. During the impact-management phase, the firm is forced to suspend the shipments already planned, while rolling out upgrades for the compromised devices. In order to prevent the complete loss of revenue, they accelerate their device launch by two months, so as to capture the market before the counterfeit product arrives. Amid all this, a loss in investor confidence leads to the suspension of a key contract from the government, as the company has displayed a lack of ability to showcase the protection of their infrastructure, hence becoming a liability. This project was supposed to contribute 5 percent to earnings; this also led to a 5 percent loss, due to a pullback from clients. As part of its long-term management, the firm is forced to undertake a myriad of policy measures, such as an enterprise-wide assessment for strengthened cyber-risk management and a plan of action. This included an intellectual property rights inventory, a sector classification, and prevention strategies, which end up deepening the cost.

The costs of a breach of this nature are ancillary and nonobvious, prima facie. They include the loss of the compromised intellectual property, a disruption on the operational front, lost contracts, increased insurance premiums, a tarnished trade name, and lost investor confidence. Our analysis, based on the 14 impact-based qualitative factors of Deloitte [18], estimates the cost at $3.2 billion, as identified in Table 5.2. The question we face after this case study is whether the elements given are credible or not. In fact, we have a litany of examples about security breaches in or via IoT devices; major cyberattacks have been orchestrated. The report, "The Big Hack," featured in Bloomberg Businessweek in 2018 [19], highlighted how a China-based motherboard manufacturer compromised the entire security infrastructure of a major US firm, through the use of a malicious chip masked as a signal conditioning unit. That chip had the power to change the system's commands and receive instructions from an external source. This event has triggered heated debate regarding IP breaches and cybersecurity, while underscoring the vulnerabilities we have in our current devices [22]. Furthermore, this is not an isolated event. An Annual Cybercrime Report by the company Cybersecurity Ventures estimated that cybercrime-related damages will reach $6 trillion by 2021 [22]. This figure has nearly doubled since 2015, when it was approximated as

TABLE 5.2
Estimated Losses of Things of Things

Cost Factor	Cost (in $ millions)	Percentage of Total Cost (%)
Technical Investigation	1	0.03
Customer Breach Notification	Not Applicable (not PII breach)	0.00
Post Breach-Customer Protection	Not Applicable (not PII breach)	0.00
Regulatory Compliance	Not Applicable (not PII breach)	0.00
Public Relations	1	0.03
Attorney Fees and Litigation	11	0.35
Cyber Security Improvements	13	0.40
Insurance Premium Increase	1	0.03
Increased Cost-to-Debt Ratio	Not Applicable	0.00
Operational Disruption	1200	36.83
Lost Value of Customer Relationship	Not Applicable	0.00
Value of Lost Contract Revenue	1600	49.11
Devaluation of Trade Name	280	8.59
Loss of Intellectual Property	151	4.63
Total	3258	100

$3 billion. Ransomware damages alone amount to $20 billion a year globally. These figures reflect the scale of the threat we face in the context of cybersecurity.

5.4 EXISTING CYBER-ATTACK DETECTION SOFTWARE AND SECURITY SCHEMES

5.4.1 CONVENTIONAL CYBER-SECURITY SCHEMES

This subsection covers various conventional cybersecurity schemes that are being currently employed in conjunction with data to ensure confidentiality and integrity, in conjunction with IoT technologies.

5.4.1.1 Access Control Technologies

Access control technologies are a segment of technology that is at the intersection of physical security and virtual security schemes. The three key tenets of access control technologies are boundary protection, authentication, and authorization.

These technologies aim to prevent unauthorized parties from viewing or accessing data that is outside their security clearance. This forms the basis for all of the layered security models implemented around the world that segment data based on the degree of confidentiality and control accessibility. Boundary protection schemes encompass methods to separate the information of distinctive magnitude by establishing tangible borders or logical boundaries between protected data and the users. The zones are known as demilitarized zones, in common parlance. Examples of boundary protection technologies include host-side firewall systems impeding illicit access via a

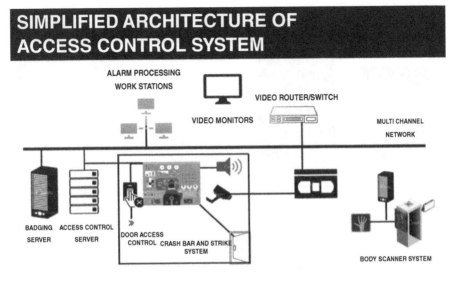

FIGURE 5.8 Simplified Architecture of Access Control Systems.

private server, content management systems, and traffic control for inappropriate content including, but not limited to, spam files or classified information. Figure 5.8 illustrates an example of boundary protection technology and its associated architecture. Technology dependent on authentication works to identify and associate an identity with an individual based on three qualitative types: individual identity, such as biometric data, or iris scan; individual possession, such as smart cards and a token system; and last, individual privilege type, such as a password or code. A two-factor authentication seems to have become the industry norm, so as to reinforce security with access control.

5.4.1.2 System Integrity

Integrity encompasses system reliability or, in other words, an integrity-check mechanism that ensures the system maintains integrity and a malicious payload or attack has not affected it. Antivirus and antispyware software are common examples of technical software for this purpose. Essentially, a system integrity checker is tasked to ensure that malware has not modified, destroyed, or corrupted a system. The malware in question could be a virus, a Trojan horse, a worm, spyware, adware, etc. [23]. This software guards system gateways so as to impede any incoming malware and repair any damage malware may have caused. Figure 5.9 below depicts a simplified architecture for a system integrity checker [24].

5.4.1.3 Cryptography

Cryptography is an indispensable tool for protecting information in computer systems, entailing the cryptographic system and the principle of a shared key. The origin of cryptography can be traced back to the development of the RSA algorithm, which

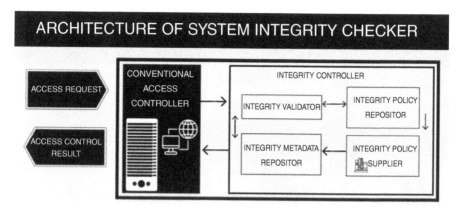

FIGURE 5.9 Simplified Architecture of System Integrity Checker.

was eventually granted a US patent [25]. "Cryptography" is defined as the study of the modification of data in a manner such that it achieves a form that hides its true nature, essentially making it a secret. Cryptography can be divided into three classes of algorithms. First is asymmetric algorithms, which use two keys, one public and one private. A public key enables the conversion of a plain message into cipher text, and the private key enables the decryption algorithm. As the name suggests, the private key is stored on a secure server and is not known to everyone. The second category is symmetric algorithms that involve a single key, which enables the conversion of a plain message into cipher text and cipher text conversion back into a plain message. The last category is hashing, which converts a plain message via a hashing function into a fixed length. This ensures integrity, as the value in the hashing function matches on the sender and receiver sides. VPN, TLS, PPTP, and SSL are a few examples of its implementation.

5.4.1.4 Audit and Monitoring

Audit and monitoring tools record the activities of a system, mapping responses for investigation purposes. Furthermore, they assess the status of security of devices, performing an analysis for attacks that are in progress or have concluded. The primary two classes of software in audit and monitoring are: intrusion detection system, intrusion protection system, S-E correlation, and cyber-forensics.

Intrusion detection can be further bifurcated into misuse detection and anomaly detection. Misuse detection (MD) consists of in-depth information about detected attacks and the weak points of the system, supplied by experts in a manner similar to a knowledge system. MD rummages around for attackers that decide to execute these attacks or gain an advantage based on system vulnerabilities. Though MD is often correct in detecting well-known attacks, these techniques cannot identify cyber-threats that are unknown to the system's knowledge base. Anomaly detection (AD) depends on the assessment of profiles that exhibit conventional behavior of connections in the network, users of the system, and the host. AD identifies

Security Vulnerabilities, Challenges, and Schemes 103

TABLE 5.3
Misuse and Anomaly Detection Tools

Misuse Detection Tools	Anomaly Detection Tools
Data mining techniques	Statistics based
Rule-based approach	Rule-based approach
Algorithms based on state-transition analysis	Distance-based technique
Signature methods	Profiling methods

conventional authorized cyberactivity by employing a plethora of methods and then employs a range of quantitative and qualitative indicators to identify aberrations from the outlined conventional activity, as a prospective anomaly. Here, the advantage is that AD can detect unknown attacks, with the drawback of having a high false-notification rate. It may be noted that the aberrations identified by AD algorithms may not be an instance of aberrations and may in fact be cases of legitimate but unconventional system behavior. Table 5.3 highlights various techniques under the ambit of anomaly detection and misuse detection software.

5.4.1.5 Configuration Management and Assurance Tools

Configuration management and assurance tools concern methods and techniques to verify whether the executed settings on a system are correct/incorrect. The various tools involved are policy enforcement tools, network management tools, continuity of operations tools, and scanners and patch management. Table 5.4 highlights various examples of security discussed here.

5.4.2 EMBEDDED-PROGRAMMING-BASED

Cyberattack detection systems (CADS) account for a crucial part of cyberattack analysis, and often these systems take up their own individual approaches. In embedded-programming-based approaches, a great deal of processing is already

TABLE 5.4
Various Classes of Cybersecurity and Their Subclasses

Class of Security Scheme	Subclasses and Examples
Access Control	Boundary Protection: Firewall and Content Management Authentication: Biometrics, Smart Token, Authorization: User Rights and Privilege
System Integrity	Integrity Checkers and Anti-Virus and Anti-Spams
Cryptography	VPN, Digital Certificates
Audit and Monitoring	IDS, IPS, Correlation Tools, Forensics Tools
Configuration, Management, and Assurance Tools	Policy, Network Management, Continuity of Operations Tools, Scanners, and Patch Management

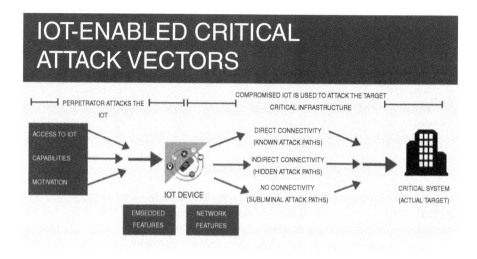

FIGURE 5.10 Critical Attack Vectors in IoT and Associated Infrastructure.

done before the information reaches the CADS, in order to reduce the latter's processing load. This same approach has been implemented in NICs [26], which reduces the traffic for computations and thus achieves a higher processing capacity for the central processor. Figure 5.10 explains the various attack vectors associated with an IoT infrastructure. One of the most common vulnerabilities in an IoT device is a lack of processing capacity, which results in the DoS attack, as explained in Section 2 of this chapter. To avoid such attacks, embedded programming proves to be quite useful.

5.4.3 AGENT-BASED APPROACH

Under the aegis of analysis methodologies, another popular method implemented in a CADS system is an agent-based approach. The working principle behind this paradigm is the ability of servers to exchange information among themselves and inform and alert each other about a possible breach or malicious activities. It may be possible to contain a breach if the infected subnet is disconnected from the main network, essentially limiting the damage. This strategy can work with compromised servers, routers, switches, and other elements of a network. The drawback of this approach is the added work and processing of the detection system to enforce these measures on the network. This approach can be bifurcated into autonomous distributed systems, where they manage and perform the necessary communications with other entities in the environment, and multi-agent systems, which entail four elementary agents: the basic agent, the coordination agent, the global coordination agent, and the interface agent [27]. These individual agents each take up a certain task in the system, so as to efficiently divide the work in CADS.

Security Vulnerabilities, Challenges, and Schemes 105

TABLE 5.5
Examples of Various CADS Systems

S. No	CADS System
1.	Haystack
2.	MIDAS. Expert systems in cyberattack detection
3.	IDES/NIDS.A real-time intrusion-detection expert system
4.	Wisdom & Sense Detection
5.	NADIR An automated system for detecting network attack and misuse
6.	Hyperview. A neural network component for cyberattack detection.
7.	DIDS Distributed Intrusion Detection Systems
8.	ASAX Architecture and rule-based language for audit trail analysis
9.	USTAT State transition analysis
10.	GrIDS. A graph-based intrusion detection system for large networks
11.	Honey Pot
12.	EMERALD: Event monitoring enabling responses to anomalous live disturbance

5.4.4 Software-Engineering- and Artificial-Intelligence-Based

The software used in CADS systems is usually the key function and backbone of the system, hence there is reinvigorated interest among CADS developers to update the technology so as to better the system. A myriad of papers exists in the literature discussing various systems that implement novel programming tools for efficiency [28]. This system implements the signature-based approach, which comes under the aegis of misuse detection and anomaly detection. Further, as artificial intelligence and machine learning gain traction, there is active research in this domain. Techniques include fuzzy logic, genetic algorithm, and artificial neural network. Table 5.5 gives an example of various CADS systems currently in use.

5.5 CONCLUSION AND OPEN CHALLENGES

The coming of IoT devices and technology was followed by the growing threats to privacy, security, and the confidentiality of personal data. These growing technologies, especially those in cyberspace, like machine learning, together with the IoT, are ready to create an everlasting mark on how humans perceive the virtual realm [29]. This comes along with the wide range of applications of the IoT in intelligent transportation systems across metropolitan cities [30], to nanotechnology, biomedical engineering, and bioinformatics [31] Taking into account the extensive amounts of metadata entered into the architecture, and also the ubiquitous nature of the IoT network, it is nearly impossible to avoid or neglect the significance of the IoT in human life. On the one hand, the IoT weaves together a fabric of data based on the objects connected in the infrastructure, which has significantly reduced human effort. But on the other hand, the loopholes left by emerging architecture, the lack of a standard international framework guiding corporations and researchers to mitigate the differences, communication barriers, and address of data privacy require viable and appropriate solutions in the near future [32]. These conditions are worsened even further by the industries that are readily advocating and adopting this advanced technology to better human lives. Such industries can become human nurseries for attackers to get hold of data. This chapter thus emphasizes the architectural challenges with reference to the health care industry and focuses on the need to update IoT network systems and infrastructure to keep up with the challenges of the modern virtual world.

REFERENCES

1. Al-Fuqaha, A., Guizani, M., Mohammadi, M., Aledhari, M., and Ayyash, M., 2015. Internet of Things: A survey on enabling technologies, protocols, and applications. *IEEE Communications Surveys & Tutorials*, 17(4), pp. 2347–2376.
2. Van Kranenburg, R., 2008. *The Internet of Things*. Amsterdam: Institute of Network Cultures.
3. Sharma, D. K., Ajay Kumar, K., Aarti, G., and Saakshi, B., 2020. Internet of Things and Blockchain: Integration, need, challenges, applications, and future scope. In Krishnan, S., Balas, V. E., Golden, J., Robinson, Y. H., Balaji, S., and Kumar, R. (eds.) *Handbook of Research on Blockchain Technology*. London: Elsevier, pp. 271–294. doi:10.1016/b978-0-12-819816-2.00011-3.
4. Pal, S., García Díaz, V., and Le, D., 2020. *IoT*, 1st ed. Boca Raton: CRC Press, pp. 134–157.
5. Jindal, M., Gupta, J., and Bhushan, B., 2019. *Machine learning methods for IoT and their future applications*. In *2019 International Conference on Computing, Communication, and Intelligent Systems (ICCCIS)* (pp. 430–434). Greater Noida, India: IEEE.
6. Babbar, G., and Bhushan, B., 2020. Framework and methodological solutions for cyber security in industry 4.0. *SSRN Electronic Journal*. doi:10.2139/ssrn.3601513.
7. Vashishth, V., Chhabra, A., and Sharma, D., 2019. GMMR: A Gaussian mixture model based unsupervised machine learning approach for optimal routing in opportunistic IoT networks. *Computer Communications*, 134, pp. 138–148.
8. Chhabra, A., Vashishth, V., and Sharma, D., 2017. A fuzzy logic and game theory based adaptive approach for securing opportunistic networks against black hole attacks. *International Journal of Communication Systems*, 31(4), p. e3487.
9. Chhabra, A., Vashishth, V., and Sharma, D., 2017. *A game theory based secure model against Black hole attacks in Opportunistic Networks. In 2017 51st Annual Conference on Information Sciences and Systems (CISS)*(pp. 1–6). Baltimore, MD: IEEE.
10. Gilchrist, A., 2017. *Iot Security Issues*, 1st ed. Boston: De/G Press.
11. Bhardwaj, K. K., Anirudh, K., Deepak Kumar, S., and Chhabra, A., 2019. Designing energy-efficient iot-based intelligent transport system: Need, architecture, characteristics, challenges, and applications. In Mittal, M., Tanwar, S., Agarwal, B., and Goyal, L. M. (eds.) *Energy Conservation for Iot Devices*, pp. 209–233. Springer, Singapore. doi:10.1007/978-981-13-7399-2_9.
12. Varshney, T., Sharma, N., Kaushik, I., and Bhushan, B., 2019. *Architectural model of security threats & their countermeasures in IoT*. In *2019 International Conference on Computing, Communication, and Intelligent Systems (ICCCIS)* (pp. 424–429). Greater Noida, India: IEEE. doi:10.1109/CCCIS48478.
13. Li, S., Xu, L., and Zhao, S., 2014. The internet of things: A survey. *Information Systems Frontiers*, 17(2), pp. 243–259.

Security Vulnerabilities, Challenges, and Schemes

14. Choi, J., Li, S., Wang, X. and Ha, J. 2012. *A general distributed consensus algorithm for wireless sensor networks. In 2012 Wireless Advanced (WiAd)* (pp. 16–21). London: IEEE.
15. Fielding, R., and Taylor, R., 2002. Principled design of the modern Web architecture. *ACM Transactions on Internet Technology (TOIT)*, 2(2), pp. 115–150.
16. van Kranenburg, R., and Bassi, A., 2012. IoT challenges. *mUX: The Journal of Mobile User Experience*, 1(9). doi:10.1186/2192-1121-1-9.
17. Peris-Lopez, P., Hernandez-Castro, J., Estevez-Tapiador, J., and Ribagorda, A., 2009. Cryptanalysis of a novel authentication protocol conforming to EPC-C1G2 standard. *Computer Standards & Interfaces*, 31(2), pp. 372–380.
18. Deloitte United States. 2020. Deloitte identifies 14 business impacts of a cyberattack: Press release. https://www2.deloitte.com/us/en/pages/about-deloitte/articles/press-releases/deloitte-identifies-14-business-impacts-of-a-cyberattack.html Accessed on April 1, 2020.
19. Robertson, J., and Riley, M., 2020. The big hack: How China used a tiny chip to infiltrate America's top companies. Bloomberg.com. https://www.bloomberg.com/news/features/2018-10-04/the-big-hack-how-china-used-a-tiny-chip-to-infiltrate-america-s-top-companies. Accessed on April 1, 2020.
20. Cybercrime Magazine, 2019. Cybercrime damages $6 trillion by 2021. https://cybersecurityventures.com/cybercrime-damages-6-trillion-by-2021/. Accessed on September 5, 2019.
21. Ning, H., Hong, L., and Yang L. T., 2013. Cyberentity security in the Internet of Things. *Computer*, 46(4), pp. 46–53. doi:10.1109/mc.2013.74.
22. Sridhar, S., and Manimaran, G., 2010. *Data integrity attacks and their impacts on SCADA control system.* In *IEEE PES General Meeting*, Providence, RI: IEEE. pp. 1–6. doi: 10.1109/PES.2010.5590115.
23. Quisquater, J., and Couvreur, C., 1982. Fast decipherment algorithm for RSA public-key cryptosystem. *Electronics Letters*, 18(21), p. 905.
24. Otey, M., Parthasarathy, S., Ghoting, A., Li, G., Narravula, S., and Panda, D., 2003. *Towards NIC-based intrusion detection.* In *Proceedings of the Ninth ACM SIGKDD International Conference on Knowledge Discovery and Data MiningKDD'03* (pp. 723–728). Washington, D.C.: Association for Computing Machinery.
25. Ran, Z., Depei, Q., Chongming, B., Weiguo, W., and Xiaobing, G., n.d. *Multi-agent based intrusion detection architecture.* In *Proceedings 2001 International Conference on Computer Networks and Mobile Computing* (pp. 494–504). Beijing: IEEE.
26. Vigna, G., Valeur, F., and Kemmerer, R., 2003. Designing and implementing a family of intrusion detection systems. *ACM SIGSOFT Software Engineering Notes*, 28(5), p. 88.
27. Abraham, A., Jain, R., Thomas, J., and Han, S., 2007. D-SCIDS: Distributed soft computing intrusion detection system. *Journal of Network and Computer Applications*, 30(1), pp. 81–98.
28. Bhardwaj, K. K., Khanna, A., Sharma, D. K., and Chhabra, A., 2019. Designing energy-efficient IoT-based intelligent transport system: Need, architecture, characteristics, challenges, and applications. In Mittal M., Tanwar S., Agarwal B., Goyal L. (eds.) *Energy Conservation for IoT Devices: Studies in Systems, Decision and Control*, vol. 206. Springer, Singapore, pp. 209–233.
29. Bhardwaj, K., Banyal, S., and Sharma, D., 2019. Artificial intelligence based diagnostics, therapeutics and applications in biomedical engineering and bioinformatics. In Balas, V. E., Son, L. H., Jha, S., Khari, M., and Kumar, R. (eds.) *Internet of Things in Biomedical Engineering*. London: Elsevier, pp. 161–187.
30. Jain, A., Crespo, R., and Khari, M., 2020. *Smart Innovation of Web of Things*, 1st ed. Boca Raton: CRC Press, pp. 21–51.

31. Banyal, S., Bhardwaj, K., and Sharma, D., 2020. Probabilistic routing protocol with firefly particle swarm optimisation for delay tolerant networks enhanced with chaos theory. *International Journal of Innovative Computing and Application*, 12(2), pp. 25–37.
32. Goel, A. K., Rose, A., Gaur, J., and Bhushan, B., 2019. *Attacks, countermeasures and security paradigms in IoT*. In *2019 2nd International Conference on Intelligent Computing, Instrumentation and Control Technologies (ICICICT)* (pp. 875–880). Kannur, Kerala, India: IEEE.

6 Advanced Security Using Blockchain and Distributed Ledger Technology

M. Al-Rawy
IkubINFO, Tirana, Albania

A. Elci
Hasan Kalyoncu University

CONTENTS

6.1 Introduction 110
6.2 Blockchain and Distributed Ledger Technology (DLT)..............................111
    6.2.1 Blockchain Types...112
        6.2.1.1 Permissionless Blockchain..112
        6.2.1.2 Permissioned Blockchain 113
        6.2.1.3 Hybrid Blockchain 114
        6.2.1.4 Classification Scheme 114
        6.2.1.5 Blockchain Mechanism...114
        6.2.1.6 Cryptography...114
        6.2.1.7 Peer-to-Peer Distributed Network.....................................116
        6.2.1.8 Smart Contracts...117
    6.3 Structural Difference between Traditional Databases and
        Blockchain 118
    6.3.1 Immutability of Blockchain 119
    6.3.2 Performance 120
    6.3.3 Robustness and Disintermediation..120
    6.3.4 Blockchain Technology as the Future of Cybersecurity 120
6.4 Future Trends of Blockchain..122
6.5 Blockchain beyond Bitcoin 122
6.6 Threat Management and Defense..123
6.7 Discussion 126
6.8 Conclusion...126
References..127

6.1 INTRODUCTION

Computer-use has been widespread since the late 1970s, with IBM's first personal computers appearing in the early 1980s. Initially, hacking was a behavior that seemed to be just a distraction for technology-crazed teenagers. Since then, however, it has evolved into international warfare. Nowadays, the internet has become the field for new types of conflicts, with all electronic destruction tools, such as espionage and malware dedicated to the destruction of governmental and nongovernmental websites, as well as efforts to control and manipulate the sensitive data of vital facilities, such as power plants and nuclear reactors, not to mention the destruction and theft of bank and state deposits. Such adverse outcomes have led researchers to develop electronic methods to adopt cryptography, to safeguard and protect data during transportation and storage; encryption is a scientific response in that line. In 2008, Satoshi Nakamoto[1] introduced his by-now famous white paper [1], which revolutionized the world of cybersecurity. He brilliantly combined encryption algorithms to achieve ultimate immutability and provide sufficient protection to transfer and store data in a distributed manner, using a well-known approach called distributed ledger technology (DLT) [2]. Also known as the cryptocurrency transaction log, this keeps data resistant to tampering with an ever-growing blockchain. DLT allows the secure exchange of transactions of valuables such as funds, stocks, or data access rights. Unlike conventional trading systems that require a broker or central control node to follow the exchange of transactions, blockchain enables all parties to communicate and transport transactions directly to each other.

Blockchain is an immutable database that is consensually shared and synchronized across a community of willing participants, all equally capable of extending it. Bitcoin was the first real-world application of the blockchain, which is used to this day. The invention of Ethereum [3] followed Bitcoin, and also other electronic currencies that emphasized peoples' needs in the financial sector for transparency and decentralization in today's rapidly evolving world. Not limited to the financial industry, blockchain technology is also utilized by many different industries, such as digital copyright [4,5], voting [6], healthcare [7,8], electronic governance [9,10,11], and the Internet of Things (IOT) [12,13], while many others [14,15] have shown interest in it and made efforts to utilize it, to achieve full trust in safeguarding data.

This chapter contributes a systematic review of the emergence of blockchain, the DLT, and how they have been applied to tackle security issues (e.g., trust, immutability, availability, transparency, etc.) in many real-world industries. A set of key references are surveyed, as well as potential implementations of blockchain in nonfinancial domains. A great deal of research focuses on the security of the blockchain and its applications, for example [16,17,18,19]; however, this research focuses on the specific structure of the blockchain and its superior features, which have the potential to combat the vulnerabilities of the networks of traditional databases. This research guides the reader to understand the mechanism of different types of blockchains, as well as the structural differences between traditional databases and blockchain that can enhance the security of systems.

---

[1] Nakamoto is a pseudonym used by the blockchain innovator to identify himself/herself in the original white paper for bitcoin, released in 2008.

Advanced Security Using Blockchain and Distributed Ledger Technology

In this chapter, we introduce what blockchain and distributed ledger technologies are, their types, the differences between them, and their features, in Section 6.2. Section 6.3 discusses the differences between the traditional database and blockchain-based data storage. The importance of the blockchain technology in the future of cybersecurity is analyzed in Section 6.4. The likely future trends of blockchain, and its role beyond Bitcoin, are discussed in Sections 6.5 and 6.6, respectively. Security aspects, especially threat management, and defenses are taken up in Section 6.7. Sections 6.8 and 6.9 present a discussion and conclusions.

6.2 BLOCKCHAIN AND DISTRIBUTED LEDGER TECHNOLOGY (DLT)

Nakamoto introduced blockchain as a linear series of packages (called "blocks"), as a ledger that can preserve data stored within it and prevent modification [20]. A blockchain ledger is entirely replicated over fully distributed nodes of a peer-to-peer network, which is called a DLT [2]. Blockchain is also defined as a digitized decentralized public ledger for all cryptocurrency transactions. Currently and primarily, blockchain is used with digital currency, although it is possible to code, digitize and insert any kind of data token into a blockchain.

The emergence of DLT, with the full-distribution principle, has enabled a new level of protection by doing away with the central node or a dominant party (i.e., the nonexistence of a single data center). Put more simply, every node contains a full copy of the entire data of any software built on top of the blockchain, and all work as data centers, which is the essence of this technology. Figure 6.1 shows different types of networks: centralized, decentralized, and distributed.

A blockchain network relies on a group of nodes, instead of a single data center, where every node performs many tasks, storing transactions, verifying them, and communicating and transmitting processes that occur inside the system. In this section, we illustrate blockchain types and the mechanism behind this novel technology.

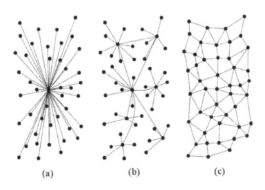

FIGURE 6.1 Network types: (a) centralized, (b) decentralized, and (c) distributed.

6.2.1 BLOCKCHAIN TYPES

The open-source idea of this undeniably genius invention (DLT) has enabled the development of many iterations of distributed ledgers, and consequently, cryptocurrency communities have striven to design a one-size-fits-all solution to blockchain's shortcomings. For instance, Ethereum formed its blockchain by using different properties than Bitcoin's blockchain, to be more practical, in their view [21]. Likewise, many so-called cryptocurrencies have emerged to compete with Bitcoin, in terms of anonymity. States and stakeholders started using the core principles of distributed ledger technology after changing its nature from being open network to being private (permissioned), for example, at banks. These efforts and endeavors have led to the emergence of many designs, such as Hashgraph [22], Tangle [22], the directed acyclic graph (DAG) [23], and so on. In this section, we classify blockchains in three different classes: permissionless, permissioned, and hybrid, which combines both aforementioned types. Figure 6.2 provides a simple illustratration for the classification of blockchain types.

Every type of blockchain has advantages and disadvantages, based on its usage. More accurately, this depends on user needs, along with system requirements, cost, transparency, speed, and scalability. The following illustrates those types and the differences between them.

6.2.1.1 Permissionless Blockchain

Permissionless (or "public") blockchain is the original type of blockchain, which is being used as a base technology for Bitcoin. It is known as permissionless, because it allows anonymous participants to join the network when the system is democratic. More simply, any participant is authorized to create blocks, verify transactions, and transfer or contribute data, as long as the outcome becomes validated through the consensus protocol [24]. This protocol ensures that the nodes agree on a unique order in which entries are appended to the chain [25]. The permissionless blockchain is:

Public: Any person/organization (known or unknown to the network) can join the network, access it, extend it, verify transactions in it, and create blocks within it, keeping in mind that the block created will only be accepted by consensus, after it is validated.

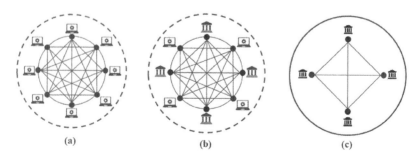

FIGURE 6.2 Blockchain types: (a) public blockchain, (b) hybrid blockchain, and (c) private blockchain.

Advanced Security Using Blockchain and Distributed Ledger Technology **113**

Transparent: Blockchain keeps blocks' data transparent/open to the public in a way that has not existed in financial and nonfinancial systems, as a new standard for transparency.

Available: A public blockchain network is a compound of many groups of nodes, all connected (peer to peer) while being unknown to each other; as the number of nodes increases, the strength and availability of the system also increase. This structure allows the network to keep working, unaffected by the failure of one or many nodes, regardless of the reason for the failure, whether it be a malfunction or a falsification. Thus, the network continues its work, either by extracting the failed node or by reconstructing a sound copy of the data and forcing it on the corrupted node, if available.

Distributed: As mentioned earlier, a blockchain is a decentralized/distributed digital ledger that does not depend on a central authority; instead, it replicates itself across many nodes, enforcing infinite immutability on the records involved. Thus, those records cannot be modified retroactively, without modifying/altering all subsequent blocks within the chain.

Disintermediated: This means, appending blocks to the blockchain does not require any third-party intermediary or dominant node to agree to the process. Acting alone, anyone can validate and append blocks to the blockchain. However, via the consensus protocol, the new blockchain must earn 51 percent of network participants' votes regarding its validity, to be accepted in the network as the longest chain (the latest valid version).

The Bitcoin and Ethereum cryptocurrencies are examples and live proof of the credibility of these characteristics regarding permissionless blockchain.

6.2.1.2 Permissioned Blockchain

Private blockchains were designed by taking advantage of public blockchain technology, but in a converse manner: they work by setting up certain groups or entities to participate internally, in a closed network of allowed nodes. This means the network parties are known to each other and proven to be trusted, unlike the public blockchain, which allows anyone to join the network.

There are some major differences between public and private blockchains, as regards validating transactions, and managing and auditing privileges that are confidential and could only be effected by the owner of the permissioned, a.k.a. "private," blockchain. As for the reading authority, it is also subjected to the owner's/company's decision. It could be open to the public or restricted to an authorized person or group of people. Special access and privileges must be given to the person or party who wants to make a change, in such a blockchain.

A private blockchain is appropriate to more traditional business and governance models, due to its nature, which requires it to be more centralized. This technique could also enhance the world of cybersecurity, by leveraging its scalability and cryptography standards to gain the necessary confidentiality. Private blockchains provide better scalability, by limiting the number of network participants. This also helps in achieving a higher processing speed [26]. Hyperledger Fabric is one of the best-known private blockchains. The Hyperledger Fabric platform was launched in 2015, by the Linux foundation, as an open-source blockchain that aims to improve performance and reliability, besides supporting global business transactions in financial and nonfinancial large enterprises [27].

6.2.1.3 Hybrid Blockchain

On the one hand, public blockchains are considered slow and expensive, and they require high computational power. On the other hand, private blockchains have a limited number of participants, with limited privileges, while also having a more centralized nature, which leaves them prone to manipulation by malicious actors. However, both types also have efficient characteristics. This has led researchers to invent a mixture of public and private blockchain, known as a hybrid blockchain. This type of blockchain combines the efficient qualities of both types of blockchains, while attempting to reduce their flaws. Thus, state institutions, enterprises, and stakeholders can use such a technology to limit access privileges to trusted authorities, for example, to perform modifications and maintenance on some information or data within the private/concealed section, while exposing unconcealed information and transactions to the public, to read and verify it, in terms of transparency. XinFin is a real-world example of a hybrid blockchain [28].

6.2.1.4 Classification Scheme

In this section, we propose a scheme of private and public infrastructures classified by access, speed, security, identity, asset, cost, energy consumption, and approval frequency, as illustrated in Table 6.1.

6.2.1.5 Blockchain Mechanism

This section dives deeper into the blockchain's architecture and operational mechanism, to show how it works and how it could enhance data security.

6.2.1.6 Cryptography

Blockchain is composed of interconnected blocks concatenated cryptographically to each other in sequence. As illustrated in Figure 6.3, each block consists of two parts: (1) a block header and (2) block contents:

1. Block header
   a. Version
   b. Timestamp

TABLE 6.1
Permissionless and Permissioned Blockchains: Comparison

Characteristic	Public	Private
Access	Does not require Read/Write privilege to participate (public/permissionless)	Requires permission to Read/Write (permissioned/restricted)
Speed	Slower	Faster
Security	Proof of Work, Proof of Stake, etc.	Preapproved participants
Identity	Anonymous/Pseudonymous	Known/identified identities
Asset	Native asset	Any asset
Cost	High cost	Lower cost
Energy consumption	Large energy consumption	Lower energy consumption
Approval time	Long transaction-approval time	Long transaction-approval time

Advanced Security Using Blockchain and Distributed Ledger Technology

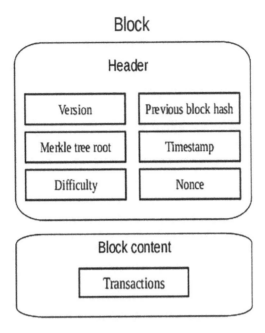

FIGURE 6.3 A single block of a blockchain.

  c. Previous block hash, which is the link that connects a block to the previous one in the chain, as shown in Figure 6.3.
  d. Nonce, a term for a "number only used once," which is a number added to a block incrementally, until a challenge is found that gives the block's specified hash zero bits [1].
  e. Difficulty
  f. Merkle tree root, which is a hash of the selected transactions to be part of the block, combined computationally in a Merkle root tree.).
2. Block contents, i.e., a list of transactions.

A blockchain's structure provides assurance of the integrity of the stored transactions and the entire chain to the very first block (the "genesis block"). A genesis block (block 0) is the first block of a blockchain, which is almost hardcoded into every application that employs the blockchain.

A block's contents of header and body are encrypted, to aggregate a unique hash value, "block hash," for that block, to achieve block integrity, as well as being stored in the next block, as shown in Figure 6.4.

The adoption of the SHA256 encryption algorithm provides an excellent advantage, where the length of the hash is constant and unique regardless of the input, any miner change leads to an alteration in the hash value. Besides, any change in the hash values of the blocks leads to the collapse of the chain. This remarkable structure, one of the masterpieces of this technology, is also one of the most important protection features that have been contributed to the cybersecurity world.

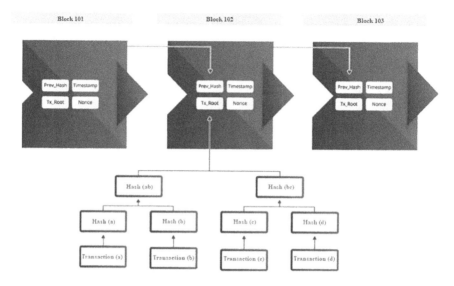

FIGURE 6.4 Blockchain hashing.

6.2.1.7 Peer-to-Peer Distributed Network

The term peer-to-peer (P2P) refers to the exchange of values between peers via a distributed network. A P2P platform enables network participants to execute trades directly between each other, without requiring intermediaries. P2P network architecture is an important quality of distributed ledger technology. A peer, also known as a node, is any device connected voluntarily to the network (i.e., computer, mobile, etc.), which is responsible for verifying transactions and can create blocks. Once a device participates in the network, it receives a synchronized replica of the blockchain. Each node is an administrator that performs the same work, like all the other nodes. Each node has an incentive for joining the network; for example, considering cryptocurrencies, it will have an equal opportunity to mine Bitcoins or Ethereum, according to the protocol of the competition.

In P2P networks, distributing transactions initiates by defining consensus protocol [25]. There are many types of competition protocols used in permissionless and permissioned blockchains, such as PoW (Proof of Work), PoS (Proof of Stake), DPoS (Delegated Proof of Stake), PBFT (Practical Byzantine Fault Tolerance), and so on [25]. Proof of Work and Proof of Stake are the protocols best known today for mining cryptocurrencies in the financial industry.

Proof of Work [29]: PoW is a mining algorithm used to validate/confirm transactions in the original blockchain, to produce new blocks into the chain. The concept of PoW has existed since the mid-1990s. It was submitted in 1996 by Adam Back under the name of "Hashcash" [30]. This mechanism enables the anonymous peers of a network to reach an agreement, while ensuring security. In this mechanism, a miner has to find a block hash result that starts with a number of zeros, which reflects the

Advanced Security Using Blockchain and Distributed Ledger Technology   **117**

number of miners on the network. The greater the number of zeros, the more difficult it is for the miner to find the result. This mechanism is also applied in Bitcoin's blockchain, to force miners to compete against each other to decide who wins the competition, according to the mining process speed. The winner, i.e., the first to create the block, is rewarded for achieving the distributed consensus to extend the blockchain.

The term consensus refers to the process of gaining approval votes from network participants to decide whether the newly created block is valid, to accept the new (extended) blockchain as the longest version. In permissionless blockchain, anyone can create a block, whether is it valid or not, regardless of its legitimacy. Thus, this protocol serves to warrant the integrity and validity of the chain, by verifying the legitimacy of the new block, along with the transactions included in it, at the same time preventing double-spending attempts. This algorithm is currently being used by many cryptocurrencies, starting with Bitcoin, Ethereum, and others. However, the disadvantage of this protocol is that it requires tremendous computational time and high-power computing hardware.

Proof of Stake [31]: The PoS algorithm achieves the same objective as the PoW distributed consensus; however, the mechanism of the miners' competition is different from that of PoW. This algorithm also forces miners to compete against each other by betting a certain amount of coins as their stake in the network. The more a miner bets, the greater his chances of winning the competition to create blocks and receive rewards. In order to prevent the wealthiest node or miner in the network from always winning, the process implements "Randomized Block Selection" and "Coin Age Selection" methods.

This peer-to-peer network is based on strong synchronization, i.e., any change occurring in one node notifies the neighboring nodes, and hence 51 percent of network nodes must agree that the update is valid, using a consensus algorithm. Then, every node within the network overwrites its replica with the latest copy of the updated blockchain. In case the update is manipulated or incorrect (i.e., the chain is broken), the network will exclude the corrupted node or destroy the modified version, to overwrite it with the distributed confirmed version of the blockchain.

6.2.1.8 Smart Contracts

The smart contract (also known as a crypto-contract) was first introduced by Nick Szabo in 1994 [32]. In his paper, he defined smart contracts as computerized transaction protocols self-executing the terms of a contract. Later on, in 2008, Nakamoto introduced smart contracts also as computer code, but running on top of a blockchain, directly controlling the transfer of digital assets and enforcing an agreed-upon performance between the two parties under the terms of the contract [33]. This is the simplest form of decentralized automation. Smart contracts are very complex electronic contracts, where the functionality extends beyond the transfer of assets, to carry out transactions in a wide range of domains, such as premiums, property, legal processes, collective financing agreements, and so on [34]. Smart contracts enable efficient characteristics for the executed transactions, such as traceability, transparency, and irreversibility. In addition, this type of contract is executed directly by the contracting parties themselves, which facilitates the need to run a lot of routine

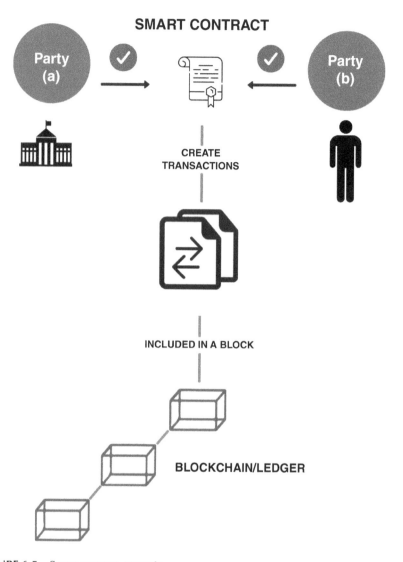

FIGURE 6.5 Smart contract, example.

processes and saves the expense of arranging for third-party services, such as those provided by lawyers or banks, or incurring large fees, as shown in Figure 6.5.

6.3 STRUCTURAL DIFFERENCE BETWEEN TRADITIONAL DATABASES AND BLOCKCHAIN

DLT is a form of a distributed database. Both blockchain and traditional databases store data, but there are fundamental differences in their structures. The traditional database has a central authority (administrator) to maintain, control, replicate/

Advanced Security Using Blockchain and Distributed Ledger Technology

distribute the data, and distribute read/write privileges to other authorities. Also, traditional databases run over a centralized network. Often, databases run in private networks behind a firewall in data centers, where control, security, and trust are handled by big enterprises that provide storage services. By contrast, blockchain is a digital ledger that receives, encrypts, distributes, and stores transactions, without requiring a trusted third party (i.e., disintermediation). For example, today's Bitcoin, for cryptocurrency [35]. The blockchain storage method is formed by a linear series of blocks, each connected to the previous one in a sequence that provides high data immutability and transparency. In this section, we highlight features that make blockchain superior, as a base for data.

6.3.1 Immutability of Blockchain

In contrast to centralized databases, stored data in the blockchain is not erasable and is almost impossible to modify. Blockchain is unlike a traditional database, which allows users to perform create, read, update, and delete operations (C.R.U.D); blockchain allows only insert and read operations. Blockchain stores transactions in a block, then links them to each other by a Merkle root algorithm [36], as shown in Figure 6.6. Then, blockchain connects all the blocks by encrypting the blocks' information, including the previous block's "hash" to form an interconnected chain.

A Merkle root hash is the ultimate encryption result for the transactions included in a block, to provide the immutability of the transaction. Therefore, any change, no matter how simple, in the transaction within the block leads not only to changing the Merkle root hash but also to changing the block hash, leading to a break in the chain.

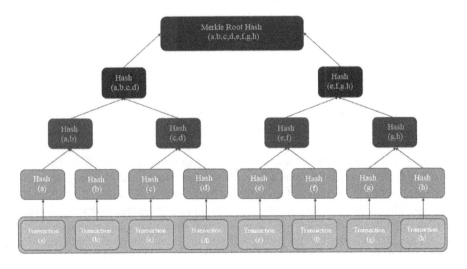

FIGURE 6.6 Merkle root tree, example.

6.3.2 Performance

Traditional databases are very fast compared to the blockchain, because the latter uses cryptography to link its blocks and also employs the consensus principle to provide full distribution, by allowing the majority of peers to agree on the outcome of the transaction, to accept it into the chain. Every node in a blockchain network contains a copy of the full data ledger, so any node experiencing failure will be shut down, not affecting the network that completes its work over the other nodes. In the case of distributed databases, the trust factor between the parties must be significantly guaranteed to preserve data integrity. Blockchain operationally repeals the trust factor, to allow an independent transfer of transactions regardless of the identity of the parties. In conclusion, traditional databases are fast but not fully distributed. Permissionless blockchains use the principle of consensus, which makes them slow; however, permissioned blockchains are relatively fast, while not exactly fully distributed.

6.3.3 Robustness and Disintermediation

Blockchain transactions contain their proof of validity and their proof of authorization, instead of requiring some centralized application logic to enforce those constraints. Transactions can, therefore, be verified and processed independently by multiple "nodes," with the blockchain acting as a consensus mechanism to ensure those nodes stay in sync. This mechanism is the essence of the blockchain, which enables the sharing of ledgers across boundaries of trust, without requiring a central dominator node.

6.3.4 Blockchain Technology as the Future of Cybersecurity

Nowadays, current technological developments facilitate the generating and accessing of data everywhere. This creates difficulties as people try to maintain their privacy, especially their personal and sensitive data stored and transmitted over internet networks by smart phones and computers. Data is the most valuable asset in today's world (the information era), and it is the tool that helps the world's largest known companies to grow bigger, from Wikipedia, Facebook, and Amazon to other social networking sites. Thus, protection methods should have evolved, to combat the currently developed methods of intrusion, penetration, and destruction.

Before the emergence of distributed ledger technology, organizations used to privilege a limited number of trusted employees to access sensitive data. However, trustworthiness is insufficient as a factor to preserve data security. There are many questions about the concept of trust in this area, such as whether an employee is reliable. The answer will always be vague and is likely to change over time. Trusting an unworthy intermediary leads to the loss or theft of sensitive data or intrusions into it. In the last decade, the blooming of blockchain and its first implementation (Bitcoin) have encouraged the world to consider the possibility of dispensing with the trusted party, whether an organization or a person. Blockchain provides the following key benefits to improve cybersecurity, due to its intrinsic design features:

Advanced Security Using Blockchain and Distributed Ledger Technology

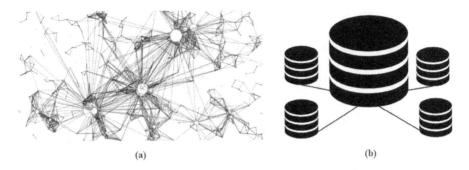

FIGURE 6.7 (a) Blockchain network structure, (b) Traditional database network structure.

1. Distribution Aspect to Secure Data: Blockchain distributes data across the entire network, to be stored in every node in a replicable manner, instead of using a single repository of data or cloud storage. Figure 6.7 shows the differences between the structures of a blockchain network and a traditional database network. Hence, data loss is almost reduced to zero, because, as mentioned earlier, if one or more nodes go down, the data is not affected, unless the entire network of all nodes fails, which has an extremely low probability. Cutting out the middle man, the central authority, or the trusted third party in processing data achieves a distributed immutable ledger of data records.
2. Encryption and Validation: Blockchain is built on two foundations, the distribution and serialization of blocks, a mechanism that provides data credibility and integrity (irrevocable data). Let us take a scenario of file integrity verification, by generating a digital signature for a particular file and then storing it in a blockchain. The integrity of the file can be confirmed by matching the stored signature with the validated file signature, to verify that no alteration has affected the file as long as the blockchain network has been running. On this basis, the encryption and verification of the blockchain help to increase the protection of data from unwarranted reading, manipulation, and forgery.
3. Difficulties in Hacking: Traditional database networks are often vulnerable to malicious attacks that allow tampering and cause data corruption. A blockchain network can counter hacking attempts, as proven in its first application, Bitcoin. This technology encrypts data, appending it to blocks and then connecting those blocks, after which data is distributed across the peer-to-peer network, to obtain sufficient data privacy. Therefore, attackers must break the majority of the network nodes at once, which is practically impossible, because it would require infeasible computing power and time.
4. Blockchain Diversity: The blockchain "fever" that began in the last decade led to the creation of various versions, such as private, public, and hybrid blockchains. Every type provides beneficial features that can be utilized in various domains. For example, public blockchain provides transparency and full distribution (decentralization), which in turn is very beneficial for many kinds of transactions,

for example, involving money, contracts, ownership, and copyrights. Private blockchain also allows the use of a P2P network, but is partially distributed among known parties, which in turn is suitable for banks, transportation companies, and health care companies. Hybrid blockchain would cater to a mixture of both private and public blockchain operations, depending on its design.

6.4 FUTURE TRENDS OF BLOCKCHAIN

The Gartner Research Company has announced that the second most popular word searched for on their website is "blockchain" [37]. Similarly, distributed ledger technology (DLT) continues to gain great prominence in many industries. However, blockchain technology is still in its early stages of prominence, as the potential of this technology has yet to be discovered by big brands. Walmart, IBM, and Amazon are exploring ways of adopting the technology. Aforementioned are only a few examples of how wide-ranging the application of this technology will be, moving forward. Who knows what the 2020s will bring? Organizations will start to build their applications on top of blockchain technology to achieve enhanced security, since this technology has proved an ultimate ability to protect sensitive data from penetration and counteract common cyberattacks [38]. Blockchain-based application development will revolutionize technological advancement.

In addition, blockchain will improve the Internet of Things (IoT), by utilizing decentralized communication models to achieve reliability, reduce risk, decrease costs, and speed up transactions. The dependency on centralized communication models to interact with the system is the major drawback of the IoT, where devices are identified, validated, and connected through centralized cloud servers. In the next section, we will see how blockchain is moving beyond cryptocurrency, and how it is revolutionizing many industries.

6.5 BLOCKCHAIN BEYOND BITCOIN

Many security concerns have been addressed by blockchain technology, such as privacy, ownership, trust, forgery, and so on. Blockchain provides features to support the cybersecurity world, such as audit trails, distributed storage, confidentiality, and disintermediation, as a solution for these security issues. The real difference between centralized systems and blockchain-based systems is that the latter utilizes code and algorithms to do away with the need for a central authority. Fundamentally, any central authority could be unreliable, noncooperative, prone to error, or expensive. By utilizing blockchain when transferring funds around the world, for example, there would be no need for intermediary banks. Similarly, when buying a house or any property, parties can transact business directly between themselves, without the need for a real estate agent to complete the transaction. Below, we discuss different use cases that take advantage of blockchain.

1. Smart Contracts: Self-executing smart contracts currently are the most popular application of blockchain technology. This case is similar to a normal real-world contract, but it replaces the third party who legalizes and monitors

Advanced Security Using Blockchain and Distributed Ledger Technology **123**

contracts with cryptographic code, facilitating agreements between two or more parties without requiring a middleman [39,40].

2. Money Transfers across Borders: Blockchain enables transferring money overseas 24/7, without the intervention of a bank or any third-party broker to impose possibly high fees and take a long time to process the transaction. Using blockchain, money can be transferred between two or more people in a way that is easier, faster, low cost, and affords absolute transparency [41].

3. Digital IDs: Research and experiments on building an electronic identity on top of blockchain have gone a long way in recent years. Almost one billion people around the world do not have an electronic identity. As an attempt to overcome this problem, Microsoft has begun to create electronic identities based on blockchain [42]. Digital IDs would serve to recognize refugees and impoverished people, for example, to bring them into the formal financial sector [43,44].

4. Supply Chain Management: Blockchain technology has enormous potential for the supply chain industry. This technology dispenses with paper and traditional databases, by adopting blockchain technology for trackable, transparent, immutable, verifiable, and accurate records of the life cycle of products, allowing both customers and businesses to track a product's trail back to the source of origin [45,46,47].

5. Digital Voting: Claims of election rigging and tally fraud are increasing every year worldwide. By integrating electronic identity into elections systems and storing votes in blockchains' immutable ledgers to eliminate tally fraud, sound and fair elections could be achieved [6,44,48]. The Sierra Leone government was the first to employ blockchain in the voting system for vote verification purposes; other countries have followed.

6. Decentralized Applications (DApps): DApps are applications built on top of a P2P network of servers, rather than a single server, where everyone can maintain their data and deal with others, without the need for a middleman. This is unlike today's apps (think Amazon or Facebook) that require a third party to function or manage users' data or information [49].

7. Financial Markets: Blockchain is often associated with the smart contract. The global market is exploring means to utilize this concept for financial assets, such as securities, derivative contracts, and currencies. Utilizing this technology in this market protects market participants and societies at large [50,51].

Blockchain technology is revolutionizing many industries beyond Bitcoin, as shown by the examples listed in Table 6.2.

6.6 THREAT MANAGEMENT AND DEFENSE

As blockchain is a relatively recent technology, it is an active research and practice area [16]. There are some inherent conceptual problems for which different solutions are being developed, according to the consensus algorithm and the network features. In this section, we describe some limitations, grouped by attack risk, that can affect the overall efficient functioning of blockchain technology.

TABLE 6.2
Blockchain-Based Financial and Nonfinancial Applications

Category	Type	Apps	Running	Developing	Contemplating
Nonfinancial	Cybersecurity	Guardtime [52]			
		REMME [53]			
	Media	Kodak [54]			
		Ujomusic [55]			
	Real Estate	Ubiquity [56]			
	Health Care	Gem [57]			
		Everledger [58]			
		MedRec [59]			
		SimplyVital Health [60]			
	Manufacturing	Provenance [61]			
		Hijro [62]			
		Blockverify [63]			
		STORJ.io [64]			
	Government	Dubai [65]			
		South Korea [66]			
		Estonia [67]			
	Transportation and Tourism	Arcade City [68]			
		Webjet [69]			
	Smart Contracts	Slock.it [70]			
		Neo [71]			
	Voting	Follow My Vote [72]			
		BlockVotes [6]			
		Blockchain-based digital voting system [44]			
		eVoting [41]			
	Decentralized Internet of Things (IoT)	Filament [73]			
Financial	Cryptocurrencies	Ethereum [21]			
		Ripple [74]			
		Monero [75]			
		Litecoin [76]			

1. Fifty-One Percent Majority Attack: Theoretically, the consensus mechanism of blockchain has 51 percent vulnerability, which can be exploited by attackers to compromise at least 51 percent of the nodes of a network of blockchain. Exploiting this vulnerability may be possible by joining some nodes of the network to aggregate a hash power that is equal to or more than 51 percent of the blockchain's total hash power, therefore dominating the competition. This vulnerability may be exploited by a malicious attacker to carry out the following attempts:

Advanced Security Using Blockchain and Distributed Ledger Technology 125

    a. Run a double-spending attack, which means spending the same credit twice, in the Bitcoin case.

    b. Change transactions' order inside blocks.

    c. Destroy the network by preventing fair mining competition, by dominating block generation for themselves.

However, in a PoW, while a majority attack is theoretically possible, it is practically infeasible.

2. Selfish Mining Attack: This is another fundamental problem regarding blockchain network security. Selfish miners can generate a private fork of the chain, and after that, make it public, as an attempt to gather more rewards, indifferent to the honest miners or fair competition. Qianlan Bai et al. [77] establish a novel Markov chain model to characterize all the state transitions of public and private chains. The proposed chain allows profitable delay increases with the decrease of the hashrate of selfish miners, forcing mining pools to be more cautious about performing selfish mining.

3. Race Attack: This attack is another way around double-spending, requiring the victim to accept unconfirmed transactions as payment [78]. An attacker could leverage a race attack by supplying an unconfirmed transaction to the recipient as payment, while the attacker broadcasts a conflicting transaction to the network. The attacker creates an illusion for the recipient, by showing the transaction in the recipient node before network confirmation. The network notices the double-spending and cancels the recipient transaction. So, as a countermeasure, it is recommended to wait for a minimum of six posterior confirmed transactions, to check that the participant did not receive falsified tokens.

4. Sybil Attack: The term Sybil comes from the name of Sybil Dorsett, a psychiatric patient with dissociative identity disorder, also known as multiple personality disorder. Theoretically, a Sybil attack is a security breach on online systems, where one person or node tries to take over the network by claiming multiple identities, accounts, or computers, behaving similarly to a person with multiple social media accounts. In the world of decentralized systems, attackers might control multiple nodes on the blockchain network. Attackers can then refuse to transmit or receive blocks, effectively blocking honest users/miners from a network. Even though a Sybil attack is still theoretical, researchers have made many attempts over the last decade to achieve full trust between strangers. In "TrustChain: A Sybil-Resistant, Scalable Blockchain," [79], the authors nearly created a blockchain that enables the broadcasting of trusted transactions among anonymous users without central control, by building trust between individuals. They offered scalability, transparency, and Sybil-resistance by replacing PoW with a mechanism to establish the validity and integrity of transactions.

5. Balance Attack: This attack is active against blockchain systems that use a PoW consensus algorithm to enable double-spending in Ethereum and Bitcoin blockchains. Attackers could leverage a delayed communication with balanced mining power in the blockchain network between multiple subgroups of nodes. In "The Balance Attack against Proof-of-Work Blockchains" [80],

the authors provide a theoretical study based on the configurations and other related statistics that are similar to the blockchain infrastructure utilized by the R3 consortium.

6.7 DISCUSSION

The internet has changed our lives over the past 20 years, although many pitfalls have been faced since then, i.e., opening the door to hacking and cyberattacks. However, the continuous success of Bitcoin currency in the last ten years, especially the recognition of this cryptocurrency in many developed countries as an official currency, proves the robustness and efficiency of the technology behind it (blockchain). We believe that blockchain is a philosophical opening for how we can rearrange society and make the world better, as much as it is a technology. This is due to the ability of blockchain first, to decrease systemic risk, second, to facilitate overall transactions, and finally, to help minimize fraud.

Blockchain has some pitfalls as well, such as:

- Blockchain has an environmental cost;
- Its complexity means end users find it hard to appreciate the benefits;
- Blockchains can be slow and cumbersome.

Even so, blockchain has great advantages to be exploited compared to its disadvantages, which could be overlooked or mitigated in the future.

6.8 CONCLUSION

In this chapter, we demonstrate blockchain and distributed ledger technology as cybersecurity features that can enhance systemic security levels. We also mention that blockchain use cases are not only associated with financial markets but go beyond encrypted currency to government, health records, property, voting, and so on. The innovative factors of blockchain technology are the decentralization of data and the elevated level of encryption, which conceals sensitive information from unauthorized reading or alteration. Those factors have revolutionized the technology of storing data over the internet. Peoples' health records, votes in elections, data assets, and unquestionably, money, all involve very important data that affect our lives. Maintaining such data has always imposed the need for a reliable third party, to whom to entrust our money or the provenance of data. Even if the third party is reliable, there is always the possibility of the manipulation or alteration of the data stored in traditional databases or networks. Blockchain technology enhances the security of many domains by providing unlimited immutability and disintermediation, which will revolutionize many financial and nonfinancial industries to facilitate our lives:

1. In the financial domain, money can be easily transferred from one person to another without the need for a third party, such as banks that charges high fees and may require a long time to execute the transaction.

2. In the law domain, the sale, rental, or even the sharing of a property could be easily accomplished between two people by using smart contracts, without the need for an intermediary to ratify and codify the agreement.

3. In the voting domain, the credibility of the process has often been questioned, including whether electronic elections are fair, given the vulnerabilities to many kinds of cyberattacks, or the possibility of manipulated election tallies. Employing blockchain secures the storage of votes and voters' identities, facilitates the election process for voters, and keeps the results intact and immutable.

4. Big data and AI are the essences of today's online services; big companies sit on the huge amount of the data they have collected. For example, the more data Google has, the better search results it can provide to users; the same thing goes for Amazon, Facebook, and so on. All of those organizations have worked independently up to now, with hardly any collaboration between them. Blockchain can enable them to collaborate, to provide better service for people by sharing data in a secure, privacy-assured and decentralized form.

5. Last but not least, data provenance and data tracking records between databases and the cloud are crucial matters for forensics and the credibility of data in many domains, as well as for knowing the intermediate sources that aid in composing the data. Frequently, it is difficult to determine historical records for data, due to the absolute possibility of altering, forging, hacking, and manipulating the original evidence of digital assets. By taking advantage of the immutability and proven protection characteristic of distributed ledger technology, data sources could be protected, along with every single user action on the data. Blockchain can enhance the privacy and availability of data provenance, to provide a decentralized and transparent tamper-proof historical record, with robust data liability.

REFERENCES

1. Nakamoto, S., 2008. Bitcoin: A peer-to-peer electronic cash system. https://citeseerx.ist.psu.edu/viewdoc/summary?. doi:10.1.1.221.9986.
2. Wattenhofer, R., 2017. *Distributed Ledger Technology: The Science of the Blockchain.* South Carolina: CreateSpace Independent Publishing Platform.
3. Dannen, C., 2017. *Introducing Ethereum and Solidity-Foundations of Cryptocurrency and Blockchain Programming for Beginners.* New York: Apress.
4. Mehta, R., Kapoor, N., Sourav, S., and Shorey, R., 2019. *Decentralised image sharing and copyright protection using blockchain and perceptual hashes.* In *2019 11th International Conference on Communication Systems & Networks (COMSNETS)* (pp. 1–6). Bangalore, India: IEEE.
5. Liang, W., Lei, X., Li, K.-C., Fan, Y., and Cai, J., 2019. *A dual-chain digital copyright registration and transaction system based on blockchain technology.* In *International Conference on Blockchain and Trustworthy Systems* (pp. 702–714). Guangzhou, China: Springer.
6. Wu, Y., 2017. *An E-voting system based on blockchain and ring signature.* University of Birmingham. https://www.dgalindo.es/mscprojects/yifan.pdf. Accessed on December 20, 2018

7. Witchey, N. J., 2019, July 02. Healthcare transaction validation via blockchain, systems and methods. https://patents.google.com/patent/US10340038B2/en. Accessed on August 14, 2019.
8. Shieber, J., 2017. Gem looks to CDC and European giant Tieto to take blockchain into healthcare. https://techcrunch.com/2017/09/25/gem-looks-to-cdc-and-european-giant-tieto-to-take-blockchain-into-healthcare/. Accessed on May 2, 2019.
9. Petkova P., and Jekov, B., 2018. *Blockchain in e-governance. Anniversary International Scientific Conference*, (pp. 149–156).
10. Markusheuski, D., Rabava, N., and Kukharchyk, V., 2017. Blockchain technology for e-governance. http://sympa-by.eu/sites/default/files/library/blockchain_egov_brief_eng.pdf. Accessed on May 5, 2019.
11. Pal, S. K., 2019. *Changing technological trends for E-governance.* In *E-Governance in India*, pp. 79–105. doi: 10.1007/978-981-13-8852-1_5.
12. Sharma, T., Satija, S., and Bhushan, B., 2019. *Unifying blockchian and IoT: Security requirements, challenges, applications and future trends.* In *2019 International Conference on Computing, Communication, and Intelligent Systems (ICCCIS)* (pp. 341–346). Greater Noida, India: IEEE
13. Huh, S., Cho, S., and Kim, S., 2017. *Managing IoT devices using blockchain platform.* In *2017 19th International Conference on advanced Communication Technology (ICACT)* (pp. 464–467). Bongpyeong, South Korea: IEEE.
14. Arora, D., Gautham, S., Gupta, H., and Bhushan, B., 2019. *Blockchain-based security solutions to preserve data privacy and integrity.* In *2019 International Conference on Computing, Communication, and Intelligent Systems (ICCCIS)* (pp. 468–472). Greater Noida, India: IEEE.
15. Zīle, K., and Strazdiņa, R., 2018. Blockchain use cases and their feasibility. *Applied Computer Systems,* 23(1), pp. 12–20.
16. Syed, T. A., Alzahrani, A., Jan, S., Siddiqui, M. S., Nadeem, A., and Alghamdi, T., 2019. A comparative analysis of blockchain architecture and its applications: Problems and recommendations. *IEEE Access*, 7, pp. 176838–176869.
17. Soni, S., and Bhushan, B., 2019. *A comprehensive survey on blockchain: Working, security analysis, privacy threats and potential applications.* In *2019 2nd International Conference on Intelligent Computing, Instrumentation and Control Technologies (ICICICT)* (Vol. 1, pp. 922–926). Kannur, India: IEEE.
18. Al-Jaroodi, J., and Mohamed, N., 2020. Blockchain in industries: A survey. *IEEE Access*, 7, pp. 36500–36515.
19. Zhang, J., Zhong, S., Wang, T., Chao, H.-C., and Wang, J., 2019. Blockchain-based systems and applications: A survey. *Journal of Internet Technology*, 21(1), pp. 1–14.
20. Nofer, M., Gomber, P., Hinz, O., and Schiereck, D., 2017. Blockchain. *Business & Information Systems Engineering*, 59(3), pp. 183–187.
21. Wood, G., 2014. Ethereum: A secure decentralised generalised transaction ledger. *Ethereum Project. Yellow Paper*, 151, pp. 1–32,.
22. Schueffel, P., 2017, December 15. Alternative distributed ledger technologies Blockchain vs. Tangle vs. Hashgraph: A high-level overview and comparison. *Computer Science*. http://dx.doi.org/10.2139/ssrn.3144241. Accessed on May 18 2019.
23. Benčić, F. M., and Žarko, I. P., 2018. *Distributed ledger technology: Blockchain compared to directed acyclic graph.* In *2018 IEEE 38th International Conference on Distributed Computing Systems (ICDCS)* (pp. 1569–1570). Vienna: IEEE.
24. Baliga, A., 2017. Understanding blockchain consensus models. *Persistent*, 2017(4), pp. 1–14.

25. Cachin, C., and Vukolić, M., 2017. *Blockchain consensus protocols in the wild*. In *31st International Symposium on Distributed Computing (DISC 2017)* (pp. 1–16). Vienna: Dagstuhl Publishing.

26. Lu, N., Zhang, Y., Shi, W., Kumari, S., and Choo, K.-K. R., 2020. A secure and scalable data integrity auditing scheme based on hyperledger fabric, *Computer Security*, 92, p. 101741.

27. Cachin, C., 2016. *Architecture of the hyperledger blockchain fabric*. In *Workshop on Distributed Cryptocurrencies and Consensus Ledgers* (Vol. 310).

28. West, A., and Fin, X., 2019. *An enterprise-ready hybrid blockchain platform that delivers secure andeEfficient international transactions*. https://cointelegraph.com/explained/proof-of-work-explained. Accessed on May 22, 2019.

29. Tar, A., 2018. *Proof-of-work, explained*. https://cointelegraph.com/explained/proof-of-work-explained. Accessed on May 28, 2019.

30. van Wirdum, A., 2018. The genesis files: Hashcash or how Adam Back designed Bitcoin's Motor Block. https://bitcoinmagazine.com/articles/genesis-files-hashcash-or-how-adam-back-designed-bitcoins-motor-block. Accessed on August 10, 2019.

31. Kiayias, A., Russell, A., David, B., and Oliynykov, R., 2017. *Ouroboros: A provably secure proof-of-stake blockchain protocol*. In *Annual International Cryptology Conference* (Vol. 10401 pp. 357–388). Cham: Springer.

32. Szabo, N., 1997. The idea of smart contracts. *Nick Szabo's Papers and Concise Tutorials* 6,.

33. Cong, L. W., and He, Z., 2019. Blockchain disruption and smart contracts. *Review of Financial Studies*, 32(5), pp. 1754–1797.

34. Wang, S., Ouyang, L., Yuan, Y., Ni, X., Han, X., and Wang, F.-Y., 2019. Blockchain-enabled smart contracts: Architecture, applications, and future trends. *IEEE Trans. Syst. Man, Cybern. Syst.*, 49(11), pp. 2266–2277.

35. Dinh, T. T. A., Liu, R., Zhang, M., Chen, G., Ooi, B. C., and Wang, J., 2018. Untangling blockchain: A data processing view of blockchain systems. *IEEE Transactions on Knowledge and Data Engineering*, 30(7), pp. 1366–1385,.

36. Szydlo, M., 2004. *Merkle tree traversal in log space and time*. In *International Conference on the Theory and Applications of Cryptographic Techniques* (Vol. 3027 , pp. 541–554). Berlin: Springer.

37. Gartner, Inc., 2020. *Blockchain technology: What's ahead?* Be ready for the next phase of the blockchain revolution. https://www.gartner.com/en/information-technology/insights/blockchain. Accessed on January 15, 2020.

38. Saini, H., Bhushan, B., Arora, A., and Kaur, A., 2019. *Security vulnerabilities in information communication technology: Blockchain to the rescue (A survey on Blockchain Technology)*. In *2019 2nd International Conference on Intelligent Computing, Instrumentation and Control Technologies (ICICICT)* (Vol. 1, pp. 1680–1684). Kannur, Kerala, India: IEEE

39. Zhang, Y., Kasahara, S., Shen, Y., Jiang, X., and Wan, J., 2018. Smart contract-based access control for the internet of things. *IEEE Internet Things J.*, 6(2), pp. 1594–1605.

40. Karamitsos, I., Papadaki, M., and Al Barghuthi, N. B., 2018. Design of the blockchain smart contract: A use case for real estate. *Journal of Information Security*, 9(3), pp. 177–190,

41. Adams, R., Parry, G., Godsiff, P., and Ward, P., 2017. The future of money and further applications of the blockchain. *Strategic Change*, 26(5), pp. 417–422.

42. Microsoft, 2020. *Own your digital identity.*. https://www.microsoft.com/en-us/security/business/identity/own-your-identity. Accessed on February 7, 2020.

43. Chalaemwongwan, N., and Kurutach, W., 2018. *A practical national digital id framework on blockchain (NIDBC)*. In *2018 15th International Conference on Electrical Engineering/Electronics, Computer, Telecommunications and Information Technology (ECTI-CON)* (pp. 497–500). Chiang Rai, Thailand: IEEE.

44. Al-Rawy, M., and Elci, A., 2018. A design for Blockchain-based digital voting system. In Antipova T., Rocha A. (eds.) *Digital Science. DSIC 2018: 32nd International Symposium on Distributed Computing.* Advances in Intelligent Systems and Computing, vol. 850, pp. 397–407. Cham: Springer.

45. Korpela, K., Hallikas, J., and Dahlberg, T., 2017. *Digital supply chain transformation toward blockchain integration.* In *Proceedings of the 50th Hawaii International Conference on System Sciences.* (Vol 50, pp. 4182–4191). University of Hawaii at Manoa: HICSS.

46. Abeyratne, S. A., and Monfared, R. P., 2016. Blockchain ready manufacturing supply chain using distributed ledger. *International Journal of Research in Engineering and Technology,* 5(9), pp. 1–10.

47. Tian, F., 2016. *An agri-food supply chain traceability system for China based on RFID & blockchain technology.* In *2016 13th International Conference on Service Systems and Service Management (ICSSSM)* (pp. 1–6). Kunming: IEEE.

48. Al-Rawy, M., and Elci, A., 2019. Secure i-voting scheme with Blockchain technology and blind signature. *Journal of Digital Science,* 1(1) , pp. 3–14.

49. Khan, S., Al-Amin, M., Hossain, H., Noor, N., and Sadik, M. W., 2020. *A pragmatical study on Blockchain empowered decentralized application development platform.* In *Proceedings of the International Conference on Computing Advancements* (pp. 1–9). Dhaka, Bangledesh: Assocation for Computing Machinery.

50. Lewis, R., McPartland, J., and Ranjan, R., 2017. Blockchain and financial market innovation. *Economic Perspectives,* 41(7), pp. 1–17.

51. Nguyen, Q. K., 2016. *Blockchain: A financial technology for future sustainable development.* In *2016 3rd International Conference on Green Technology and Sustainable Development (GTSD)* (pp. 51–54). Kaohsiung, Taiwan: IEEE.

52. Guardtime Cyber, 2020. Guardtime cyber helping to achieve cyber resilience. Available: https://cyber.guardtime.com/. Accessed on July 9, 2019.

53. Yasin, D., 2019. Remme is revolutionizing password protection through the Blockchain. https://cryptopotato.com/remme-revolutionizing-password-protection-blockchain/. Accessed on July 15, 2019.

54. Businesswire, 2018. KODAK and WENN Digital partner to launch major blockchain initiative and cryptocurrency. Accessed on July 9, 2019. https://www.businesswire.com/news/home/20180109006183/en/KODAK-WENN-Digital-Partner-Launch-Major-Blockchain. Accessed on July 10, 2019.

55. ConsenSys, 2018, September 13. Ujo and capitol records bring blockchain innovation to music.https://media.consensys.net/consensys-ujo-and-capitol-records-bring-blockchain-innovation-to-music-319f2c649790. Accessed on June 1, 2019.

56. Ubitquity, 2020. One block at a time. https://www.ubitquity.io/. Accessed on July 12, 2019.

57. Gem, 2020. The best crypto portfolio tracker does the work for you. https://gem.co/. Accessed on March 15, 2019.

58. Everledger, 2020. Everledger: Traceability. Provenance. Authenticity. https://www.everledger.io/. Accessed on March 15, 2019.

59. MedRec, n.d. What is MedRec? https://medrec.media.mit.edu/. Accessed on March 15, 2019.

60. Damiani, J., 2020. SimplyVital health is using blockchain to revolutionize healthcare. https://www.forbes.com/sites/jessedamiani/2017/11/06/simplyvital-health-blockchain-revolutionize-healthcare/#75f9cd1a880a. Accessed on March 15, 2019.

61. Provenance, 2020. Provenance: Every product has a story. https://www.provenance.org/. Accessed on March 15, 2019.

Advanced Security Using Blockchain and Distributed Ledger Technology 131

62. Businesswire, 2020. Fluent rebrands as Hijro, announces blockchain trade asset marketplace. https://www.businesswire.com/news/home/20161117005566/en/Fluent-Rebrands-Hijro-Announces-Blockchain-Trade-Asset. Accessed on March 15, 2019.
63. Hulseapple, C., 2020. Blockverify. https://cointelegraph.com/news/block-verify-uses-blockchains-to-end-counterfeiting-and-make-world-more-honest. Accessed on May 05, 2019.
64. Storj Labs, 2020. Decentralized cloud storage is here.. https://storj.io/. Accessed on March 15, 2019.
65. Radcliffe, D., 2017. Could blockchain run a city state? Inside Dubai's blockchain-powered future. https://www.zdnet.com/article/could-blockchain-run-a-city-state-inside-dubais-blockchain-powered-future/. Accessed on March 15, 2019.
66. Buck, J., 2017. Samsung wins public sector Blockchain contract for Korean Government. https://cointelegraph.com/news/samsung-wins-public-sector-blockchain-contract-for-korean-govt. Accessed on March 15, 2019.
67. Guardtime, 2020. Blockchain-enabled cloud: Estonian government selects Ericsson, Apcera and Guardtime. https://guardtime.com/blog/blockchain-enabled-cloud-estonian-government-selects-ericsson-apcera-and-guardtime. Accessed on March 15, 2019.
68. Arcade City, Inc., 2020. Connect freely. https://arcade.city/. Accessed on March 15, 2019.
69. Foxley, W., 2020. Digital travel firm webjet has launched its booking verification Blockchain. https://www.coindesk.com/digital-travel-firm-webjet-has-launched-its-booking-verification-blockchain. Accessed on March 15, 2019.
70. Slock.It, 2020. Slock.it connects devices to the blockchain, enabling the economy of things. https://slock.it/. Accessed on March 15, 2019.
71 Neo, 2020. An open network for the smart economy. https://neo.org/. Accessed on March 15, 2019.
72. Followmyvote, 2020. Introducing a secure and transparent online voting solution for the modern age. https://followmyvote.com/. Accessed on March 10, 2019
73. Filament, 2020. Filament. https://www.iotone.com/supplier/filament/v2122. Accessed on February 25, 2019.
74. Ripple, 2020. Move money to all corners of the world. https://ripple.com.
75. MONERO, n.d. A reasonably private digital currency. https://www.getmonero.org/.
76. Litecoin Project, 2020. The cryptocurrency for payments based on blockchain technology. https://litecoin.org/. Accessed on February 30, 2019.
77. Bai, Q., Zhou, X., Wang, X., Xu, Y., Wang, X., and Kong, Q., 2019. *A deep dive into blockchain selfish mining.* In *ICC 2019-2019 IEEE International Conference on Communications (ICC)* (pp. 1–6). Shanghai: IEEE.
78. Park, J., and Park, J., 2017. Blockchain security in cloud computing: Use cases, challenges, and solutions. *Symmetry (Basel)*, 9(8), p. 164.
79. Otte, P., de Vos, M., and Pouwelse, J. A., 2017. TrustChain: A Sybil-resistant scalable blockchain *Future Generation Computer Systems*. 107, pp. 770–780.
80. Natoli, C. and Gramoli, V. 2016. The balance attack against proof-of-work blockchains: The R3 testbed as an example. https://www.semanticscholar.org/paper/The-Balance-Attack-Against-Proof-Of-Work-The-R3-as-Natoli-Gramoli/b6b291da2871920510367a89c1fbf63f534fc8dc. Accessed on February 17, 2019.

7 Blockchain Technology and Its Emerging Applications

N. Rahimi, I. Roy, B. Gupta
Southern Illinois University

P. Bhandari
Southeast Missouri University

Narayan C. Debnath
Eastern International University

CONTENTS

7.1 Introduction 135
7.2 Terminology 136
    7.2.1 Block 136
    7.2.2 Chain 136
    7.2.3 Blockchain Ledger 136
    7.2.4 Node 136
    7.2.5 Proof of Work 136
    7.2.6 Key 137
    7.2.7 Input 137
    7.2.8 Output 137
    7.2.9 Hash Function 137
7.3 History and Working of Blockchain Technology 137
    7.3.1 History of Blockchain Technology 137
    7.3.2 Working of Blockchain Technology 138
        7.3.2.1 Cryptography of Private and Public Keys 138
        7.3.2.2 Public Distributed Ledger (Peer-to-Peer Network)............138
        7.3.2.3 Program or Protocol (Rules) 138
7.4 Types of Blockchain 140
    7.4.1 Public Blockchain 140
    7.4.2 Semiprivate Blockchain 141
    7.4.3 Private Blockchain 141
    7.4.4 Consortium Blockchain 141
7.5 Advantages and Disadvantages of Blockchain Technology..............................141
    7.5.1 Key Characteristics/Advantages of Blockchain Technology 141

		7.5.1.1	Decentralization	141
		7.5.1.2	Immutability/Persistence	141
		7.5.1.3	Higher Capacity	142
		7.5.1.4	Higher Security	142
		7.5.1.5	Anonymity	142
		7.5.1.6	Auditability	143
		7.5.1.7	Faster Processing	143
		7.5.1.8	Cheaper Transactions	143
		7.5.1.9	Transparency	143
	7.5.2	Disadvantages of Blockchain Technology		143
		7.5.2.1	Blockchain Technology Is New	143
		7.5.2.2	Blockchain Technology Can Be Expensive	143
		7.5.2.3	Blockchain Technology Is Unregulated	144
		7.5.2.4	Security Risk	144
		7.5.2.5	Technological Risk	144
7.6	Issues/Limitations of Blockchain Technology			145
	7.6.1	Complexity		145
	7.6.2	Necessity for Large Networks		145
	7.6.3	Necessity of High-Quality Input		145
	7.6.4	Security Flaw		145
	7.6.5	Behavioral Change		145
	7.6.6	Scaling		145
	7.6.7	Bootstrapping		146
	7.6.8	Government Regulations		146
	7.6.9	Fraudulent Activities		146
7.7	Applications of Blockchain Technology			146
	7.7.1	Elections/Electronic Voting		146
	7.7.2	Traditional Financial Institutions: Improvements		146
	7.7.3	Health Care Technology		147
	7.7.4	Cross-Border Payments/Remittance		147
	7.7.5	Smart Contracts		147
	7.7.6	Copyrighted Contents		148
	7.7.7	Identity of Things		148
	7.7.8	Internet of Things		148
7.8	Real-World Examples of Blockchain Technology Applications			148
	7.8.1	Cybersecurity		148
		7.8.1.1	GuardTime	148
		7.8.1.2	REMME	149
	7.8.2	Health Care		149
		7.8.2.1	Gem	149
		7.8.2.2	MedRec	149
		7.8.2.3	SimplyVital Health	149
	7.8.3	Financial Services		149
		7.8.3.1	Abra	150
		7.8.3.2	Bank Hapoalim	150

	7.8.3.3	Barclays	150
	7.8.3.4	Maersk	150
	7.8.3.5	Aeternity	150
	7.8.3.6	Augur	150
7.8.4	Manufacturing and Industrial Sectors		150
	7.8.4.1	Provenance	151
	7.8.4.2	JioCoin	151
	7.8.4.3	Hijro	151
7.8.5	Governmental Services		151
	7.8.5.1	Dubai	151
	7.8.5.2	Estonia	151
	7.8.5.3	South Korea	151
	7.8.5.4	GovCoin	151
	7.8.5.5	Followmyvote.com	152
7.8.6	Charity		152
7.8.7	Retail Services		152
	7.8.7.1	OpenBazaar	152
	7.8.7.2	Loyyal	152
7.8.8	Real Estate Services		152
7.8.9	Transportation and Tourism Sectors		153
	7.8.9.1	IBM Blockchain Solution	153
	7.8.9.2	Lazooz	153
7.8.10	Media Services		153
	7.8.10.1	Kodak	153
	7.8.10.2	Ujomusic	153
7.9	Conclusion		153
References			154

7.1 INTRODUCTION

Blockchain is an intangible public record in the digital world. It can be considered as a ledger that contains all transactions, all of which can be accessed by anyone. Even though blockchain does not give anything in hand, it does provide a sense of ownership. This ownership can be transferred to anyone else. Any transaction performed in blockchain stays in blockchain forever, which is the beauty of this technology, making it more secure than others. Another factor that makes this technology safer is that self-regulation is easily carried out using blockchain technology, which requires multiple confirmations before a transaction is made [1,2].

The chapter begins by developing a common ground with respect to blockchain terminology. Section 7.3 summarizes the history and working of blockchain technology. In Sections 7.4 and 7.5, respectively, types of blockchain are discussed, with their advantages and disadvantages. Section 7.6 investigates the limitations of blockchain technology. Section 7.7 presents applications for blockchain technology. Section 7.8 gives several real-world examples of blockchain technology applications. Section 7.8 concludes.

7.2 TERMINOLOGY

Blockchain combines the terms "block" and "chain." A block can be seen as a file that contains information about all the transactions that have been processed. Every transaction contains information about the sender and receiver, and some form of identification that makes the transaction unique and connected to the others. For a chain, the blocks are arranged in linear sequence. In blockchain technology, every block or information about the transaction relates to other blocks by adding the previous block's information. This is the reason why, once placed in blockchain technology, a transaction cannot be reversed or manipulated, because the block is already connected with others [3].

7.2.1 BLOCK

A block is a list of transactions containing information processed on the network. Every piece of information processed by the network is stored one block at a time.

7.2.2 CHAIN

When a transaction is processed, every computer in a network tries to solve a puzzle of algorithms. Once the calculation is completed, cryptography creates the chain of blocks, placing them in order. The placement of a block in the chain is verified and secured by most computers in network, which causes the chain to keep growing longer with each transaction.

7.2.3 BLOCKCHAIN LEDGER

This is like a traditional accounting ledger, keeping transactions for all accounts in a centralized banking system. However, in the case of blockchain, it is digital and stores all of the accounts and transactions processed in the network. A typical blockchain ledger stores information such as account numbers, transactions, and balances.

7.2.4 NODE

A node is made up of every computer on a blockchain network participating in the processing of transactions. These nodes connect with each other in a network and help verify transactions. They can also view all the transactions processed in a blockchain.

7.2.5 PROOF OF WORK

For any transaction to be processed, there is a method of verification used in block-chain technology that verifies that a transaction is real and not processed automatically, trying to hack or do something wrong. This method or information algorithm used for verification or processing transactions is called proof of work. Every

Blockchain Technology and Its Emerging Applications 137

computer in a network has its proof of work, and when this has been verified, a block is added to the blockchain.

7.2.6 KEY

The security of data transferred in a blockchain is very important. Hence, in this technology, a key is used to encrypt or decrypt data. These keys are very strong, providing a high level of security, while encrypting data and passing it on in a network and/or decrypting it and providing it to the public, for informational purposes [4]. Private keys and public keys are used in this technology. A private key should always be kept secret, as it is used as the digital signature of a user in a network, whereas a public key can be shared publicly. Private keys are therefore very important, as this is where most of the hacking occurs in blockchain technology, and because people can transfer ownership using their private keys [5].

7.2.7 INPUT

Input refers to any values that any user has received within blockchain through incoming transactions.

7.2.8 OUTPUT

When a user wants to contribute in blockchain, they distribute values for the outgoing transactions to all other members of the blockchain. These are called "output."

7.2.9 HASH FUNCTION

A hash function is a computer program used to store large amounts of transactional information processed in blockchain. Because of the hash function, even though blockchain processes large quantities of transactions, the memory is never overwhelmed [6]. Hash functions convert all input values into a string of alphanumeric values, to produce a digital fingerprint. The important characteristic for blockchain is that the two separate input values never create the same hash values using the hash function.

7.3 HISTORY AND WORKING OF BLOCKCHAIN TECHNOLOGY

As a young technology, blockchain is having a rippling effect in various sectors, from finance to manufacturing and education.

7.3.1 HISTORY OF BLOCKCHAIN TECHNOLOGY

The idea of blockchain technology emerged back in the 1990s, when an electronic ledger was used to digitally sign documents, making sure that the documents signed had not been changed. This idea is the basis of blockchain technology, which was then implemented by Bitcoin, the first digital cash, and was presented in an initial paper describing Bitcoin electronic cash solutions [7]. This paper was published pseudonymously by Satoshi Nakamoto; the actual author(s), or the owner of the first Bitcoin, is still is a mystery. There are various other currencies now, which are all based on Nakamoto's blueprint, with variations. After the successful start of Bitcoin as the first use case or application of blockchain, the technology became linked with Bitcoin, and, as a result, people believe that the two terms are same. Also, some people believe blockchain technology is used only for monetary transactions, which is incorrect and changing only slowly. Other currencies or schemes used blockchain technology before Bitcoin, but none was successful enough to become popular. One big reason that Bitcoin used blockchain was the direct transaction between users, without the need to involve a third party. Bitcoin is therefore a distributed currency, rather than being controlled by a single entity. Blockchain also enables users to be pseudonymous, meaning they are anonymous, but their accounts are publicly observable, with all their transactions. This has enabled complete transparency for transactions. As people's trust in this system keeps increasing, more currencies have started entering the digital world. Businesses have also started using digital currencies, to deal with untrusted and unknown users and to avoid sending the same digital asset to many users.

7.3.2 Working of Blockchain Technology

Blockchain combines various technologies; three of them are [8,9]:

7.3.2.1 Cryptography of Private and Public Keys

For blockchain technology to work safely, every transaction needs to have a digital identity. These identities are created by using a combination of private keys and public keys. This combined identity is the digital signature used in blockchain [10,11,12,13,14].

7.3.2.2 Public Distributed Ledger (Peer-to-Peer Network)

Blockchain technology works with a peer-to-peer network. When a transaction is started, it goes through all the computers in a network and is only processed with the consensus of a majority of the computers in the network [15,16].

7.3.2.3 Program or Protocol (Rules)

Rules are required before computers are allowed to become involved in blockchain. These rules are called the protocol or program, with reference to a blockchain network. In the digital world, this protocol is a mathematical puzzle that a computer solves in order to accept or deny blockchain transactions. It is also known as "data mining." Every blockchain transaction is a piece of data, in the form of a block that contains an individual user's digital signature and all the other information about a transaction. As illustrated in Figure 7.1, an easier way to explain the workings of blockchain technology is with an example: person A, sending money to person B.

Blockchain Technology and Its Emerging Applications

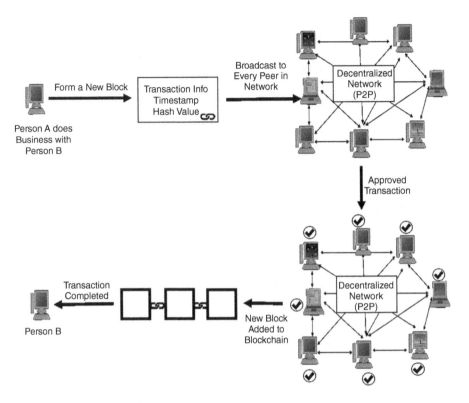

FIGURE 7.1 Simple Working of Blockchain Technology.

Every transaction in blockchain is represented as block, so when a person A sends money, the block is broadcasted to every node (peer) in the network. The block is always timestamped, so that users can verify the authority of the data. The hash value present in the block is very important information, when verifying transactions by nodes. Nodes can be any computers on the network participating in blockchain. They might be participating in blockchain for various reasons, and one important reason is mining, where the computers participating in blockchain solve a mathematical puzzle, to participate in the transaction. If successful, they receive a share of the transaction. This process is called the proof of work, which can change depending on the technology used and vary from time to time [17]. The nodes verify varied information contained in the block and make sure that the transaction is valid. The protocols used in proof of work must be validated by each node. If the protocols are not met, then the transactions are denied by the node. Once enough nodes approve the transaction, the transaction is then approved, and the block is added to the blockchain. The processing of a transaction might take from a few seconds to minutes or even longer, depending on how the blockchain technology is set up. The hash value stored in each block is very important. In fact, it is critical, because this is what helps maintain network security. In the digital currency world, when a hash is solved, it generates coins, so each computer participating in the blockchain network try to solve as many hashes as possible, so they can mine more currencies. But because of the proof of work, the balance of currencies mined in blockchain is maintained [18,19].

7.4 TYPES OF BLOCKCHAIN

There are various types of blockchain, depending on the way the blockchain has evolved, as shown in Figure 7.2. Table 7.1 shows the main types. They are as follows:

7.4.1 PUBLIC BLOCKCHAIN

"Public" blockchain means that anyone in the world can contribute to the blockchain, participating in the transaction process as a node. This type of blockchain is also called "permissionless." Economic incentives for cryptographic verification may or may not be present. Some popular examples include Bitcoin, Ethereum, and so on.

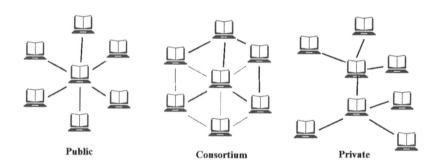

FIGURE 7.2 Types of Blockchain Network.

TABLE 7.1
Comparison between Main Types of Blockchain

Property	Public	Consortium	Private
Consensus	All peers	Designated peers	One organization
Security	Immutable	Could be altered	Could be altered
Efficiency	Low	High	High
Centralized	No	Partial	Yes/hierarchical

Blockchain Technology and Its Emerging Applications 141

7.4.2 Semiprivate Blockchain

This type of blockchain is semiprivate in nature and is run by single organization or group of individuals. These organizations or individuals can grant access to any user for organizational purposes. This blockchain is semiprivate, because it has a public part that is exposed to everyone for participation.

7.4.3 Private Blockchain

"Private" blockchain gives one organization, or a certain group of individuals, the permission to write. The read permissions are public or restricted to large set of users, and transactions are verified by a very few nodes in system. Some popular examples include the Gem health network, Corda, etc. [20,21,22,23].

7.4.4 Consortium Blockchain

"Consortium" blockchain is consensus based, where consensus power is restricted to a set of people or nodes. This type of blockchain is also known as permissioned private blockchain. There are no economic rewards for mining in this type of blockchain, but approval time is fast, because there are fewer nodes in the chain. Some examples include financial institutions such as Deutsche Boerse and R3 [24].

7.5 ADVANTAGES AND DISADVANTAGES OF BLOCKCHAIN TECHNOLOGY

The following are major advantages and disadvantages of blockchain.

7.5.1 Key Characteristics/Advantages of Blockchain Technology

Blockchain has various characteristics:

7.5.1.1 Decentralization

The first and foremost characteristic of blockchain technology is decentralization, meaning data is spread across the network without the need for a central authority, as compared to conventional centralized transactions, which are validated by a central trusted agency, such as a bank. Blockchain mainly uses a long random string of numbers, called a public key. However, it may also use private keys, which could be used as passwords, so the owners can access digital data. The data in blockchain technology is very safe and incorruptible, and with blockchain's decentralization characteristics and omission of a central authority, no individual is preferred over another. The basic characteristics of centralized and decentralized networks are illustrated in Figure 7.3, and their main differences are summarized in Table 7.2.

7.5.1.2 Immutability/Persistence

Another important characteristic of blockchain technology is that it can create immutable ledgers. Transactions performed in blockchain are not only validated quickly but are nearly impossible to alter, as a majority of the nodes/computers involved must accept changes in the ledger, as compared to a centralized database, which can easily be

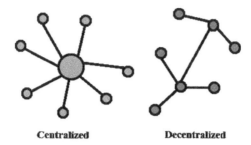

FIGURE 7.3 Centralized vs. Decentralized Network.

TABLE 7.2
Differences between Centralized and Decentralized Networks

Centralized	Decentralized
A core authority is present to dictate to and control other participants in the network.	There is no core authority present.
The authority needs to provide access for the users, so the participants can access transaction history and/or new transactions.	Every participating user can access transaction history and/or new transaction [25].

corrupted. All the parties involved in blockchain transactions go through the mathematical calculations, and hence transactions are validated very quickly. Only those transactions are added to the chain that succeed in the process of validation. This way, invalid transactions are discovered immediately and are never accepted into blockchain.

7.5.1.3 Higher Capacity

Because blockchain technology uses peer-to-peer technology, it combines the power of all the computers connected in the network and hence has a greater capacity than traditional centralized servers do [26,27].

7.5.1.4 Higher Security

Blockchain technology is considered a well-secured form of technology. Reportedly, it has never been hacked. While there are situations where keys used in the technology are sometimes hacked, since the technology uses a network of computers that confirm transactions, the blockchain is impossible to break into and is always safe and secure. By comparison, this is unlike traditional servers, which could be hacked because all the attacker needs to do is to hack one server [28,29].

7.5.1.5 Anonymity

Each user in blockchain interacts with others with a generated address that has no name associated with it. Therefore, the transactions performed are anonymous.

Blockchain Technology and Its Emerging Applications 143

7.5.1.6 Auditability

Another important key characteristic of blockchain technology is its auditability. Every transaction in blockchain refers to previous transactions and stores data about user balances. The transaction status is unspent and spent, so any transaction is easily verifiable and tracked.

7.5.1.7 Faster Processing

Blockchain technology is very fast in terms of processing, unlike traditional banking, which can take a long time to settle transactions, depending on the institution's policy. With the use of blockchain, processing time is much faster, and money transfers can even be settled instantly, which ultimate benefits financial institutions, saving time and resources.

7.5.1.8 Cheaper Transactions

Blockchain technology utilizes the computing power in a network and hence is distributed between lots of users, with a much simpler transaction verification process, compared to centralized systems.

7.5.1.9 Transparency

With blockchain technology, all transactions are performed with public addresses and encrypted via complex algorithms. Identities are hidden and represented by the public addresses. Because the addresses are public, all transactions can be seen directly, which provides great transparency in financial transactions. This level of transparency within a financial system has never existed before. It encourages more trust by users.

7.5.2 Disadvantages of Blockchain Technology

There are various disadvantages of using blockchain technology, some of which are explained here:

7.5.2.1 Blockchain Technology Is New

Even though blockchain was introduced back in the 1990s, its popularity increased only after Bitcoin was introduced in 2009. So, the technology is still considered new, which can be a disadvantage. Unlike other technologies, blockchain needs the participation of a majority of users. Everyone should be encouraged to understand this, so they can contribute to improving blockchain. Because the technology is new, it is unfamiliar to people. Also, since blockchain is computer related, the public needs to be updated about new developments, or in time, the technology could fail.

7.5.2.2 Blockchain Technology Can Be Expensive

As with any other technology, moving to blockchain is somewhat expensive in the initial stage, because businesses must switch from what they are currently using, which could require improvements and investments to accomplish. This might not be the case every time, however, depending on how the business is planning to use the technology.

7.5.2.3 Blockchain Technology Is Unregulated

While the lack of regulation has been discussed as a key feature of this technology, it could also be considered a disadvantage. If for any reason we lose from using blockchain, there is no way to go and ask any authority or administrator for help, a fact that might scare some people. Even though people have a full control on this technology, everyone needs to take care, on their own, as to how best to use it.

7.5.2.4 Security Risk

Blockchain is considered a secure technology, because it is a large network of computers and because it has the requirement that a majority of computers accept transactions in order for the transactions to be validated. However, this might not always be the case. As blockchain keeps developing, attackers also are constantly working on ways to hack it. As with any other technology, this system is computer-based, so there is always some form of risk involved. Even though there is no record yet of the system being hacked, there is also no guarantee that people or attackers will be unable to penetrate the system in one way or another. One possible approach, for example, would be through the hash used in each block.

7.5.2.5 Technological Risk

Like any other technologies, blockchain is always at technological risk, meaning there might be situations where this technology could be replaced by another, better technology that can override the advantages or characteristics of this one. Since technology always keeps changing, one that is currently doing well might not stay trouble-free forever. Table 7.3 presents in brief some of the advantages and disadvantages of blockchain.

TABLE 7.3

Summary of Advantages and Disadvantages of Blockchain Technology

Advantages	Disadvantages
Decentralization	New Technology
Immutable/Persistence	Expensive in Some Cases
Higher Capacity	Unregulated
Higher Security	Security Risk
Anonymity	Technological Risk
Auditability	
Faster Processing	
Cheaper Transactions	
Transparency	

Blockchain Technology and Its Emerging Applications | **145**

7.6 ISSUES/LIMITATIONS OF BLOCKCHAIN TECHNOLOGY

There are issues/limitations currently in existence with this technology:

7.6.1 COMPLEXITY

Because it is new, this technology uses various terms, some of which might make it difficult for people to understand. There are various resources available to facilitate the study of this technology. However, at first, it might be very difficult for some.

7.6.2 NECESSITY FOR LARGE NETWORKS

This technology works with a network of computers. A large network is required, to make the blockchain more powerful and effective. It does work with small networks, but it is very limited, and the full benefits of the technology cannot thereby be experienced.

7.6.3 NECESSITY OF HIGH-QUALITY INPUT

This technology requires high-quality input information. Since the user is not allowed to make changes, once entered, information is assumed to be correct.

7.6.4 SECURITY FLAW

Blockchain is based on the 51 percent attack concept. It assumes that if 51 percent of users approve a transaction or change, it is appropriate, which might not be the case every time and could be a limitation. Even though overriding more than 51 percent of users is very difficult, it might still be possible.

7.6.5 BEHAVIORAL CHANGE

Like any other technology, this approach requires a change in people's behavior. Since this technology depends mostly on users, people need adapt to using blockchain.. Resistance to change is a part of the real world. Instead of using current technology, to ask people to change and start using blockchain requires strong motivation, and blockchain needs to improve the public's sense of trust and security.

7.6.6 SCALING

One issue that mainly concerns the digital currency part of this technology is scaling. The process of validating and downloading the entire set of existing blocks can be very difficult for anyone who wants to start doing transactions right away. It takes hours to validate and download all the blocks, and since the number of blocks is increasing exponentially, this will be harder in days to come.

7.6.7 BOOTSTRAPPING

If an organization decides to move to this new technology, the transfer of current records to blockchain technology is very difficult, as it requires many migration tasks, which involves a great deal of time and money.

7.6.8 GOVERNMENT REGULATIONS

Countries' regulations regarding blockchain will always be an issue, because regulations will always exist to monitor blockchain technology, which is the opposite of the key characteristics of this technology. Many laws and regulations are being introduced around the world to monitor and regulate blockchain.

7.6.9 FRAUDULENT ACTIVITIES

Since blockchain is pseudonymous, there is always the possibility of fraudulent activities using this technology. Managing these activities and allowing law enforcement agencies to monitor and prosecute them will be a challenge and an issue for blockchain.

7.7 APPLICATIONS OF BLOCKCHAIN TECHNOLOGY

Blockchain has been very successful so far with digital coins, and it gained in popularity with the use of the platform for Bitcoin and other digital coins. Now, however, there is much discussion of where this technology can best be used. Programmers have found innovative ways to use the technology, because of its secure characteristics. The areas where blockchain technologies could be applied include:

7.7.1 ELECTIONS/ELECTRONIC VOTING

Throughout the world, voting systems are very time consuming, as people must wait in line for their turn to vote in elections. Previously, and still in some countries, paper voting was the only option. Now some countries have developed electronic voting, but still people have to go to election centers to vote or send in their votes through the postal system. Blockchain can probably solve this problem in the near future. With this technology, votes can be recorded as individual transactions, and since the technology keeps a record of each transaction, they are very safe. This technology can also reduce the marginal error in elections and reduce the possibility of fraud, as well, because each user in the network can easily see the summary of all votes as transactions. More important, the increased ease of voting could also increase voter participation [30].

7.7.2 TRADITIONAL FINANCIAL INSTITUTIONS: IMPROVEMENTS

With the introduction of digital coins, a big challenger appears to traditional financial institutions, mostly banks. To stay competitive, traditional banking has slowly been

Blockchain Technology and Its Emerging Applications

understanding the technology and trying to use it to better serve their clients and provide greater security. Research is ongoing by these traditional financial institutions to better understand how to use blockchain technology. Some primary ways it could be utilized include retail payments, consumer lending, and reference data systems. At this point, financial institutions take time to process every transaction, from sending money online or making purchases with cards. By using blockchain technology, these institutions could improve their processing time. Similarly, consumer lending by financial institutions takes a long time, due to all the paperwork and intermediate tasks, both of which could be substantially improved by using blockchain technology. As a further point, given the data involved with financial institutions, sharing information is not easy. But with the real-time information sharing technique used in blockchain, data can be shared very easily, as it is easily visible to anyone in the blockchain network.

7.7.3 HEALTH CARE TECHNOLOGY

Health care is another field where blockchain technology could be used, mostly to maintain patient electronic medical records and/or facilitate insurance claims. Medical records can be updated or entered in blockchain, which is very safe, and no one would be able to change the records, once entered. This also helps insurance companies, as every transaction has a timestamp, which could make insurance claims much easier, since the insurance company would have records of each diagnosis and when it was made. Health insurance companies could also use this technology to determine patients' preconditions very easily. Also, using the smart contract aspect of blockchain technology would automatically determine payments and claims, thus reducing administrative costs and other costs associated with manual work [31,32].

7.7.4 CROSS-BORDER PAYMENTS/REMITTANCE

At present, the transfer of money internationally is expensive, error-prone, and an easy method for money-laundering. In addition, the whole process takes a long time. The use of blockchain could resolve all these issues, and transactions could be faster, cheaper, and more secure. One example of a remittance service provider using this technology is Abra, but many other businesses and organizations will probably take advantage of this technology in the near future [33].

7.7.5 SMART CONTRACTS

"Smart contracts" are better ways to handle contracts using blockchain technology. Traditional contracts require a third party in an intermediary position. Smart contracts would eliminate the need for these third parties. Both parties signing a smart contract are always aware of the details in the contract and know when and how it will be implemented. These contracts not only govern the rules and penalties for each party, but also enforce them as appropriate. They have a good use scenario for individuals, businesses, and IoT devices [34].

7.7.6 Copyrighted Contents

Copyrighted contents could benefit quite a bit from blockchain technology. Right now, it is very confusing to know who are the owners of copyrighted materials and whenever a royalty must be paid to the author or the publishing company or the artist, for copyrighted materials. If the smart contract aspects of blockchain were used, all material would carry information about its owners, making it easy to pay royalty fees directly when needed [35].

7.7.7 Identity of Things

Blockchain technology can help in providing a seamless identity. In blockchain, identity is pseudonymous, meaning there is an identity but no need of name. Whenever a transaction is made, it is associated with the identity, but without disclosing the name, unless the individual wants to do so. This way, everyone can have an identity in blockchain, which would enhance the global identity for every person in the world. This same identity concept can be used for products, packages, machines, and many other things. This is called the identity of things (IDoT). The IDot concept could play a very important role in supply chain management, or infrastructure. It could also make the shopping and retail experiences easier for everyone [36,37].

7.7.8 Internet of Things

With the invention of billions of devices in the world, the information collected by each device is scattered in various systems. Slowly, however, a need is developing to integrate all of this information, which is the Internet of Things (IoT). There is a need to save all this information, and blockchain technology is best suited for the task. As a distributed technology, using a large number of computers in a network, it provides extensive storage and bandwidth, and enhanced security, along with enhanced processing power. Traffic lights and rain gauges are examples of where this could be used.

7.8 REAL-WORLD EXAMPLES OF BLOCKCHAIN TECHNOLOGY APPLICATIONS

Different fields have different ways of using blockchain technology to improve people's lives and make them easier and safer. According to *Forbes Magazine*, there are various real-world examples that offer amazing promise, in the fields of finance, health care, government, media, and various other innovative sectors.

7.8.1 Cybersecurity

Examples from the cybersecurity sector include:

7.8.1.1 GuardTime

The GuardTime Company was started in 2007, with the goal of eliminating the need for trusted authorities within Estonian government networks. Headquartered in Lausanne, Switzerland, the company has offices in Estonia, the United States, the

Blockchain Technology and Its Emerging Applications

Netherlands, the United Kingdom, and Singapore. Currently, GuardTime has more than 40 patents, granted since 2007, devoted to building technologies and applications in cryptography and computer science. Among the various technologies and applications, this company's use of the keyless signature blockchain technology to store Estonian citizens' health records is very significant [20,38,39].

7.8.1.2 REMME

Headquartered in the United States, Remme is a company that provides blockchain-based public key infrastructure (PKI), which resolves various problems with cyberattacks and focuses on preventing the hacking of central systems. This company is also working on a decentralization authentication system, aimed to resolve the problems of logins and passwords with SSL certificates. The company plans to store SSL certificates on blockchain [21,40].

7.8.2 HEALTH CARE

Examples from the health care sector include:

7.8.2.1 Gem

Gem is one a software company in California, which is working in blockchain technology. Their focus is on health care solutions, and hence they are partnered with Philips, to securely store, access, and share health care data, with full HIPAA compliance. Their blockchain-based architecture provides full efficiency, transparency, and cost-reduction to managing health care data. They are working with the U.S. Centers for Disease Control, looking at ways to manage health and disaster responses through data collection and analysis.

7.8.2.2 MedRec

MedRec is a technology project supported by a grant from the Robert Wood Johnson Foundation and with additional support from the MIT Media Lab. This project was first implemented and designed in August 2016, and another version is currently being developed. The blockchain technology used in this project stores patient data and maintains authorization data [41].

7.8.2.3 SimplyVital Health

Based in Brighton, Massachusetts, SimplyVital Health was founded in 2017. They have developed an open source Health Nexus protocol with blockchain infrastructure, which provides tools to increase data access and decrease the cost of health care globally. This company is a fork of Ethereum and has features such as a distributed database for storage, secured key pair system, and a new validation and governance period [42].

7.8.3 FINANCIAL SERVICES

Blockchain has been utilized extensively by financial service providers. Examples from the financial services sector include:

7.8.3.1 Abra

Abra is a cryptocurrency wallet based on blockchain technology. There are about 30 cryptocurrencies and 50 fiat currencies in Abra, which allows people to start investing easily, using bank transfers, credit/debit transactions, or cryptocurrencies. It is easy to use and a safe way to manage cryptocurrencies in a single place [33].

7.8.3.2 Bank Hapoalim

Bank Hapoalim, in Israel, is trying to utilize blockchain technology with the help of Microsoft, to manage digital bank guarantees. This would make various bank processes, such as signing up guarantors, simpler and quicker and would also eliminate people's need to be physically present in the bank. This technology will also enable banks to provide documents to its clients in digital format, securely.

7.8.3.3 Barclays

Barclays is a world-renowned bank headquartered in London. The bank uses blockchain technology initiatives for various financial activities, such as identification, verification, and various transactional activities. Since the bank needs to be very careful with the identities of its staff and customers, blockchain could be a very important technology, moving forward [43].

7.8.3.4 Maersk

Maersk is another company using blockchain technology in the field of financial services. It utilizes the keyless signature blockchain technology developed by GuardTime, to work with shipping data. Its partners include EY, Willis Towers Watson, Microsoft, and various insurance companies. Maersk hopes to make shipping insurance more efficient and to facilitate the shipping supply chain with real-time data sharing [44].

7.8.3.5 Aeternity

Aeternity is another innovative blockchain technology focused on creating smart contracts to allow the execution of credible transactions. Third or intermediate parties are not required, which provides businesses and individuals with very private contracts and makes sure they are well secured. Aeternity also plans to make low-cost transactions by leveraging private state channels between the parties involved in the contracts. [45].

7.8.3.6 Augur

A blockchain technology, Augur is based on prediction market protocol and claims to be operated and owned by the people using this. It allows users to create blockchain-based predictions for various trading markets and financial institutions [46].

7.8.4 Manufacturing and Industrial Sectors

The manufacturing and industrial sectors have also been taking advantage of blockchain technology, as follows:

Blockchain Technology and Its Emerging Applications

7.8.4.1 Provenance

This blockchain technology aims to provide solutions for transparency in product supply chains. According to the whitepaper for Provenance, presented by social enterprise Project Provenance Ltd., people have little information about the products we use. Our products travel through large networks of retailers before finally reaching us, and hence, there is the need for transparency in product supply chains [47].

7.8.4.2 JioCoin

JioCoin is a project being developed by Reliance Industries in India. This project utilizes blockchain technology to work on supply chain management and smart contracts, along with their own digital currency, Jiocoin [48].

7.8.4.3 Hijro

Hijro is a company based in Lexington, Kentucky, working on connecting users to global networks through blockchain technology, to provide financial and supply chain solutions [49].

7.8.5 GOVERNMENTAL SERVICES

Blockchain has applications in governmental services, including for:

7.8.5.1 Dubai

Dubai is currently planning to be the first blockchain-powered government, by 2020. There are various projects currently running in Dubai based on blockchain technology, which has gained support from businesses, the government, and entrepreneurs. A global blockchain council was formed in 2016, with more than 30 government entities and international companies. The global council members have announced projects utilizing blockchain technology in various possible fields, such as health, business registrations, shipping, and so on [50].

7.8.5.2 Estonia

The Estonian government is also utilizing blockchain technology. They have partnered with companies and other organizations to move their citizens' health records into blockchain technology and have been involved in creating data centers to move other public records into blockchain [51].

7.8.5.3 South Korea

Even though South Korea has banned initial coin offerings (ICOs), it is hoping to utilize blockchain technology and hence has an agreement with Samsung SDS to create a platform based on this technology for welfare, public safety, and transportation. Scheduled for completion by 2022, the project aims to increase transparency in the government services provided by South Korea [51].

7.8.5.4 GovCoin

The distribution and recording of benefit payments are among the most important government processes, and this is where the UK Department of Work and Pensions is investigating the use of blockchain technology. This was a matter of discussion for the UK government in a budget speech presented in 2017 [52].

7.8.5.5 Followmyvote.com

This is a blockchain technology-based innovation to allow secure transparent online voting. The voting system is transparent and increases voter turnout by reducing voter fraud [53].

7.8.6 CHARITY

Even charitable organizations benefit from blockchain. For instance, BitGive, founded in 2013, is the first Bitcoin nonprofit organization in the United States, a status that allows federal tax exemption status. BitGives is a well known nonprofit organizations, in addition to Save the Children, the Water Project, Medic Mobile, TECHO, and several other. BitGive provides greater transparency, so donors can easily see what their donations are used for [54].

7.8.7 RETAIL SERVICES

Blockchain has been utilized by retail service providers, such as:

7.8.7.1 OpenBazaar

This is a free online marketplace, which offers selling anything without any third parties or intermediaries involved. There is no platform fee for selling and no restrictions. OpenBaazar connects people directly, via a peer-to-peer network, and people can pay with 50+ cryptocurrencies. The marketplace follows blockchain technology, and hence there is no one with control over it. Everyone shopping online on this platform contributes to the network [55].

7.8.7.2 Loyyal

As with the current trend, companies offer their users various incentives and rewards for using their services or being a part of the company in some way. A similar concept is used by Loyyal, a California-based company that has created a platform based on blockchain technology to allow companies to offer loyalty packages. It allows consumers to combine and trade their loyalty rewards earned. It uses proprietary blockchain, as well as smart contract technology [56].

7.8.8 REAL ESTATE SERVICES

Ubitquity LLC, founded in 2016 and based in Delaware, is focused on enterprise real estate organizations, government municipalities, and resellers. It is one of the first real estate companies to incorporate blockchain into its services. The company is working on building and maintaining a platform based on blockchain technology, to serve the real estate industry. This also includes helping governments with real estate activities, such as legal process. This technology is expected to smooth out the complicated legal process and make real estate transactions easier for everyone [57].

Blockchain Technology and Its Emerging Applications

7.8.9 TRANSPORTATION AND TOURISM SECTORS

Here are some examples of blockchain usage in the transportation and tourism sectors:

7.8.9.1 IBM Blockchain Solution

Among the various solutions offered by blockchain, IBM is incorporating the technology for the vehicle leasing industry. The company is working on making all of a vehicle's history available to potential clients, so they can easily navigate what they are going to lease. This would benefit everyone associated with the chain in leasing, from a government office registering vehicle information, to insurance agencies trying to set insurance rates. Manufacturers also could use the technology to determine recalls costs, and so on [58].

7.8.9.2 Lazooz

Based in Israel, Lazooz is a company working on systems for sustainability. As with the concept of sustainability, this technology encourages smart transportation by using existing resources fully before adding a new transportation. The company believes that real-time ridesharing with the use of blockchain technology will help in maintaining sustainability, by enabling people with private cars to share a journey with others traveling in same direction. The company believes this will not only save money but also enhance social relationships [59].

7.8.10 MEDIA SERVICES

Media services have also benefited from blockchain technology, for example:

7.8.10.1 Kodak

A technology company focused on imaging, Kodak has also begun working with WENN Digital on blockchain technology: KODAKOne (an image rights management platform) and KODAKCoin (photo-centric cryptocurrency). These technologies utilize blockchain and will empower photographers and agencies. For example, image registration and rights management will be handled in blockchain, to allow transparent accounts [60,61].

7.8.10.2 Ujomusic

Ujomusic is a platform for singer-songwriters and musicians, to allow them to create records of their ownership, thus facilitating the payment of royalties. The platform uses blockchain technology to bring this about [62].

7.9 CONCLUSION

Blockchain is a very fast-growing technology, with countless possibilities for growth, in various fields. Even though there are challenges with this new technology, and despite the fact that many people just consider blockchain as cryptocurrency, it can be utilized in various fields. There are many projects underway, and large-scale companies are engaged in researching blockchain's proper utilization. Time will tell how this is going to benefit all of us; but present impacts suggest that it is going to stay around for quite a long time, because of its characteristics of decentralization, security, transparency, and so on. As discussed, this technology needs participation by as many people as possible, to make it more secure. The study of this field is very important, as is people's awareness of the technology.

REFERENCES

1. Narayanan, A., Bonneau, J., Felten, E., and Goldfeder, S., 2016. *Bitcoin and Cryptocurrency Technologies*. Princeton: Princeton University Press.
2. Norman, A., 2017. *Cryptocurrency Investing Bible: The Ultimate Guide About Blockchain, Mining, Trading, ICO, Ethereum Platform, Exchanges, Top Cryptocurrencies for Investing and Perfect Strategies to Make Money*. Scotts Valley, CA: CreateSpace Independent Publishing, LLC.
3. Norman, A., 2017. *Blockchain Technology Explained: The Ultimate Beginner's Guide About Blockchain Wallet, Mining, Bitcoin, Ethereum, Litecoin, Zcash, Monero, Ripple, Dash, IOTA and Smart Contracts*. Scotts Valley, CA: CreateSpace Publishing LLC.
4. Hill, B., Chopra S., and Valencourt, P., 2018. *Blockchain Quick Reference: A Guide to Exploring Decentralized Blockchain Application Development*. Birmingham, UK: Packt Publishing Ltd.
5. James, J., 2018. *Blockchain: The Ultimate Beginner's Guide to Understanding Blockchain and Blockchain Technology*. Karnataka, India: Independently Published.
6. Sebastian, L., 2018. *Blockchain: 3 Books – The Complete Edition on Bitcoin, Blockchain, Cryptocurrency and How It All Works Together in Bitcoin Mining, Investing and Other Cryptocurrencies*. Cleveland, OH: Positive Impact Books.
7. Watney, M., 2017. *Blockchain for Beginners: The Complete Step by Step Guide to Understanding Blockchain Technology*. Scotts Valley, CA: CreateSpace Publishing LLC.
8. Nakamoto, S., 2008. *Bitcoin: A Peer-to-Peer Electronic Cash System*. https://bitcoin.org/bitcoin.pdf. Accessed on April 13, 2019.
9. Rahimi, N., Reed, J. J., and Gupta, B., 2018. On the significance of cryptography as a service. *Journal of Information Security*, 9(4), pp. 242–256.
10. Fernández-Caramès, T. M., and Fraga-Lamas, P., 2020. Towards post-quantum blockchain: A review on blockchain cryptography resistant to quantum computing attacks. *IEEE Access*, 8, pp. 21091–21116.
11. Arora, D., Gautham, S., Gupta, H., and Bhushan, B., 2019. *Blockchain-based security solutions to preserve data privacy and integrity*. In *2019 International Conference on Computing, Communication, and Intelligent Systems (ICCCIS)* (pp. 468–472). Greater Noida, India: IEEE.
12. Rahimi, N., 2020. Security consideration in peer-to-peer networks with a case study application. *International Journal of Network Security & Its Applications (IJNSA)*, 12(2). pp. 1–16
13. Fleming, S., 2017. *Blockchain technology: Introduction to Blockchain technology and its impact on business ecosystem*. Scotts Valley, CA: CreateSpace Publishing.
14. Singh, V., (2017). Understand Blockchain Technology: Your quick guide to understand blockchain concepts. https://www.google.com/books/edition/Understand_Blockchain_Technology/3yiCDwAAQBAJ?hl=en&gbpv=0. Accessed on November 10, 2019.

Blockchain Technology and Its Emerging Applications

15. Rahimi, N., Sinha, K., Gupta, B., Rahimi, S., and Debnath, N. C., 2016. *LDEPTH: A low diameter hierarchical p2p network architecture.* In *2016 IEEE 14th International Conference on Industrial Informatics (INDIN)* (pp. 832–837). Emden, Germany: IEEE.
16. Rahimi, N., Gupta, B., and Rahimi, S., 2018. *Secured data lookup in LDE based low diameter structured P2P network.* In *Proceedings of the 33rd International Conference on Computers and Their Applications (CATA).*(pp. 56–62). Las Vegas: ISCA Publishing.
17. Blockgeeks, 2019. What is blockchain technology? A step-by-step guide for beginners. https://blockgeeks.com/guides/what-is-blockchain-technology. Accessed on November 17, 2019.
18. Marr, B., 2018. 35 Amazing real world examples of how blockchain is changing our world. Forbes.com. https://www.forbes.com/sites/bernardmarr/2018/01/22/35-amazing-real-world-examples-of-how-blockchain-is-changing-our-world/#3c43c0dd43b5. Accessed on November 17, 2019.
19. Sharma, T., Satija, S., Bhushan, B., 2019, October. *Unifying blockchian and IoT: Security requirements, challenges, applications and future trends.* In *2019 International Conference on Computing, Communication, and Intelligent Systems (ICCCIS)* (pp. 341–346). Greater Noida, India: IEEE.
20. Guardtime.com, 2019. https://guardtime.com/. Accessed on April 28, 2019.
21. Blockchain.cioreview.com, 2019. REMME: Delivering effective data security with decentralization. https://blockchain.cioreview.com/vendor/2018/remme. Accessed on April 28, 2019.
22. Remme.io, 2019. Remme: Distributed PKI and apps for the modern web. https://remme.io/. Accessed on April 28, 2019.
23. Gem, 2019. *Home.* https://enterprise.gem.co/ Accessed on November 17, 2019.
24. Shieber, J., 2017. Gem looks to CDC and European giant Tieto to take blockchain into healthcare – TechCrunch. https://techcrunch.com/2017/09/25/gem-looks-to-cdc-and-european-giant-tieto-to-take-blockchain-into-healthcare/. Accessed on April 28, 2019.
25. Gupta, B., Rahimi, N., Rahimi, S., and Alyanbaawi, A., 2017. *Efficient data lookup in non-DHT based low diameter structured P2P network.* In *Proceedings of the 2017 IEEE 15th International Conference on Industrial Informatics (INDIN)* (pp. 944–950). Emden, Germany: IEEE.
26. Gupta, B., Rahimi, N., Hexmoor, H., and Maddali, K., 2018. *Design of a new hierarchical structured peer-to-peer network based on Chinese remainder theorem.* (pp. 944–950). *Proceedings of the 33rd International Conference on Computers and Their Applications (CATA)*, Las Vegas: ISCA Publishing.
27. Rahimi, S., 2017. A novel linear diophantine equation-based low diameter structured peer-to-peer network. *PhD Diss.*, Southen Illinois University, Carbondale, IL.
28. Zhaofeng, M., Xiaochang, W., Jain, D. K., Khan, H., Hongmin, G., and Zhen, W., 2019. A Blockchain-based trusted data management scheme in edge computing. *IEEE Transactions on Industrial Informatics.* 16(3), pp. 2013–2021.
29. Saini, H., Bhushan, B., Arora, A., and Kaur, A., 2019. *Security vulnerabilities in Information communication technology: Blockchain to the rescue (A survey on Blockchain Technology).* In *2019 2nd International Conference on Intelligent Computing, Instrumentation and Control Technologies (ICICICT)* (Vol. 1, pp. 1680–1684). Kannur, India: IEEE.
30. Follow My Vote, 2019. The online voting platform of the future: Follow my vote. https://followmyvote.com. Accessed on April 28, 2019.
31. Simplyvitalhealth.com, 2019. https://www.simplyvitalhealth.com/. Accessed on November 17, 2019.

32. Hendren, L., 2017. What is health nexus?. https://medium.com/simplyvital/health-nexus-the-overview-e8deb57bdd07. Accessed on April 28, 2019.

33. Abra, 2019. Home: Abra. https://www.abra.com. Accessed on November 17, 2019.

34. Meyer, D., 2017. Microsoft's blockchain experiments expand to digital bank guarantees. http://fortune.com/2017/09/07/microsoft-bank-hapoalim-blockchain-bank-guarantees/. Accessed on April 28, 2019.

35. Solomon, S., 2017. Bank Hapoalim, Microsoft join forces on blockchain technology. https://www.timesofisrael.com/bank-hapoalim-microsoft-join-forces-on-blockchain-technology/. [Accessed on April 28, 2019].

36. Friese, I., Heuer, J., Kong, N., 2014. *Challenges from the Identities of Things: Introduction of the Identities of Things discussion group within Kantara initiative.* In *2014 IEEE World Forum on Internet of Things (WF-IoT)* (pp. 1–4). Seoul, Korea: IEEE.

37. Lam, K. Y., and Chi, C. H., 2016. *Identity in the Internet-of-Things (IoT): New challenges and opportunities.* In *International Conference on Information and Communications Security* (pp. 18–26). Cham: Springer.

38. Guardtime, 2019. Blockchain-enabled cloud: Government, ericsson, Apcera. https://guardtime.com/blog/blockchain-enabled-cloud-estonian-government-selects-ericsson-apcera-and-guardtime. Accessed on April 28, 2019.

39. Rahimi, N., Nolen, J., and Gupta, B., 2019. Android security and its rooting: A possible improvement of its security architecture. *Journal of Information Security*, 10(2), pp. 91–102.

40. Attaran, M., and Gunasekaran, A., 2019. Blockchain and cybersecurity. In Attaran, M., and Gunasekaran, A. (eds.) *Applications of Blockchain Technology in Business*, pp. 67–69. Cham: Springer.

41. Medrec.media.mit.edu, 2019. MedRec. https://medrec.media.mit.edu/. Accessed on November 17, 2019.

42. Monteil, C., 2019. Blockchain and health. *Digital Medicine*. Cham: Springer, pp. 41–47.

43. Barclays Corporate, 2019. What does Blockchain do? https://www.barclayscorporate.com/insights/innovation/what-does-blockchain-do/. Accessed on April 28, 2019].

44. Hackett, R., 2017. Maersk and Microsoft tested a blockchain for shipping insurance. http://fortune.com/2017/09/05/maersk-blockchain-insurance/. Accessed on April 28, 2019.

45. Aeternity Blockchain, 2019. A Blockchain for scalable, secure, and decentralized apps. https://www.aeternity.com/. Accessed on April 28, 2019.

46. Augur, 2019. Home page. https://www.augur.net/. Accessed on April 28, 2019.

47. Provenance, 2017. Blockchain: The solution for transparency in product supply chains. https://www.provenance.org/whitepaper. Accessed on April 28, 2019.

48. Das, S., 2018. JioCoin: India's biggest conglomerate to launch its own cryptocurrency. https://www.ccn.com/jiocoin-indias-biggest-conglomerate-launch-cryptocurrency. Accessed on April 28, 2019.

49. Hijro, 2019. Hijro: Trade asset marketplace. https://hijro.com/. Accessed on April 28, 2019.

50. Radcliffe, D., 2017. Could blockchain run a city state? Inside Dubai's blockchain-powered future | ZDNet. https://www.zdnet.com/article/could-blockchain-run-a-city-state-inside-dubais-blockchain-powered-future/. Accessed on April 28, 2019.

51. Buck, J., 2017. Samsung wins public sector blockchain contract for Korean Government. https://cointelegraph.com/news/samsung-wins-public-sector-blockchain-contract-for-korean-govt. Accessed on April 28, 2019.

Blockchain Technology and Its Emerging Applications

52. Herian, R., 2017. Why a blockchain startup called Govcoin wants to "disrupt" the UK's welfare state. Research at The Open University. http://www.open.ac.uk/research/news/blockchain-startup-called-govcoin. Accessed on April 28, 2019.
53. Abuidris, Y., Kumar, R., and Wenyong, W., 2019. *A survey of blockchain based on E-voting systems*. In *Proceedings of the 2019 2nd International Conference on Blockchain Technology and Applications* (pp. 99–104). Xi'an China: ACM.
54. BitGive Foundation, 2019. About Us: BitGive foundation. https://www.bitgivefoundation.org/about-us/. Accessed on April 28, 2019.
55. OpenBazaar, 2019. Home page. https://openbazaar.org/. Accessed on April 28, 2019.
56. Loyyal, 2019. The internet of loyalty. https://loyyal.com/. Accessed on April 28, 2019.
57. Ubutquity, 2019. UBITQUITY: The enterprise ready blockchain-secured platform for real estate recordkeeping|One block at a time. https://www.ubitquity.io/. Accessed on April 28, 2019.
58. Youtube, 2017. IBM blockchain car lease demo. https://www.youtube.com/watch?v=IgNfoQQ5Reg. Accessed on April 28, 2019.
59. LaZooz, 2019. A value system designed for sustainability. http://lazooz.org/. Accessed on April 28, 2019.
60. Kodabone, 2019. Kodak and WENN Digital partner to launch major blockchain initiative and cryptocurrency. https://www.kodak.com/US/en/corp/press_center/kodak_and_wenn_digital_partner_to_launch_major_blockchain_initiative_and_cryptocurrency/default.htm. Accessed on April 28, 2019.
61. Holding I. Ryde, 2019. KODAK One: Image rights management platform. https://kodakone.com/. Accessed on April 28, 2019.
62. Ujo Music, 2019. Ujo Music: Get played, get paid. https://ujomusic.com. Accessed on April 28, 2019.

8 Emergence of Blockchain Technology

A Reliable and Secure Solution for IoT Systems

A.K.M. Bahalul Haque
School of Engineering and Physical Sciences,
North South University, Bangladesh

Bharat Bhushan
School of Engineering and Technology, Sharda University, India

CONTENTS

8.1 Introduction 160
8.2 Blockchain Methodology 162
    8.2.1 Blockchain Characteristics.. 162
    8.2.2 Digital Wallet 163
    8.2.3 Block Structure 163
8.3 Types of Blockchain... 164
    8.3.1 Public Blockchain 164
    8.3.2 Private Blockchain 164
    8.3.3 Consortium Blockchain 165
    8.3.4 Hybrid Blockchain 165
8.4 Consensus Algorithms.. 165
    8.4.1 Proof of Work (PoW) 167
    8.4.2 Proof of Stake (PoS) 167
    8.4.3 Delegated Proof of Stake (DPoS) 167
    8.4.4 Proof of Burn (PoB).. 168
    8.4.5 Practical Byzantine Fault Tolerance (PBFT) 168
    8.4.6 Ripple.. 168
8.5 Securing the IoT using Blockchain 168
    8.5.1 IoT Infrastructure 169
        8.5.1.1 Stage One: Sensors/Actuators 169
        8.5.1.2 Stage 2: The Internet Gateways.. 169
        8.5.1.3 Stage 3: Edge IT and the Cloud 169
        8.5.1.4 Stage 4: Applications Domain.. 170
    8.5.2 Security Issues in the IoT and Solutions Using Blockchain 170
        8.5.2.1 Data Privacy and Integrity (Unauthorized
              Access to Devices) 170
        8.5.2.2 IoT Device Authentication 170

	8.5.2.3	Insecure Devices 171
	8.5.2.4	Sybil Attacks 171
	8.5.2.5	Software Attack (Malware and Ransomware) 171
	8.5.2.6	RPL Routing Attack 171
	8.5.2.7	Sinkhole and Wormhole Attack...................................... 171
	8.5.2.8	Malicious Code Injection 172
	8.5.2.9	Congestion/Jamming... 172
	8.5.2.10	Spoofing Attack... 172
	8.5.2.11	Deviation and Disruption of Protocols........................... 172
	8.5.2.12	The Exploitation of Misconfigurations (OS, Servers, Frameworks, etc.)..................................... 173
	8.5.2.13	Single Point of Failure 173
8.6	Applications of Blockchain.. 173
	8.6.1	Smart Power Utilization 173
	8.6.2	Smart City.. 175
	8.6.3	Health Care 176
	8.6.4	Smart Contract... 176
	8.6.5	Electronic Voting System... 176
	8.6.6	Smart Identity 176
	8.6.7	Cryptocurrency 177
	8.6.8	Warranty and Insurance Claims...................................... 177
	8.6.9	Transparent Supply Chains... 177
	8.6.10	Document Verification 177
8.7	Conclusion and Future Scope... 177
References.. 178

8.1 INTRODUCTION

The Internet of Things (IoT) is one of the most prominent and impactful technologies in today's information and communication technology world. It is used everywhere in different forms. From home usage to industry usage, the IoT has appeal and importance. It has widespread usage in smart cities [1], smart farming, smart grids, and smart meters, agriculture, home automation security, etc. [2]. Due to its small decentralized architecture, easy installation, and ease of usage, it has attracted all types of users. However, as the application vector increases, the attack vector has also increased. The primary purpose of a perpetrator is to collect as much data as possible from the environment. IoT devices, including various sensors, are connected with the physical environment. If the attacker can get their hands on the network, he can collect a considerable amount of information.

Blockchain, a distributed digital ledger technology, emerged back in 1991 with a group of enthusiastic researchers in an attempt to timestamp digitally embedded documents, so that they would not become alterable or obsolete. Later, in 2008, a group of anonymous people known by the name Satoshi Nakamoto [3] invented the first Blockchain-based cryptocurrency for digital transactions. Known as Bitcoin, it consists of an ordered series of blocks for recording digital transactions. It is based on a peer-to-peer [4] and decentralized architecture that promotes transparency in the

Emergence of Blockchain Technology

transaction. Since the transaction inside this type of currency network takes place without the interaction of a third party, users have some benefits, e.g., reduced costs, less time delay, and avoiding mishaps caused by third parties. Blocks are the heart of the blockchain network, which comprises transaction data, cryptographic hash, and many other components. The first block in a blockchain is named the genesis block/ foundation block. It is linked with other blocks, creating an interlinked data structure for founding a blockchain network. The respective blocks are each connected to the previous block in the chain by storing the digital hash code of the last block. With each new transaction, a new block is appended to the chain of blocks [5].

IoT devices generally are connected in a centralized environment. In some cases, various types of wireless or wired sensors are connected, too. In a centralized network, the modules in the network are connected to a centralized node. This node controls all the other nodes, in terms of connection, data collection, data flow, maintenance, etc. Data collected by the nodes are stored or flow through this central node. In the case of a home automation network that is to design a smart home, the sensors and other devices are connected to a home server. This server controls the data collection and storage from various other devices. If we consider a smart agriculture environment too similar to a centralized structure is seen. A centralized structure always has a single point of failure, which can cause significant damage to the network itself [6,7]. There are other security issues involved that make the IoT network vulnerable to attacks and threats.

On the other hand, because blockchain is based on a decentralized structure, it does not have a single point of failure. The transactions are transparent to every user by consensus, which makes blockchain persistent, secure, and immutable, as no one can change the data inside the block without the other blocks acknowledging the changes. Any new transactions made in the ledger are validated by all of the participants in the network and are shared and made available to all the users in the blockchain network. This provides maximum security and permanency to the blocks of aggregated data. These features make blockchain a possible solution for various platforms, including the IoT.

This chapter comprises a holistic overview of blockchain methodology, including its application in the different arenas of information technology and its possible use in leveraging IoT security issues. The chapter includes a detailed methodology of blockchain and its various types, the simplified architecture of the IoT, and IoT security issues in brief, as follows:

- Characteristics and methodology of blockchain, in detail
- Application of blockchain in different sectors, with explanations
- Architecture of the Internet of Things, with a simplified summarizing figure
- Brief survey of security issues and challenges of the IoT
- Tabular representation of proposed blockchain solutions for IoT security issues
- Current challenges and future direction of the usage of blockchain

The chapter is organized as: Section 8.2 provides an overview of blockchain methodology. Section 8.3 describes various types of blockchain in detail, including a comparative analysis of its different types. Section 8.4 consists of consensus algorithms,

162   Blockchain Technology for Data Privacy Management

e.g., Proof of Work and Proof of Stake, etc. Section 8.5 consists of a detailed simplified architecture of the IoT, security issues of the IoT in detail,and proposed solutions, with blockchain addressing the security issues of the IoT. Section 8.6 consists of the application of blockchain. Section 8.7 outlines possible future research directions and usages of blockchain.

8.2 BLOCKCHAIN METHODOLOGY

Blockchain creates a secure, transparent, and trustworthy environment. For this reason, it is ideal for financial transactions (cryptocurrency, e.g., Bitcoin), confidential data storage, etc. To protect integrity, blockchain uses a hash inside each block. The length of the hash is the same for any amount of data stored inside a block. Trying to tamper with any data residing in a block would ultimately result in a complete change of hash [8].

8.2.1 BLOCKCHAIN CHARACTERISTICS

Blockchain has following characteristics [9,10]:

- *Decentralization:* Blockchain is based on distributed ledger technology. Unlike traditional networks, it does not store transaction history in a single central node of the network. The transaction history of each node is shared in the blockchain network. So, each node in the network has the same information about the transaction.
- *Transparency:* Every node in the blockchain network shares the digital ledger. As a result, an attempt to add a transaction needs to be validated by all the users. Only if a majority of the users give their consent can a block be added to the network, thus preserving transparency within the system.
- *Open Source:* Blockchain code is available online. Anyone can collect it and modify it, according to their individual needs. The open-source development scheme of blockchain enables more users to participate in the blockchain network and makes the system more versatile.
- *Autonomy:* Blockchain can execute transactions without any central authority, due to its autonomous design. So, it relies on the users themselves for functioning. Users keep the network alive and functional.
- *Immutability:* Once a transactional block is added in the digital ledger, it cannot be altered. Each block has a unique hash code, along with the hash of the previous block, forming a chain of blocks. Changing any data in one block means a change in all the previous blocks. Thus, the blockchain network is immutable.
- *Anonymity:* The anonymity of the users is provided through cryptography. It uses a public key to communicate and share information between users, and a private key to keep a user's identification hidden.

Bitcoin is a blockchain-based cryptocurrency. Since its inception in 2008, its market value has risen $300 billion, due to its reliability and security. Though bitcoin is

Emergence of Blockchain Technology

based on blockchain and implemented in a public blockchain, its reliability and security are ensured through consensus algorithms. There are several consensus algorithms in the blockchain, which are described in Section 8.4. Using a consensus algorithm along with the mining process, the nodes in a blockchain network have to agree upon every single transaction's validation. Before that, through the mining process, the nodes (miners) have to solve a complex mathematical problem. Upon completion of all the processes, a new block can be added to the blockchain [11,12].

8.2.2 DIGITAL WALLET

A digital wallet or cryptographic wallet is a virtual place where financial information, transaction information, and currency information are stored. Each person owns a digital wallet. Each digital wallet is identified by a public address, which is a combination of alphanumeric characters. This is similar to a bank account number used for monetary transactions, except that there is no user name or bank name associated with it, making the digital wallet anonymous to other users. Everyone can see the wallet ID and the amount associated with it, but not the real identity of the owner. For transaction and verification purposes, there is also a private key linked with it [13]. For example, User A intends to send digital currency to User B. Here, as the transaction is being processed, the destination address shall be checked against the private key of User B's wallet. If the checking process is successful, the amount will be transferred; otherwise, it will be declined.

8.2.3 BLOCK STRUCTURE

Blockchain consists of blocks containing information. Apart from the user data it stores, other information is stored inside the block [14]. The structure of a block formation, including a pictorial representation, is given in Figure 8.1.

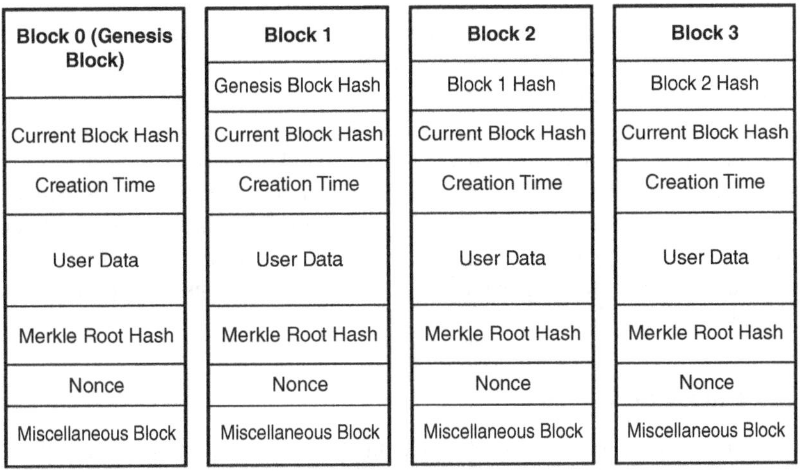

Block 0 (Genesis Block)	Block 1	Block 2	Block 3
	Genesis Block Hash	Block 1 Hash	Block 2 Hash
Current Block Hash	Current Block Hash	Current Block Hash	Current Block Hash
Creation Time	Creation Time	Creation Time	Creation Time
User Data	User Data	User Data	User Data
Merkle Root Hash	Merkle Root Hash	Merkle Root Hash	Merkle Root Hash
Nonce	Nonce	Nonce	Nonce
Miscellaneous Block	Miscellaneous Block	Miscellaneous Block	Miscellaneous Block

FIGURE 8.1 Pictorial Representation of Block Structure and Elements.

- Data: Blockchain is used in various application services, such as banking, insurance, e-commerce, etc. These application sectors define which type of data will be stored in this "Data" section. This portion stores the transaction information.
- Timestamp: Each block in a blockchain is produced at a specific time and date. This time and date data helps verify the block's existence, as it is created without human intervention and cannot be changed.
- Block Hash: A hash is created using various algorithms, e.g., SHA-256. Every block (except the genesis block) contains the hash of the previous block. A hash is unique and irreversible. Each block stores the hash of the previous block. So, changing the block's data becomes impossible, since it would change all the previous hash values [14].
- Nonce Value: Generally used for mining purposes, this value is generated using random functions. This 4-byte value helps miners to calculate the exact hash values of the block, so that it can be added to the blockchain. Here, the calculated hash value has to be less than the stipulated one.
- Merkle Root Hash: This value is used for the flexible verification and validation of application-specific block data. We have already seen that each block contains the hash of the previous block. A Merkle root hash is the hash of all previous transactions [15].

8.3 TYPES OF BLOCKCHAIN

The critical characteristics of blockchain have enabled its usage in different application services and sectors, such as smart contracts, cryptocurrency, banking, insurance, etc. Multiple types of blockchain have been developed for implementation in various types of entities. Briefly, these are [16]:

8.3.1 PUBLIC BLOCKCHAIN

In this type of blockchain, any user can participate in a transaction; that is, it is open source. No particular organization controls it. As this is open for participation and contribution, user identity is well hidden, which provides one of the best anonymities over the network. This type of blockchain also uses an incentive mechanism. That is, for mining purposes, incentives are presented to the miners. Some examples of public blockchain that are widely used and accepted are Bitcoin and Ethereum. This type of blockchain is also called permissionless blockchain. In terms of performance, public blockchain works a little bit more slowly than private blockchain does, because it has a lot of participating members. The consensus protocols used here also have some impact.

8.3.2 PRIVATE BLOCKCHAIN

This is also known as permissioned blockchain. The nodes participating in the mining and verification process are fixed earlier by the authority, meaning private blockchain imposes specific rules and restrictions in operation. A group of individuals or

Emergence of Blockchain Technology

entities participates in the controlling of the blockchain. The consensus is also based on the particular objectives of the entities or organizations. Private blockchain sometimes poses a centralized architectural model, as it is controlled and restricted to a certain number of individuals or organizations. Various consensus algorithms are used here. Depending on the consensus algorithm and the types of organization the blockchain belongs to, the authority selects the participants for transaction and verification. The usage benefits of private blockchain are organizational policy restrictions, data confidentiality, the protection of organizations, the need for less scalability, and other particular features that limit any corporate or individual organizations from taking part in a public blockchain network. Moreover, as only the selected participants can access the network, potential security breaches can be detected. Hyperledger, Hash graph, Corda, etc. are examples of popular and widely adopted private blockchain platforms.

8.3.3 Consortium Blockchain

Consortium blockchain comprises functional attributes of both public and private blockchain. However, public blockchain allows anyone to participate in the transaction and mining process. This property is absent in consortium blockchain. Consortium blockchain allows specific entities to participate in network activities. For this reason, this type of blockchain resembles more a private blockchain than a public blockchain, revealing a partially decentralized architecture. A consortium blockchain is also known as a federated blockchain. In this type of blockchain network, a permission access attribute is used, so that there is a controlled flow of participants, and organizational policies, rules, regulations can easily be incorporated.

8.3.4 Hybrid Blockchain

A hybrid blockchain comprises both public and private blockchain. It incorporates the characteristics and benefits of both types. The benefits of hybrid blockchain can range from permissioned access, modification according to the implementing organization, the use of flexible consensus protocols, etc. These benefits and features are adopted according to the requirements of the users. A hybrid blockchain is not fully permissionless; instead, it imposes specific rules while participating in the blockchain network. Ensuring the immutability and consensus procedure through particular nodes by assigned authorities, this type of blockchain increases the flexible use of the blockchain platform. It facilitates more room for verification and the detection of confidentiality breaches, and provides more control over the network. Hybrid blockchain platforms include Dragonchain. The aforementioned types of blockchains are summarized in Table 8.1 [17–20]:

8.4 CONSENSUS ALGORITHMS

This is an algorithmic process for validating blockchain transactions. It is a combined decision-making process, where the nodes or involved parties have to reach a consensus for a confident decision [21]. Blockchain is a distributed network, providing

TABLE 8.1
Summary of Attributes of Public, Private, and Consortium Blockchain

Properties	Public	Private	Consortium
Network type	Decentralized	Partially decentralized	Partially decentralized
Characteristic	No specific entity holds control of the network, anyone can enter	A specific organization holds control of the network, where access is limited	A group of predefined known entities control access of the network
Consensus determination	All miners	Miners within one organization	Selected miners within an organization
Consensus process	Permissionless	Permissioned	Permissioned
Consensus algorithms	Proof of Work (PoW), Proof of Stake (PoS), Proof of Elapsed Time (PoET), Delegated Proof of Stake (DPoS)	Practical Byzantine Fault Tolerance (PBFT), RAFT	Practical Byzantine Fault Tolerance (PBFT), Proof of Authority (PoA), Delegated Proof of Stake (DPoS)
Transaction approval frequency	Long	Short	Short
Efficiency (in terms of resources)	Less efficient, as the entire network verifies the transactions	Highly efficient, as a single organization verifies the transactions	Highly efficient, as fewer nodes are involved in the verification process
Centralization	No	Yes	Partial
Transaction cost	Costly, due to an increased number of nodes	Less costly, due to a fewer number of nodes	Less costly, due to a fewer number of nodes
Transaction speed	Slower	Faster	Faster
Write permission	Everyone	Restricted to only a single entity or organization	Restricted to authorized group
Identity	Anonymous	Known	Known
Transparency	Completely transparent	Transparent only to granted users	Transparent based on predefined protocols
Immutability	Impossible to tamper	Possible to tamper	Possible to tamper
Incentive strategy	Needs incentives to grow the network for better security	Does not need incentives, due to a single trusted node	Does not need incentives, due to selected trusted nodes
Examples	Bitcoin, Ethereum, Litecoin, etc.	Hyperledger, Corda, Hashgraph, etc.	Ripple, R3, etc.

anonymity for the users. Anonymity is vulnerable in any system and can pose threats, such as trust issues and malicious intruders. Consensus algorithms are used to address these issues and achieve reliability, by establishing a trustworthy interaction among the participants in the network. The purpose of using these algorithms is to resolve several computational complexities. There are various types of consensus algorithms, some of which are briefly discussed as follows.

Emergence of Blockchain Technology

8.4.1 Proof of Work (PoW)

According to this algorithm, a newly mined block has to provide sufficient proof of its effort to be added to the existing chain. Special nodes in the network, called miners, hold the responsibility to legitimize each transaction in the block through a process called mining. This is carried out by a computationally expensive calculation to solve mathematical puzzle problems. The complexity of the issues includes a hash function, integer factorization, and tour puzzles. On average, it takes about 10 minutes for a user node to try and calculate the correct solution [22]. All the miners in the network compete against each other to be the first to find the answer. The first miner who succeeds receives an economic reward that has been predefined by the protocol of the blockchain network.

To solve this puzzle, all the nodes need to put their verified transaction and some other information, such as previous hash and timestamp. There is a secret value that needs to be guessed by changing the nonce, and through this nonce, all the information will be inputted to an SHA-256 hash function. The value will be accepted when the output of this function is under a designated difficulty threshold. After every 2016 blocks, the difficulty is adjusted, which takes10 minutes per block [23]. Sometimes, more than one problem is solved within the given time period, which leads to the formation of several branches. However, the longest chain is considered to be the valid one in the decentralized network.

8.4.2 Proof of Stake (PoS)

Here, mining capability is determined by the number of currencies (stake) the miners already possess. The evidence of a stake follows a deterministic approach of choosing the miners (sometimes called forgers) for validating the transactions in the blockchain network. Unlike PoW, upon successful validation of a transaction, the forgers are not rewarded; instead, they obtain the transaction fees that were invested as a stake at the start of the forging process. The stake size is what determines the probability for the next forger to be selected and validate the next block. This algorithm is more energy efficient than PoW, which faces computational challenges while solving mathematical puzzles. Another concept behind this algorithm is that the nodes having higher stakes (currency) are not supposed to be involved in malicious activity with the blockchain network. So, node validation in this case entirely depends on the wealth of the nodes [24,25].

8.4.3 Delegated Proof of Stake (DPoS)

This is rather a more democratic approach than direct selection based on the amount of currency nodes have. Here, the nodes in the network elect witnesses on behalf of the stakeholders. Several witnesses select the valid block through a voting procedure. Thus, a block in the blockchain network is validated and authenticated. In this type of consensus algorithm, there are fewer nodes (witnesses) for block validation and authentication purposes. For this reason, the time needed for block validation is much less and the energy consumption is also significantly lower. As fewer nodes participate in the process of block authentication, a tendency toward centralization might arise.

Moreover, if any malicious node or a group of malicious nodes possesses the most stakes, it can threaten the network, because it can validate illicit transactions [26].

8.4.4 PROOF OF BURN (POB)

In this consensus algorithm, two different functions are used. Those function aim to burn or destroy the currencies in the validator's address. One task (function) generates a public address for cryptocurrency generation and initiates a burning procedure, in the event of a money transfer to that address. Another service (function) checks the account address's verifiability, from which the currencies cannot be spent for any transaction purpose. The block validators can validate the blocks by sending currency to the generated address. As the currency cannot be spent, that is why it is called burning the currency. The nodes receive incentives. Here, the value of the coin increases because of the burning procedure, since the total amount remains the same, concerning fewer coins [27].

8.4.5 PRACTICAL BYZANTINE FAULT TOLERANCE (PBFT)

The methodology of this replication algorithm is to make the blockchain network byzantine fault-proof. This is derived from the byzantine general problem, where an army must reach a consensus on whether to attack or not attack. In this scenario, several generals each have a group of troops and must reach a consensus about attacking or retreating. It is possible, however, that a general will not follow the agreement. That is why the fault tolerance algorithm was developed, in the case of a distributed network. This algorithm is used in a hyperledger fabric framework. The framework has fault tolerance, even if 1/3 of total nodes are malicious [28]. According to this algorithm, a node will be selected based on the network protocol. This node will be responsible for the transaction initiation. Out of the three phases of this method, the selected node needs to receive a majority of the votes. This way, the selected node is known to almost all over the network [29].

8.4.6 RIPPLE

In this type of consensus algorithm, two types of nodes are used in the system. One is a server type and another is a client. A server node takes part in the consensus for block validation and authentication. Each of them contains a list. This list comprises the number of unique nodes in the network. It is called the unique node list (UNL). A client-type node is responsible for transferring cryptocurrencies. Whenever a transaction is initiated, the server nodes ask for its validation among the UNL. If it receives an acceptance of 80 percent or more, the transaction is validated and stored in the blockchain [30,31].

8.5 SECURING THE IOT USING BLOCKCHAIN

Because it is a useful component in the modern world, the IoT has attracted a lot of attention, not only in terms of the variety of its usages but also as a potential

Emergence of Blockchain Technology

attack surface. IoT users have increased at different levels. So, the amount of valuable data stored and processed in these devices has also increased. Understanding IoT security threats first requires a comprehensive grasp of the infrastructure. This section discusses the IoT infrastructure, its security threats, and proposed solutions through using blockchain [32].

8.5.1 IoT Infrastructure

With the proliferation of advanced technologies, internet availability has become widespread. This enables devices with built-in sensors to create a network of interrelated devices for efficient data transmission and communication, giving rise to the whole new world of the IoT. In simple terms, the IoT refers to a system of interlinked devices that can communicate and exchange critical information via the internet [33,34]. The large variety of internet-connected smart devices in the IoT needs an infrastructure. To unify the different components of the IoT in one network, there are four stages, as follows:

8.5.1.1 Stage One: Sensors/Actuators

This stage is also known as the sensor/perception layer, which consists of both wired and wireless sensors and smart devices. The components of this layer perceive information from the surrounding environment The information is converted to electrical signals and passed to the actuator. The raw data is transmitted to IoT gateways. There are several types of sensors, such as body sensors, environmental sensors, home sensors, etc. For data transmission, initially, a heterogeneous network connectivity must be assured, which includes a local area network (LAN), Wi-Fi, and Ethernet. A personal area network (PAN) is another type of LAN that consists of both wired and wireless communication protocols, such as Bluetooth, ZigBee etc.

8.5.1.2 Stage 2: The Internet Gateways

Also called the network layer, this layer is responsible for processing, controlling, and managing IoT data. The architecture of this network model must maintain communication performance overall for latency, error probability, scalability, bandwidth requirements, and security, while ensuring efficient energy usage. An internet gateway is like an intermediary networking device, which aggregates data from the sensors and maintains security protocols to route it securely through Wi-Fi, the Internet, and wired LANs and transfer it to a remote server, such as the cloud.

8.5.1.3 Stage 3: Edge IT and the Cloud

Also known as the Middleware layer of the IoT architecture, the edge computing IT system receives an enormous amount of preprocessed IoT data from the previous stages. This layer is responsible for data storage, analysis, and real-time processing. For this reason, various technologies are enabled, such as databases, the cloud, and big data processing models. Edge computing leverages the power of distributed local servers to facilitate computing and manage a wide range of services. An edge IT processing system usually resides close to the end-devices, to minimize communication latency, bandwidth, and overhead traffic away from a centralized cloud [35]. The enormous amount of IoT data can quickly eat up network bandwidth, engulfing data center resources. Here, the edge comes to rescue. It reduces the burden on IT infrastructure by performing some analytics beforehand. The processed data is forwarded to the cloud.

8.5.1.4 Stage 4: Applications Domain

The application layer of the IoT architecture deals with the application services of the IoT in the physical world. The classification of application services can be made based on the heterogeneity and availability of networks, along with the network coverage size. Application sectors include the military, the environment, transportation, energy, health care, smart cities, etc. [36,37]. Horizontal markets include supply chain management, asset management, fleet management, surveillance, etc.

8.5.2 SECURITY ISSUES IN THE IOT AND SOLUTIONS USING BLOCKCHAIN

IoT devices and networks face several types of attacks due to a centralized structure, device capability, protocol design, and many other reasons. As security threats, attack techniques, and attack characteristics increase, researchers are trying to mitigate those attacks. These attacks breach the confidentiality, authenticity, and availability of IoT service [38]. In the case of industrial usage, this becomes a huge problem. Blockchain is a relatively new technology that possesses unique characteristics suitable for addressing IoT security issues [39]. In this section, the attacks and their proposed countermeasures using blockchain technology are discussed.

8.5.2.1 Data Privacy and Integrity (Unauthorized Access to Devices)

Data privacy and security are the most prudent issues that trip up the interconnected digitized world. The data stored in IoT devices is vulnerable to cyberattack, since it is shared, transmitted, and processed through multiple IoT devices. This can lead to unauthorized access to IoT networks, causing data theft and data manipulation by adversaries. Such events can compromise data privacy and integrity. To leverage such threats, blockchain can be a viable solution [40]. Blockchain uses a decentralized model that adopts cryptographic hash functions to prevent data tampering. Furthermore, blockchain maintains reliability and trustworthiness, due to its immutable nature [41].

8.5.2.2 IoT Device Authentication

Verified devices need to join a network to prevent malicious intruders. The diverse heterogeneity of devices and services in an IoT network requires different mechanisms for authentication. For this reason, there are no standard global security protocols. This poses difficulty for the authentication and authorization techniques of IoT devices. Blockchain can provide unique identifiers for IoT devices using cryptographic hash algorithms. It prevents any unauthorized IoT devices to gain access and attempt any malicious communication. A transaction made by a sender can be signed by using a unique public key [42].

Emergence of Blockchain Technology

8.5.2.3 Insecure Devices

The vast number of embedded devices in an IoT network are mostly devices with low cost and low power. These devices have limitations in terms of memory and computing power. Attackers can gain easy access to such physically insecure devices. Blockchain provides credentials as unique key pairs for each of the registered connected IoT devices. The absence of the intermediary in blockchain reduces the risk of device tampering and other malicious activities. Moreover, smart contracts in blockchain keep track of the node that needs to patch hardware, or update or reset IoT devices.

8.5.2.4 Sybil Attacks

A Sybil attack is a critical security attack in an IoT environment, where a malicious node attempts to masquerade as other nodes in the network, resulting in multiple fake identities. It helps an attacker to gain unauthorized access to the system and impede routing messages, disrupting the overlay of network operations. Sybil attacks can be addressed by blockchain technology. Generating a PoW for the mining process is very expensive, because this is a complex mathematical calculation. Yet, it can prevent malicious nodes from creating multiple fake identities. On the other hand, PoS provides a consensus based on the staking currency of each miner, which can limit the resource requirements for Sybil attacks [43].

8.5.2.5 Software Attack (Malware and Ransomware)

Malware (viruses, spyware, Trojan horses, etc.) is malicious software exploited by hackers to damage a computer system and steal confidential information. Ransomware is a variation of malware that can limit or completely lock down user access to files in a network by encrypting them and asking for a ransom in exchange for the decryption keys. Using blockchain, the whole database of sensitive data can be encrypted and stored in a digital ledger. The database is shared among all available nodes in the network, and consensus algorithms validate each transaction. This makes it almost impossible for an attacker to take over a single node and hold it for ransom, because multiple access points are available. This further prevents a single point of failure.

8.5.2.6 RPL Routing Attack

RPL is a routing protocol that transmits data packets between a sender node and a sink node, creating a destination-oriented directed acyclic graph (DODAG) [44]. The sink node is the centralized root node known as 6LBR (6LoWPAN Border Router). This centralized structure can be vulnerable to a potential single point of failure. As a decentralized ledger, blockchain eliminates the risk of a single point of failure, providing a distributed trust. All control frames are timestamped in the blocks of the blockchain with an encrypted hash of the previous block. This ensures secure data packet transmission and mitigates eavesdropping or man-in-the-middle attacks.

8.5.2.7 Sinkhole and Wormhole Attack

In this attack, a compromised end in the network attracts neighboring nodes with appealing fake routes for transferring their data packets toward the destination. This results in the dropping of the data packets midway, creating a sinkhole [45].

A wormhole attack is also an active attack, where two compromised nodes strategically place themselves at two ends of the network, creating a tunnel. This tunnel gives a false impression of being the active shortcut route of low-latency for data transfer [46]. Thus, nodes prefer to choose this route over any other routes, causing selective data forwarding, eavesdropping, and network disruption. The blockchain network enables controlled message flow through the network. The message is also encrypted in the blocks using cryptographic hashes and digital signature, which ensures data integrity and correct path selection.

8.5.2.8 Malicious Code Injection

This is a severe attack, where an attacker either physically inserts a harmful code via some exterior device to steal user data or injects a malicious program into the system. This compromises a node and causes the entire network to shut down [47]. To perform this attack, an external device must first be connected to the IoT network. Blockchain can ensure the authorization of such devices by providing cryptographically secured unique identifiers to each IoT device. This prevents malicious device communication. Smart contracts can be used to validate and authenticate the code transactions, preventing the network from injecting malicious programs [48].

8.5.2.9 Congestion/Jamming

Jamming is a type of attack on wireless networks that aims to disrupt operations by flooding the network with an illegitimate radio frequency (RF). For this reason, a legitimate data packet cannot be transmitted, which causes an IoT service malfunction [49]. The cryptographic features (consensus algorithms) of blockchain could be a probable solution. Legitimate data packets can be encrypted for temper-proofing and to ensure data integrity.

8.5.2.10 Spoofing Attack

Spoofing is a technique adopted by potential attackers to forge device or user identities, to gain unauthorized access and launch malicious attacks. The forms of spoofing attacks include IP spoofing, email spoofing, DNS spoofing, etc. Blockchain could maximize security against such attacks. Each sender node can use digital signatures to sign a transaction before sending it to the blockchain network, to establish legitimate access control. Furthermore, smart contracts can facilitate secure message communication, device authentication, and authorization [50].

8.5.2.11 Deviation and Disruption of Protocols

This type of attack compromises standard protocols, such as application protocols, network protocols, and key management protocols. This results in service unavailability [51]. Blockchain provides soft and hard fork improvements of the standard protocols via the cryptocurrency community. It reduces the overhead computational cost of devices having limited memory and computational power [52]. Decentralization omits the necessity of a third party, which provides autonomy, transparency, trust, and the prevention of disrupting protocols. Moreover, blockchains store cryptographic hashes that aid in the verification of each device [53].

Emergence of Blockchain Technology

8.5.2.12 The Exploitation of Misconfigurations (OS, Servers, Frameworks, etc.)

A secure application requires the security of its various OS platforms, frameworks, servers, database management systems, etc. A loophole in the configuration of any of these components leaves the application vulnerable to attacks. Examples of such scenarios include attackers with unauthorized access of disabled directory listings to exploit configuration files, and improper handling of error logs. This can expose sensitive information related to underlying application flaws, and exploitation of default account credentials (root, passworded) [54]. Since blockchain is immutable, it prevents the unauthorized modification of configuration files stored in the ledger. It also incorporates cryptographically enforced smart contracts that ensure the secure execution, monitoring, and management of configuration files automatically, in a tamper-proof environment [55].

8.5.2.13 Single Point of Failure

The heterogeneity of IoT network devices requires cloud identification and verification services for connection and data storage. The cloud is a trusted entity for the verification of data in the entire network, which makes it vulnerable to a single point of failure. This gives rise to critical data security and privacy issues. A tamper-proof environment for a sustainable network is a necessity for the IoT services envisioned. Blockchain's decentralized system prevents IoT devices from undergoing a single point of failure, since a digital ledger of transactions is open to every node for proper validation and authentication [56].

Table 8.2 summarizes various security issues and challenges in IoT systems. The table highlights the security challenges of the IoT, attack strategies adopted by attackers, the effects, and possible blockchain solutions [57] against those attacks.

8.6 APPLICATIONS OF BLOCKCHAIN

In recent years, as researchers have acknowledged the potential impact of blockchain, going beyond just cryptocurrency, there has been a surge in investment and excitement from all spheres of industry. Many countries have taken the initiative to adopt blockchain. Dubai has set up smart city initiatives. Since 2012, Estonia has adopted distributed ledger technology in many sectors, such as health care, legal services, and personal information management. The implementation of blockchain technology can potentially circumvent existing obsolete technologies and provide new permanent solutions [67]. Some application scenarios are as follows.

8.6.1 SMART POWER UTILIZATION

Fossil fuels provide 48 percent of the world's electrical energy, and their increasing scarcity brings about significant problems. Blockchain has been proposed to solve some of these problems, in terms of smart energy management [68]. The use of blockchain can potentially impact the business model and daily operation of

TABLE 8.2
IoT Security Issue Analysis and Blockchain Solutions

Security, Issues	Attack, Strategy	Effects	Affected, Layer	Blockchain, Solution
Ensure data privacy and integrity [54,58,59]	Nodes attacked physically, malicious code insertion	DoS attack, eavesdropping, MITM, privacy breach.	All four layers	A decentralized P2P network, cryptographic algorithms, and hashing techniques. Smart contracts for transaction validation
Authentication and authorization (users, devices, data) [60]	Physical attack, communication attack, software attack,	DoS attack, eavesdropping, privacy violation, unauthorized access	All four layers	Unique blockchain identifier, strong consensus algorithms, tamper-proof distributed ledger, smart contracts
Insecure devices [61]	Physical attack, communication attack	DoS attack, privacy violation	Perception layer	Unique identification credentials, no intermediary involved, smart contracts
Sybil attacks [62]	Communication attack	DoS attack, privacy violation, identity manipulation, spamming, resource exhaustion	Network layer	PoW and PoS significantly increase the cost of pseudonymous node creation
Software attack (malware and ransomware) [63]	Software attack	Zero-day attack, DoS attack, buffer overflow, data theft, privacy violation	Application layer	Decentralized digital ledger, encrypted transactions, no single point of failure
RPL Routing attack [64,65]	Communication attack	Eavesdropping, man-in-the-middle attack, single point of failure	Network layer	Decentralized distributed nodes, encrypted data packets, reduces network overhead
Sinkhole and wormhole attack [66]	Communication attack	DoS attack, energy consumption, privacy violation, dropped/altered routing information	Network layer	Decentralized distributed ledger, cryptographic techniques-hashing and digital signature
Malicious code injection [49]	Nodes attacked physically, malicious code insertion	Data theft, privacy violation, network disruption	Perception layer, network layer, application layer	Cryptographic device identifiers, smart contracts for code authentication
Congestion/ Jamming [49]	Physical attack	DoS attack, network disruption	Perception layer	Decentralized distributed ledger, cryptographic consensus algorithm (PoW)

(Continued)

Emergence of Blockchain Technology

TABLE 8.2 (*Continued*)
IoT Security Issue Analysis and Blockchain Solutions

Security, Issues	Attack, Strategy	Effects	Affected, Layer	Blockchain, Solution
Spoofing attack [49]	Communication attack, software attack	DoS attack, man-in-the middle attack, privacy violation, network disruption	Perception, network and application layers	Decentralized network, cryptographic digital signature, smart contracts
Deviation and disruption of protocols [66]	Communication attack	Man-in-the middle attack, network disruption, service unavailability	Network layer	Consensus algorithms, transparent distributed ledger, optimization of standard protocols, unique identifier, trusted authentication and authorization
Exploitation of misconfigurations (OS, servers, frameworks, etc.) [66]	Software attack	Data theft, privacy violation	Application layer	Decentralized ledger, immutability of records, cryptographic encryption, smart contract execution, no intermediary involved
Single point of failure [66]	Communication attack	Damage to the entire system, deterioration in services provided by the IoT	Middleware layer	Establishment of decentralized and distributed network

energy companies starting from accounting, sales and marketing, smart grid application, and management to alternative energy distribution, energy trading, and record keeping for auditory and regulatory compliance. The use of blockchain and smart contracts means each consumer could have an energy profile encompassing their energy consumption patterns. This could lead to companies tailoring different products to suit different consumer needs. Implementing blockchain technology in microgrids would enable the use of sustainable resources. Moreover, utilizing it could ensure the transparency, auditability, and better regulation of the distribution network.

8.6.2 SMART CITY

The smart city concept has attracted a lot of attention these days. Several countries have already taken initiatives to integrate and implement smart city technologies, connecting everything to the internet using sensors, IoT devices, and other elements. Smart city technology includes the smart-identity smart grid, smart-citizen, smart power, etc. Blockchain-based applications have been proposed for sustainable smart cities [69,70].

8.6.3 Health Care

Blockchain can also be used in health care technology. Since blockchain can store application-specific data, it can be used for patient's personal data and health record storage. Smart contracts integrated with blockchain can be used for payment services. This approach would protect the integrity and authenticity of data, provide transparency within the health sector in terms of payment, etc. In sum, it would create a transparent, trustworthy, efficient and reliable health network.

8.6.4 Smart Contract

The smart contract consists of coded instructions executed without any human or manual interaction. It can run automatically, in terms of certain protocols. A smart contract provides users the facility to digitize legal conditions and automatize the contract, which is usually done on paper. The concept was first introduced by Nick Szabo, a law scholar and cryptographer [71]. Once a smart contract starts, it cannot be stopped until the specific conditions are fulfilled. The contract is also immutable. Once stored inside blockchain, a smart contract it becomes trustworthy, immutable, and distributed over the network.

8.6.5 Electronic Voting System

Blockchain can also be implemented in electronic voting systems. A blockchain-based e-voting system would be fair and hence, introduce a strong democratic authority to the citizens of a country. All votes cast would be recorded in that immutable ledger. Votes could not be altered without consensus. In a democratic country, people need transparency for every kind of work in the state, especially elections. The 2018 midterm election in Virginia was held using blockchain technology. The event received a lot of hits, according to cybersecurity experts. In the voting procedure, the voters were identified using fingerprints, and then they were given electronic ballot papers. The voting procedure was carried out in a private blockchain environment, and eight nodes were used for the purpose.

8.6.6 Smart Identity

Blockchain is a possible solution for creating an immutable, trustworthy, authentic, easily traceable smart ID. Since August 2017, almost 650,000 persecuted Rohingya Muslims who fled from Myanmar to Bangladesh to take shelter, deprived of citizenship and identity, have been enrolled in a pilot project to provide them with digital identities using blockchain. The smart identifiers help them obtain services from the UN and other government bodies, in a transparent manner. Identifications created and managed in blockchain can have a greater impact in e-governance, smart contract-based payment, and blockchain-based health care systems, also.

Emergence of Blockchain Technology

8.6.7 Cryptocurrency

Cryptocurrency is another technology that uses blockchain. Cryptocurrency, like Bitcoin, was introduced in 2008 by Satoshi Nakamoto. In his paper, he described a peer-to-peer transfer of digital currency. The transaction data was stored in the blocks of blockchain. The data is encrypted with a pair of keys. Each block contains the hash of the previous one, so the smallest change in the previous block causes an enormous change in the hash, which ensures the authenticity and security of the transaction. Nowadays, there are other cryptocurrencies, such as Ethereum, Litecoin, Libra, Monero, etc.

8.6.8 Warranty and Insurance Claims

Blockchain also has potential applicability in warranty and insurance claims. Using smart contracts, both vendor and user can benefit when warranty claim issues arise. A warranty will be activated under certain conditions and void otherwise. This condition can be implemented in a smart contract and stored in blockchain, for transparency and trustworthiness. There is no need for human intervention here. Rather than human effort, as in the traditional system, a smart contract can execute warranty procedures in an effective way.

8.6.9 Transparent Supply Chains

A transparent product supply chain can be a reality, using blockchain. This can track products' minute details throughout the entire shipping process. This would help prevent waste, inefficiency, the illegal transport of unethical products, and fraudulence by any means [72]. Even though today's supply chains can handle large amounts of data, many processes are still reliant on paper. Introducing blockchain in the supply chain could help replace these manual processes and improve record keeping. Blockchain technology in the supply chain could also help improve tracking strategy, inventory management, the supply chain payment process, and even insurance and warranty claim procedures.

8.6.10 Document Verification

Blockchain could potentially be used in document verification in several sectors. In the educational sector, certificates could be verified using blockchain. Moreover, in the corporate sector, blockchain could have potential uses in crucial contract verification, using smart contracts. Document verification has possible uses in immigration, such as medical test report verification. The medical test report requested by an embassy/high commission could be stored in a blockchain and shared with the authorities. This would help immigration authorities to verify whether a report has been altered.

8.7 CONCLUSION AND FUTURE SCOPE

The Iot has become an essential tool, with the usage of smart devices such as smart homes, intelligent weather detection, etc. Apart from the security threats posed to its infrastructure and data, IoT has its appeal, due to other characteristics such as low cost, a smaller size, an integration capability with sensors and other devices, etc. Blockchain technology is relatively new. It has potential use in almost every sector of our lives. As the usage vector increases, however, issues arise. For this reason, while discussing the future scope, it is also imperative to discuss potential future application challenges. Blockchain's future application ranges are as follows:

- Big data is a buzzword in today's world. It is a collection of various types of data in large volume. Blockchain technology could be used as efficient and cost-effective online storage, to help record big data information in its original form, without distortion.
- Among the various types of blockchain, private and consortium blockchain use a centralized architecture that can lead to vulnerability issues for the network. For example, in the case of private or consortium blockchain, if the majority of or selected nodes collectively betray the organization, this could create a serious issue.
- Fraud detection in the banking and financial sectors could be another crucial future direction for blockchain technology.
- Storing data on a centralized server, such as cloud storage, is susceptible to hacking and data loss. Implementing blockchain technology would omit the centralized structure of cloud servers, making the system more robust and protected from cyberattacks. A decentralized cloud storage system can also enable participants to create a virtual marketplace.
- Document verification is another potential future application for blockchain. Educational certificates can be easily verified over a blockchain network. Moreover, storing educational data in a blockchain could also provide benefits for a future recruitment process. The human resource department could easily review blockchain data and use smart contracts to select suitable candidates.
- Integrating AI with blockchain can open doors to numerous opportunities by limiting the many challenges that blockchain holds. An example would be blockchain oracles linked with smart contracts that externally send outside information in the form of electronic data to trigger the execution of smart contracts. This introduces the intervention of a trusted third party. Here, the implementation of a smart oracle using AI that automatically learns from occurrences in the outside world and trains itself, could be a significant improvement in smart contracts. Moreover, blockchain smart contracts could be used to define rules that prevent the miscommunication of AI-based applications, such as autonomous cars.

REFERENCES

1. Eckhoff, D., and Wagner, I., 2017. Privacy in the smart city: Applications, technologies, challenges, and solutions. *IEEE Communications Surveys & Tutorials*, 20(1), pp. 489–516. doi:10.1109/COMST.2017.2748998.
2. Hassija, V., Chamola, V., Saxena, V., Jain, D., Goyal, P., and Sikdar, B., 2019. A survey on IoT security: Application areas, security threats, and solution architectures. *IEEE Access*, 7, pp. 82721–82743. doi:10.1109/ACCESS.2019.2924045.

Emergence of Blockchain Technology

3. Nakamoto, S., 2019. Bitcoin: A peer-to-peer electronic cash system. *Manubot.*
4. Bahri, L., and Girdzijauskas, S., 2019, June. *Blockchain technology: Practical P2P computing (Tutorial).* In *2019 IEEE 4th International Workshops on Foundations and Applications of Self-Protecting Systems (FAS*W)* (pp. 249–250). Umea, Sweden: IEEE. doi:10.1109/FAS-W.2019.00066.
5. Mohanta, B. K., Jena, D., Panda, S. S., and Sobhanayak, S., 2019. Blockchain technology: A survey on applications and security privacy challenges. *Internet of Things*, 8, p. 100107. doi:10.1016/j.iot.2019.100107.
6. Hamdan, O., Shanableh, H., Zaki, I., Al-Ali, A. R., and Shanableh, T., 2019, January. *IoT-based interactive dual mode smart home automation.* In *2019 IEEE International Conference on Consumer Electronics (ICCE)* (pp. 1–2). Las Vegas, NV: IEEE. doi:10.1109/ICCE.2019.8661935.
7. Glaroudis, D., Iossifides, A., and Chatzimisios, P., 2020. Survey, comparison and research challenges of IoT application protocols for smart farming. *Computer Networks*, 168. doi:10.1016/j.comnet.2019.107037.
8. Madaan, L., Kumar, A., and Bhushan, B., 2020. *Working principle, application areas and challenges for blockchain technology.* In *IEEE 9th International Conference on Communication systems and Network Technologies (CSNT)* (pp. 254–259). Gwalior, India: IEEE. doi:10.1109/CSNT48778.2020.9115794.
9. Gupta, S., Sinha, S., and Bhushan, B., 2020. *Emergence of blockchain technology: Fundamentals, working and its various implementations.* In *Proceedings of the International Conference on Innovative Computing & Communications (ICICC).* Delhi, India: Springer. doi:10.2139/ssrn.3569577.
10. Haque, A. B., and Rahman, M., 2020. Blockchain technology: Methodology, application and security issues*IJCSNS*, 20(2), pp. 21–30.
11. Tomov, Y. K., 2019, September. *Bitcoin: Evolution of blockchain technology.* In *2019 IEEE XXVIII International Scientific Conference Electronics (ET)* (pp. 1–4). Sozopol, Bulgaria: IEEE. doi:10.1109/ET.2019.8878322.
12. Ahram, T., Sargolzaei, A., Sargolzaei, S., Daniels, J., and Amaba, B., 2017, June. *Blockchain technology innovations.* In *2017 IEEE Technology & Engineering Management Conference (TEMSCON)* (pp. 137–141). Santa Clara, CA: IEEE. doi:10.1109/TEMSCON.2017.7998367.
13. Tschorsch, F., and Scheuermann, B., 2016. Bitcoin and beyond: A technical survey on decentralized digital currencies. *IEEE Communications Surveys & Tutorials*, 18(3), pp. 2084–2123. doi:10.1109/COMST.2016.2535718.
14. Varshney, T., Sharma, N., Kaushik, I., and Bhushan, B., 2019, October. *Authentication and encryption based security services in blockchain technology.* In *2019 International Conference on Computing, Communication, and Intelligent Systems (ICCCIS)* (pp. 63–68). Greater Noida, India: IEEE. doi:10.1109/icccis48478.2019.8974500.
15. Saghiri, A. M., 2020. Blockchain architecture. In Kim, S., and Deka, G. C. (eds.) *Advanced Applications of Blockchain Technology.* Singapore: Springer, pp. 161–176. doi:10.1007/978-981-13-8775-3_8.
16. Niranjanamurthy, M., Nithya, B. N., and Jagannatha, S., 2019. Analysis of blockchain technology: Pros, cons and SWOT. *Cluster Computing*, 22(6), pp. 14743–14757. doi:10.1007/s10586-018-2387-5.
17. Zheng, Z., Xie, S., Dai, H., Chen, X., and Wang, H., 2017, June. *An overview of blockchain technology: Architecture, consensus, and future trends.* In *2017 IEEE International Congress on Big Data (Big Data Congress)* (pp. 557–564). Honolulu, HI: IEEE. doi:10.1109/BigDataCongress.2017.85.

18. Dib, O., Brousmiche, K. L., Durand, A., Thea, E., and Hamida, E. B., 2018. Consortium blockchains: Overview, applications and challenges. *International Journal on Advances in Telecommunications*, 11 (1 & 2) (pp. 51–64).

19. Mingxiao, D., Xiaofeng, M., Zhe, Z., Xiangwei, W., and Qijun, C., 2017, October. *A review on consensus algorithm of blockchain.* In *2017 IEEE International Conference on Systems, Man, and Cybernetics (SMC)* (pp. 2567–2572). Banff, AB, Canada: IEEE. doi:10.1109/SMC.2017.8123011.

20. Dorsemaine, B., Gaulier, J. P., Wary, J. P., Kheir, N., and Urien, P., 2016, July. *A new approach to investigate IoT threats based on a four layer model.* In *2016 13th International Conference on New Technologies for Distributed Systems (NOTERE)* (pp. 1–6). Paris: IEEE. doi:10.1109/NOTERE.2016.7745830.

21. Tasca, P., and Tessone, C. J., 2017. Taxonomy of blockchain technologies. In *Principles of Identification and Classification*. arXiv preprint arXiv: 1708.04872.

22. Crosby, M., Pattanayak, P., Verma, S., and Kalyanaraman, V., 2016. Blockchain technology: Beyond bitcoin. *Applied Innovation*, 2(6–10), p. 71.

23. Wang, W., Hoang, D. T., Hu, P., Xiong, Z., Niyato, D., Wang, P., ... and Kim, D. I., 2019. A survey on consensus mechanisms and mining strategy management in blockchain networks. *IEEE Access*, 7, pp. 22328–22370. doi:10.1109/ACCESS.2019.2896108.

24. Saleh, F., 2020. Blockchain without waste: Proof-of-stake. *The Review of Financial Studies*, forthcoming. https://doi.org/10.1093/rfs/hhaa075 .

25. Kiayias, A., Russell, A., David, B., and Oliynykov, R., 2017, August. *Ouroboros: A provably secure proof-of-stake blockchain protocol.* In *Annual International Cryptology Conference* (pp. 357–388). Cham: Springer. doi:10.1007/978-3-319-63688-7_12.

26. Larimer, D., 2014. *Delegated proof-of-stake (dpos).* Bitshare whitepaper.

27. Bhushan, B., Khamparia, A., Sagayam, K. M., Sharma, S. K., Ahad, M. A., and Debnath, N. C. 2020. Blockchain for smart cities: A review of architectures, integration trends and future research directions. *Sustainable Cities and Society*, 61(102360). doi:10.1016/j.scs.2020.102360.

28. Hyperledger Project, 2015. Advancing business blockchain adoption through global oepn source collaboration. https://www.hyperledger.org/. Accessed on June 5, 2020.

29. Castro, M., and Liskov, B., 1999, February. Practical byzantine fault tolerance. *OSDI*, 99(1999), pp. 173–186.

30. Bach, L. M., Mihaljevic, B., and Zagar, M., 2018, May. *Comparative analysis of blockchain consensus algorithms.* In *2018 41st International Convention on Information and Communication Technology, Electronics and Microelectronics (MIPRO)* (pp. 1545–1550). Opatija, Croatia: IEEE. doi:10.23919/MIPRO.2018.8400278.

31. Schwartz, D., Youngs, N., and Britto, A., 2014. The ripple protocol consensus algorithm. *Computer Science*, 5(8), pp. 1–8.

32. Sharma, T., Satija, S., and Bhushan, B., 2019, October. *Unifying blockchian and IoT: Security requirements, challenges, applications and future trends.* In *2019 International Conference on Computing, Communication, and Intelligent Systems (ICCCIS)* (pp. 341–346). Greater Noida, India: IEEE. doi:10.1109/icccis48478.2019.8974552.

33. Nie, X., Fan, T., Wang, B., Li, Z., Shankar, A., and Manickam, A., 2020. Big data analytics and IoT in operation safety management in Under Water Management. *Computer Communications*, 154, pp. 188–196. doi:10.1016/j.comcom.2020.02.052.

34. Xu, X., Sun, W., Vivekananda, G.N. and Shankar, A., 2020. Achieving concurrency in cloud-orchestrated Internet of Things for resource sharing through multiple concurrent access. *Computational Intelligence*. Special Issue, pp. 1–16. doi:10.1111/coin.12296.

Emergence of Blockchain Technology

35. Premsankar, G., Di Francesco, M., and Taleb, T., 2018. Edge computing for the Internet of Things: A case study. *IEEE Internet of Things Journal*, 5(2), pp. 1275–1284. doi:10.1109/JIOT.2018.2805263.

36. Latif, G., Shankar, A., Alghazo, J. M., Kalyanasundaram, V., Boopathi, C. S., and Jaffar, M. A., 2019. I-CARES: Advancing health diagnosis and medication through IoT. *Wireless Networks*, 26(4), pp. 2375–2389. doi:10.1007/s11276-019-02165-6.

37. Shankar, A., Sivakumar, N. R., Sivaram, M., Ambikapathy, A., Nguyen, T. K., and Dhasarathan, V., 2020. Increasing fault tolerance ability and network lifetime with clustered pollination in wireless sensor networks. *Journal of Ambient Intelligence and Humanized Computing*, pp. 1–14. doi:10.1007/s12652-020-02325-z.

38. Sinha, P., Rai, A. K., and Bhushan, B., 2019, July. *Information security threats and attacks with conceivable counteraction*. In *2019 2nd International Conference on Intelligent Computing, Instrumentation and Control Technologies (ICICICT)* (pp. 1208–1213). Kannur, Kerala, India: IEEE. doi:10.1109/icicict46008.2019.8993384.

39. Biswal, A., and Bhushan, B., 2019, September. *Blockchain for Internet of Things: Architecture, consensus advancements, challenges and application areas*. In *2019 5th International Conference On Computing, Communication, Control And Automation (ICCUBEA)* (pp. 1–6). Pune, India: IEEE. doi:10.1109/iccubea47591.2019.9129181.

40. Arora, D., Gautham, S., Gupta, H., and Bhushan, B., 2019, October. *Blockchain-based security solutions to preserve data privacy and integrity*. In *2019 International Conference on Computing, Communication, and Intelligent Systems (ICCCIS)* (pp. 468–472). Greater Noida, India: IEEE. doi:10.1109/icccis48478.2019.8974503.

41. Makhdoom, I., Abolhasan, M., Abbas, H., and Ni, W., 2019. Blockchain's adoption in IoT: The challenges, and a way forward. *Journal of Network and Computer Applications*, 125, pp. 251–279. doi:10.1016/j.jnca.2018.10.019.

42. Reyna, A., Martín, C., Chen, J., Soler, E., and Díaz, M., 2018. On blockchain and its integration with IoT: Challenges and opportunities. *Future Generation Computer Systems*, 88, pp. 173–190. doi:10.1016/j.future.2018.05.046.

43. Alachkar, K., and Gaastra, D., 2018, August. Blockchain-based Sybil attack mitigation: A case study of the I2P network. Faculty of Physics, Mathematics and Informatics University of Amsterdam. https://homepages.staff.os3.nl/~delaat/rp/2017-2018/p97/report.pdf.

44. Raoof, A., Matrawy, A., and Lung, C. H., 2018. Routing attacks and mitigation methods for RPL-based Internet of Things. *IEEE Communications Surveys & Tutorials*, 21(2), pp. 1582–1606. doi:10.1109/COMST.2018.2885894.

45. Kaur, M., and Singh, A., 2016, September. *Detection and mitigation of sinkhole attack in wireless sensor network*. In *2016 International Conference on Micro-Electronics and Telecommunication Engineering (ICMETE)* (pp. 217–221). Ghaziabad: IEEE. doi:10.1109/ICMETE.2016.117.

46. Pirzada, A. A., and McDonald, C., 2005, May. *Circumventing sinkholes and wormholes in wireless sensor networks*. In *IWWAN'05: Proceedings of International Workshop on Wireless Ad-hoc Networks* (Vol. 71, pp. 261–266). London: Centre for Telecommunications Research.

47. Sengupta, J., Ruj, S., and Bit, S. D., 2020. A comprehensive survey on attacks, security issues and blockchain solutions for IoT and IIoT. *Journal of Network and Computer Applications*, 149, 102481. doi:10.1016/j.jnca.2019.102481.

48. Ahmed, A. W., Ahmed, M. M., Khan, O. A., and Shah, M. A., 2017. A comprehensive analysis on the security threats and their countermeasures of IoT. *International Journal of Advanced Computer Science and Applications*, 8(7), pp. 489–501.

49. Bhushan, B., and Sahoo, G., 2017. Recent advances in attacks, technical challenges, vulnerabilities and their countermeasures in wireless sensor networks. *Wireless Personal Communications*, 98(2), pp. 2037–2077. doi:10.1007/s11277-017-4962-0.

50. Varshney, T., Sharma, N., Kaushik, I., and Bhushan, B., 2019, October. *Architectural model of security threats & their countermeasures in IoT*. In *2019 International Conference on Computing, Communication, and Intelligent Systems (ICCCIS)*. Greater Noida, India: IEEE. doi:10.1109/icccis48478.2019.8974544.

51. Hossain, M. M., Fotouhi, M., and Hasan, R., 2015, July. *Towards an analysis of security issues, challenges, and open problems in the internet of things*. In *2015 IEEE World Congress on Services* (pp. 21–28). New York: IEEE. doi:10.1109/SERVICES.2015.12.

52. Sial, M. F. K., 2019. Security issues in Internet of Things: A comprehensive review. *American Scientific Research Journal for Engineering, Technology, and Sciences (ASRJETS)*, 53(1), pp. 207–214.

53. Maroufi, M., Abdolee, R., and Tazekand, B. M., 2019. On the convergence of blockchain and Internet of Things (IoT) technologies. *Journal of Strategic Innovation and Sustainability*, 14(1), pp. 1–11. doi:10.33423/jsis.v14i1.990.

54. Rizvi, S., Kurtz, A., Pfeffer, J., and Rizvi, M., 2018, August. *Securing the Internet of Things (IoT): A security taxonomy for IoT*. In *2018 17th IEEE International Conference on Trust, Security and Privacy in Computing and Communications/12th IEEE International Conference on Big Data Science and Engineering (TrustCom/BigDataSE)* (pp. 163–168). New York: IEEE. doi:10.1109/TrustCom/BigDataSE.2018.00034.

55. Košťál, K., Helebrandt, P., Belluš, M., Ries, M., and Kotuliak, I., 2019. Management and monitoring of IoT devices using blockchain. *Sensors*, 19(4), p. 856. doi:10.3390/s19040856.

56. Sultan, A., Malik, M. S. A., and Mushtaq, A., 2018. Internet of Things security issues and their solutions with blockchain technology characteristics: A systematic literature review. *American Journal of Computer Science and Information Technology*, 6(3), p. 27. doi:10.21767/2349-3917.100027.

57. Saini, H., Bhushan, B., Arora, A., and Kaur, A., 2019. *Security vulnerabilities in information communication technology: Blockchain to the rescue (A survey on Blockchain Technology)*. In *2019 2nd International Conference on Intelligent Computing, Instrumentation and Control Technologies (ICICICT)* (pp. 1680–1684). Kannur, Kerala, India: IEEE. doi:10.1109/icicict46008.2019.8993229.

58. Zheng, Z., Xie, S., Dai, H. N., Chen, X., and Wang, H., 2018. Blockchain challenges and opportunities: A survey. *International Journal of Web and Grid Services*, 14(4), pp. 352–375. doi:10.1504/IJWGS.2018.095647.

59. Cong, L. W., and He, Z., 2019. Blockchain disruption and smart contracts. *The Review of Financial Studies*, 32(5), pp. 1754–1797. doi:10.1093/rfs/hhz007.

60. Weber, R. H., 2010. Internet of Things: New security and privacy challenges. *Computer Law & Security Review*, 26(1), pp. 23–30. doi:10.1016/j.clsr.2009.11.008.

61. Pulkkis, G., Karlsson, J., Westerlund, M., and Hassan, Q. F., 2018. Blockchain-based security solutions for iot systems. In Hussan, Q. F. (ed.) *Internet of Things A to Z: Technologies and applications*. Hoboken: Wiley-IEEE Press, pp. 255–274. doi:10.1002/9781119456735.ch9.

62. Alharbi, A., Zohdy, M., Debnath, D., Olawoyin, R., and Corser, G., 2018. Sybil attacks and defenses in Internet of Things and mobile social networks. *International Journal of Computer Science Issues (IJCSI)*, 15(6), pp. 36–41. doi:10.5281/zenodo.2544625.

Emergence of Blockchain Technology

63. Zhang, Z. K., Cho, M. C. Y., Wang, C. W., Hsu, C. W., Chen, C. K., and Shieh, S., 2014, November. *IoT security: Ongoing challenges and research opportunities*. In *2014 IEEE 7th International Conference on Service-Oriented Computing and Applications* (pp. 230–234). Matsue, Japan: IEEE.

64. Airehrour, D., Gutierrez, J. A., and Ray, S. K.,(2019. SecTrust-RPL: A secure trust-aware RPL routing protocol for Internet of Things. *Future Generation Computer Systems*, 93, pp. 860–876. doi:10.1109/SOCA.2014.58.

65. Li, X., Jiang, P., Chen, T., Luo, X., and Wen, Q., 2020. A survey on the security of block-chain systems. *Future Generation Computer Systems*, 107, pp. 841–853. doi:10.1016/j. future.2018.03.021.

66. Khan, M. A., and Salah, K., 2018. IoT security: Review, blockchain solutions, and open challenges. *Future Generation Computer Systems*, 82, pp. 395–411. doi:10.1016/j. future.2017.11.022.

67. Nehaï, Z., and Guerard, G., 2017, May. *Integration of the blockchain in a smart grid model*. In *The 14th International Conference of Young Scientists on Energy Issues (CYSENI)* (pp. 127–134). Kaunas, Lithuania: CYSENI.

68. Xie, J., Tang, H., Huang, T., Yu, F. R., Xie, R., Liu, J., and Liu, Y., 2019. A survey of blockchain technology applied to smart cities: Research issues and challenges. *IEEE Communications Surveys & Tutorials*, 21(3), pp. 2794–2830. doi:10.1109/ COMST.2019.2899617.

69. N. Szabo., 1997. Formalizing and securing relationships on public networks. *First Monday*, 2(9). doi:10.5210/fm.v2i9.548.

70. Rana, T., Shankar, A., Sultan, M. K., Patan, R., and Balusamy, B., 2019, January. *An Intelligent approach for UAV and drone privacy security using blockchain methodology*. In *2019 9th International Conference on Cloud Computing, Data Science & Engineering (Confluence)*. Noida, India. IEEE. doi:10.1109/confluence.2019.8776613.

71. Saberi, S., Kouhizadeh, M., Sarkis, J., and Shen, L., 2019. Blockchain technology and its relationships to sustainable supply chain management. *International Journal of Production Research*, 57(7), pp. 2117–2135. doi:10.1080/00207543.2018.1533261.

72. Karatas, E., 2018. Developing Ethereum blockchain-based document verification smart contract for Moodle learning management system. *International Journal of Informatics Technologies*, 11(4), pp. 399–406.

73. Soni, S., and Bhushan, B., 2019, July. *A comprehensive survey on blockchain: Working, security analysis, privacy threats and potential applications*. In *2019 2nd International Conference on Intelligent Computing, Instrumentation and Control Technologies (ICICICT)* (pp. 922–926). Kannur, Kerala, India: IEEE. doi:10.1109/icicict46008.2019.8993210.

9 Internet of Things and Blockchain

Amalgamation, Requirements, Issues, and Practices

Ansh Riyal, Parth Sarthi Prasad
Department of Computer Engineering, Netaji Subhas University of Technology (formerly Netaji Subhas Institute of Technology), India

Deepak Kumar Sharma
Department of Information Technology, Netaji Subhas University of Technology (formerly Netaji Subhas Institute of Technology), India

CONTENTS

9.1 Introduction 186
9.2 Blockchain...187
    9.2.1 Currently Employed Methods..188
        9.2.1.1 Public Key Cryptography (Asymmetric Cryptography).....189
        9.2.1.2 Rivest-Shamir-Adleman Encryption Technique (RSA)......189
    9.2.2 Why Blockchain Is the Future 190
    9.2.3 Comparisons between Encryption Techniques 191
9.3 The Inner Workings/Blockchain Explained 192
    9.3.1 Components of Cryptography...192
    9.3.2 The Algorithm behind Blockchain Explained.......................................193
        9.3.2.1 Finite Fields..193
        9.3.2.2 Elliptic Curve Cryptography (Ecc) over R.........................193
    9.3.3 Putting These Together 194
        9.3.3.1 Elliptic Curves over Fp (Graphical Understanding) 194
        9.3.3.2 Signature Generation..195
        9.3.3.3 Signature Verification 196
    9.3.4 Proving the Space and Time Complexities 197
        9.3.4.1 Hash Function 197
        9.3.4.2 Traversal Time Complexity...198
        9.3.4.3 Elliptic Curve Discrete Logarithm Problem (Ecdlp) 199
    9.3.5 Distributed Consensus Mechanism Based on Proof of Work 199

9.4	Risk Analysis and Mathematical Understanding	199
	9.4.1 Satoshi Nakamoto's Analysis	199
	9.4.1.1 Satoshi's False Approximation	200
	9.4.1.2 Meni Rosenfeld's Correction	200
	9.4.1.3 Closed-Form Approach	201
	9.4.1.4 Finer Risk Analysis	201
9.5	Blockchain Limitations	202
	9.5.1 Excessive Energy Requirements	202
	9.5.2 Distribution and Duplication	202
	9.5.3 Inability to Adapt to User Base Burst	202
	9.5.4 Lack of Oversight and Corresponding Manipulations	203
	9.5.5 Lack of Abstraction/Privacy	203
	9.5.6 Proof of Work	203
	9.5.7 Hybridization Need and Complexity	203
	9.5.8 Storage	203
	9.5.9 Network Security Risk	204
	9.5.9.1 Mitigating Attacks	204
	9.5.9.2 The Sybil Attack	204
	9.5.9.3 Race Attack	204
	9.5.9.4 Finney Attack	204
	9.5.9.5 Vector76 Attack	204
9.6	Conclusion and Future Prospects	205
References		205

9.1 INTRODUCTION

The Internet of Things (IoT) is a relatively new but rapidly growing process. With concerns about the effectiveness of data security and encryption technologies against network vulnerabilities, blockchain integration as an encryption technique has found implementation in data encryption, commerce, defense, and various other fields of technology.

Taking advantage of its unique blend of large-scale network management and low power consumption, the IoT has found applications in dozens of fields and subfields, especially, focusing on the gathering and transmission of data. The IoT works with the idea of connecting every device on a common platform of communication, forming a distributed platform with centralized control, and using protocols to establish a standard language/mechanism for communication inside every node of the network. The concept of an internet of interconnected things has been gaining in popularity, by virtue of being a crucial standard for low-power lossy networks (LLN), under constrained resources. The IoT's widespread acceptance facilitates everyday activities, affecting the way people interact with their surroundings. Any real-world implementation of such an intercommunication concept needs to take information security into account. One of the most efficient mechanisms is the use of blockchain. Blockchain launched its worldwide popularity with the cryptocurrency known as Bitcoin. Briefly put, blockchain is a public, permanent, and append-only ledger. As a cryptocurrency,

Internet of Things and Blockchain

Bitcoin is based on a distributed ledger storing transactions blocks and validation reward, dependent on the encryption mechanisms used in blockchain, as discussed further in this chapter. Thus, it can be presented as one of the best possible applications for blockchain, providing a perfect platform for the implementation and explanation of the newly introduced concepts. That is why the name, "Bitcoin," is often confused with and used in lieu of "blockchain," and vice versa. However, blockchain has vast potential for the field of information security, as well, to be implemented without any form of currency or wallet, as is explained in this chapter [1].

The IoT is not a single entity, having a standard structure, but rather, devices interconnected over networks and communicating among themselves. In terms of security, since connections are established between a wide variety of physical/computing devices, gaining access to the weakest node/link can potentially be used to send malicious content to other parts of the network and thus bypass the external security protocols of the high-cost-high-sensitivity (in terms of proprietary/classified information) devices, to steal useful information, sabotage costly devices, or even crash the whole network. This situation creates a need for information security in three major ways: information security and cryptography; the prevention of access to data; and, the prevention of the readability of transmitted data. The prevention of access to data is managed with firewalls and security protocols and readability in transmission is handled with the use of cryptography. Intrusion detection systems (IDS) use pattern-recognition systems to classify anomalies and recognize external intrusions. Finally, firewalls' access to control and security protocols help prevent intrusions by utilizing encryption. Encryption works by giving identities and classifications based on the content of the file packets, without letting them be read by an attacker without the decryption key.

In this chapter, Section 9.2 describes the currently employed techniques and the relevance of blockchain to them, on the basis of key comparative parameters. Section 9.3 dives deeper into blockchain and explains the backbone of the concept and its implementation as a security mechanism. Section 9.4 describes the implementation of the explained concept in the world of the IoT. The limitations faced by blockchain in the fields of the IoT and cryptocurrency are subsequently described in Section 9.5. Section 9.6 briefly mentions blockchain's endless future possibilities and concludes the chapter.

9.2 BLOCKCHAIN

Blockchain is a linear combination of nodes developed to store interactions between users, in the form of transactions. It is a linked list, with transaction identities and originalities protected by various verification and authentication processes that involve digital signatures in the form of hashes.

Basically, blockchain is a distributed consensus ledger network, the details of which are explained in further sections. It offers a top-of-the-line method of communication and data transfer, safeguarded against any attacker that does not control at least 51 percent of the network. There is no central agency for someone to focus attacks on, and every node is given equal weight. Every transaction entry is agreed upon, and the complete history is visible to every verified node, giving blockchain a democratic setup.

Blockchain's introduction to the world was accompanied by that of Bitcoin cryptocurrency, so for most people, the two are intertwined. But the fact of the matter is that while bitcoins are specific implementations of blockchain in the creation and usage of a cryptocurrency, blockchain is an encryption and data security mechanism, which uses chains to store the transaction history of the whole group. That data can subsequently be cryptocurrency (as implemented in Bitcoin), or it can be used in internetwork communication and data transfer.

Implementing the IoT network initializes the sensor network in a way similar to the implementation of blockchain. By some minor tweaks to the software protocols, an IoT network can be relatively easily molded into a blockchain-secured network. Blockchain redefines the concept of the ownership of data. Having data means having the permission protocols to modify/share it. Blockchain boasts a centralized data string spread out across all nodes, given equal priority about all transactions of the whole network.

All the transaction records of communication are to be stored in a ledger that is simultaneously updated throughout the network. Since blockchain requires a decentralized transaction record that exists in every node, it is thus verified by the network as a collection. The method of transactions that is associated with blockchain can be loosely translated to data transfer in the IoT.

The actual implementation of data and file transfer in implementing blockchain in the IoT is a little more complex than its Bitcoin variant. In file transfer, the files cannot be stored on the blockchain, as that would create multiple copies of the whole database, wasting too much space and rendering the verification process too slow. The solution to that is something called the InterPlanetary File System (IPFS). We store the files to be transferred in a common storage database that is given a tag ID based on hashing. The sender then sends the hash/ID of the file to the receiver, who can then access it from the common IPFS database. This however creates a privacy problem, as the hash ID of the file sent as a message is recorded as a transaction in the blockchain, which is recorded for every user and accessible to all. To resolve this, pairing-based cryptography in blockchain provides the solution. The sender first encrypts the whole file with the public key of the recipient, and then the file is uploaded to the IPFS database. By the concepts of public key cryptography, the receiver has the private key to decrypt the file and read it. All data packets except for small communication messages are sent by this mechanism, so as to avoid creating an N-way copy on the server.

9.2.1 CURRENTLY EMPLOYED METHODS

In cryptography systems throughout the world, a shift from secret key sharing to public key cryptography has been observed. Besides the scope of implementing different pairing mechanisms for faster computation, and the opportunity to introduce pairing-based cryptography systems, the biggest advantage of public key systems over those using secret keys is not having to establish a guaranteed secure transmission channel whereby to share a secret key. The private key is generated by and kept by the recipient, who simply makes the public key public for everyone.

Internet of Things and Blockchain

9.2.1.1 Public Key Cryptography (Asymmetric Cryptography)

Public key cryptography systems are implemented by first generating a public key and a private key. The logic is that every user's public key will be available to everyone and can be used to encrypt data to be sent for communication. Encrypted data cannot be decrypted (and therefore understood) without the private key usage, even if the encryption key used is accessible (one-way trapdoor functions) [2]. The encrypted message is then received and decrypted by using the private key to reveal the original message. One of the cornerstones of understanding cryptography systems is to assume a malicious attacker that has access to every message that is in transmission. It is a generic standard to assume an attacker named X, a sender named B, and a receiver named A. The public key of A is assumed to be PubA and the corresponding private key to be PrivA and message M.

B encrypts message M by using PubA, which is freely available to all. The message generated is called E_m. While transmitting E_m through an unprotected transmission medium, attacker X cannot find the message M by the use of E_m and PubA, because of the trapdoor function. A then receives the message E_m, which is then decrypted using PrivA to recover M.

9.2.1.2 Rivest-Shamir-Adleman Encryption Technique (RSA)

RSA is an application of asymmetric cryptography. RSA works on the principles of exponents and mod-N mathematics.

Note: Congruence is represented by ~ in the chapter. Two values are said to be congruent in modular mathematics if their values wrapped around the modulus limit N (in a mod N system) are equal.

i.e., $a \sim b$ if a modulo $n = b$ modulo n.

We have a message M that is represented by using a number between 0 to N-1 (inclusive).

The public key PubA is called E (i.e., encryption key).

The private key PrivA is called D (i.e., decryption key).

The encrypted message Em is called C (i.e., ciphertext).

1. Key generation:

At the receiver's side, we first choose two very large random prime numbers, p and q, and calculate $N = p * q$.

Then we find the totient of N, i.e., $\phi(N)$.

$\phi(N)$ by calculation comes out to be $(p - 1)(q - 1)$.

Then we select a key E, such that it is co-prime with $\phi(N)$.

D is calculated by finding a value satisfying the following relation:

$$E \times D \sim 1 \left(\bmod N \right)$$

i.e., find a D such that ED-1 is divisible by $\phi(N)$.

E serves as the public key and D serves as the private key for the encryption.

190 Blockchain Technology for Data Privacy Management

2. Message encryption:

At the sender's side, we have the standard N, the message M, and the public key of receiver E.

We calculate C by using E, M, and N by the relation:

$$C = M^E \left(\bmod N \right) \tag{9.1}$$

The message is then sent in the form of the ciphertext.

3. Message decryption:

At the receiver's side, we use D, C, and N to find the original message by the relation:

Received message: $M_{rec} = C^D (mod\ N)$. Now, from (9.1)

$$C^D \sim \left(M^E \right)^D \left(\bmod N \right)$$

$$C^D \sim M^{E \times D} \left(\bmod N \right) \tag{9.2}$$

Applying properties of modular mathematics, we have

$$M^{E \times D} \sim M^{E \times D} \left(\bmod N \right)$$

Since $DE = 1 + K \times \phi(N)$ (for some value of K),

$$M^{D \times E} \sim M^{1+\phi(N)} \left(\bmod N \right)$$

$$M^{D \times E} \sim M \left(M^K \right)^{\phi(N)} \left(\bmod N \right)$$

$$M^{D \times E} \sim M \left(\bmod N \right)$$

So, from (9.2) we generate M from M_{rec}.
Thus we can obtain the original message for all values of M, p and q.

9.2.2 Why Blockchain Is the Future

With information security as its most desired and heavily invested-in field, the data industry wants a fault tolerant safe network. The problem with centralized security networks is their dependence on a central node. In the eventuality that the central node is hacked/not working, security is compromised, and the network stops. The IoT is in itself a decentralized concept, which acts on an interconnected web of individual nodes, which are taught a standard language for communication with nodes forming client server pairs, depending on the communication requirements. Blockchain is by far one of the most fault-tolerant systems with no central node. Every node contributes to verifying the validity of the blockchain. Even if one node fails, the network will not be disrupted, as is the requirement in the IoT.

Internet of Things and Blockchain

There is no central node, so there is no focus of attack for hijacking the network. The only way to take control of transactions is to take control of more than 51 percent of the network. Such a task is extremely difficult and becomes even more so with the randomization of miners in their physical locations in the blockchain groups formed [3].

9.2.3 COMPARISONS BETWEEN ENCRYPTION TECHNIQUES

Starting with a theoretical explanation, we understand that encryption techniques are mainly compared on two points: the speed of encryption-decryption and the minimum percentage of the network, the security of which must be breached in order to take control of the whole network. Now, these two points are generally in a multiplicative trade-off with each other, i.e., both of them in relative moderation, which produces the best overall result, whereas trying to get either one into a positive extreme results in the other one deteriorating drastically. Hashing is one of the fastest one-way trapdoor functions we have; it is virtually impossible to reverse, and there are very few approaches to illegally decrypting hashing without using brute force approaches (which are completely cost inefficient). Blockchain also ensures that in order to take control of the entire network, we need a minimum of 51 percent of the nodes in the network. As mentioned before, this ratio is very hard to achieve and thus makes intrusion detection easy and quick.

Parameters for the analysis of cryptographic algorithms include:

- Number of rounds used for encryption and decryption [4]
- Size of blocks
- Length of key to be used
- Encryption rates [5]

As seen in Figure 9.1, asymmetric algorithms are fewer, but newer. This is because, even though asymmetrical algorithms are better from a security perspective, they require greater computation processing time and higher data storage capabilities.

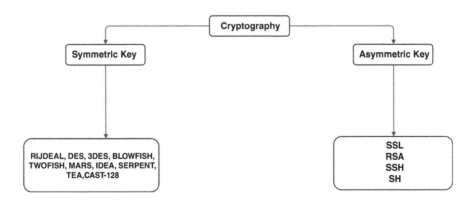

FIGURE 9.1 Classification of encryption algorithms.

Asymmetric algorithms, such as RSA, use key exchange, and symmetric algorithms use encryption/decryption [6].

9.3 THE INNER WORKINGS/BLOCKCHAIN EXPLAINED

A blockchain is identified by several parameters T = (p, a, b, G, n and h), and its digital signature algorithm is created on the basis of the same mathematics that is behind the eliptic curve digital signature algorithm (ECDSA). In an asymmetric key exchange composed of a public and private key pair (K_{priv}, K_{pub}) using equation

$$K_{pub}(\text{public key coordinate}) = k_{priv}(\text{private key coordinate}) * G \quad (9.3)$$

If two points of communication, A and B, are required to verify and authenticate the receiver signal, this can be done by generation of message signing and authentication using a digital signature [7].

9.3.1 COMPONENTS OF CRYPTOGRAPHY

In blockchain, a cryptographic hash is used on digital signatures carrying a certain amount of data stored in the Merkle tree [8]. Figure 9.2 shows the structure of a blockchain network, with header hash and a Merkle tree storing the data [9]. Various

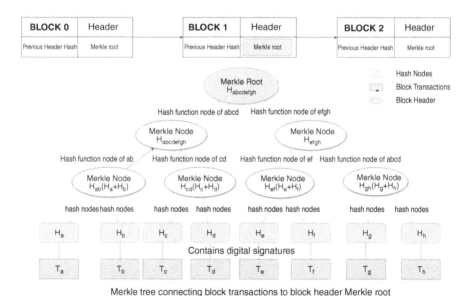

FIGURE 9.2 Merkle tree connecting representations of block header routes and transactions.

Internet of Things and Blockchain

components include the methods of data storage, the authentication and hashing mechanism, and the methodology of storage of hashed information. Its components are: header block, Merkle tree, cryptographic hash function, data authentication, and timestamping [10].

9.3.2 The Algorithm behind Blockchain Explained

9.3.2.1 Finite Fields

A finite field can be perceived as a set range of positive numbers, within which each calculation must fall. Any number that is not in this range must be wrapped around to ultimately fall within the range. Theoretically, for every finite field, there exists a number t such that $\Sigma\, t$ equals 0, i.e, $1 + 1 + 1 \ldots 1 = 0$ [5,11].

The first number a, following this property t, must belong to set of prime numbers P'' and is defined as characteristic of the field. The simplest example for this is calculating the remainder modulus (mod) with respect to a number. Figure 9.3 shows the wrapping of Secp256k1 over finite field F23.

9.3.2.2 Elliptic Curve Cryptography (Ecc) over R

Elliptic curves are defined by the function ($f(x)$) of the form:

$$f(x)^2 = x^3 - ax + b \tag{9.4}$$

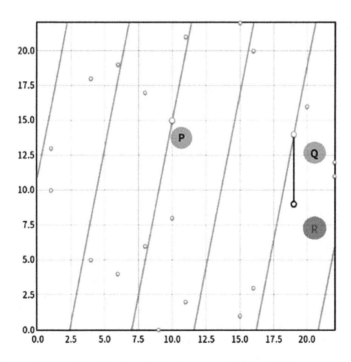

FIGURE 9.3 Finite field graph of an elliptic curve over F23 (mod 24).

This is also called the "short Weierstrass form" and is the general form to talk about elliptic curves. The other form of elliptic curve that can be discussed is the "Edward form":

$$x^2 + y^2 = 1 + dx^2 \tag{9.5}$$

This is used to "sign" data in a manner such that third parties can authenticate the signature of the signer, since the private key can only be created by the signer. Elliptic curves have several useful properties that make them usable for Bitcoin; these can be studied by plotting the Weierstrass form on the graph. Two important properties of ECDSA, which are used for the generation of points, are: point addition and point doubling.

The addition $(P + Q)$ and doubling $(2P)$ of points are used together for scalar multiplication $Z = kX$, defined as the repetitive addition of point X to itself k times. Taking an example from [12], each number can be represented as the continued sum and doubling of itself.

The important idea to be understood is that if we have point X and point Z, we cannot find preimage k. This implies that we cannot find $k = Z/X$, as no inverse for point addition or point doubling exists. Thus, the irreversibility of the ECDSA point multiplication and one-way function helps the system of asymmetric keys work to maintain a cryptographically encrypted secure system.

9.3.3 PUTTING THESE TOGETHER

9.3.3.1 Elliptic Curves over Fp (Graphical Understanding)

Floating point arithmetic has a drawback, in that integers, in time, are taken for computations, which is unavoidable from the ECDSA algorithm. For faster computation, we require input without decimal numbers. Hence, we consider an elliptic curve over finite fields [13].

Definition: an EC $(E(F_q))$ defined over F $(: F_q)$ is a set of the following points:

$P_i = (x_i, y_i)$ $y^2 = x^3 + 7$ over F_q
which can be expressed as
$y^2 = x^3 + 7 \pmod{q}$
for bitcoin: $q = 2^{256} - 2^{32} - 2^9 - 2^8 - 2^7 - 2^6 - 2^4 - 1$ (see Figure 9.4 for a finite field graph of a Secp256 curve).

As explained in [13], a cyclic subgroup of an EC is used while choosing parameters in place of the complete curve. The secondary additional parameters, such as a, b, p, k and point $G \in E(Fq)$, are used to choose the sufficient cyclic subgroup. The parameters are then shared, to generate public keys.

For their signature on any message, the sender is required to randomly choose an integer A as their private key (K_p), and compute the public key as $(K_a) = nAG$. A trusted authority (TA) is used to store all the public keys, to make them public within the network.

Internet of Things and Blockchain

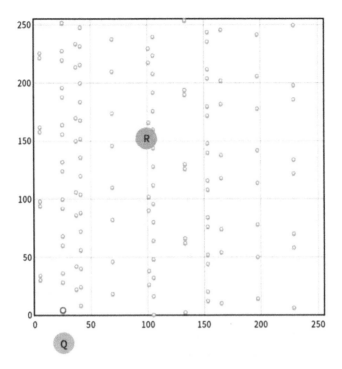

FIGURE 9.4 Secp256 curve ($y^2 = x^3 + 7$) defined over the field $Z_2{}^{256}{}_{-2}{}^{32}{}_{-977}$, where X and Y are coordinated 256-bit integer modulo large numbers.

9.3.3.2 Signature Generation

ECDSA signal generation specifies signing a message m, A as an entity and several domain parameters. $D = (q, FR, a, b, G, n, h)$ and associated pair of public keys (d, Q). The message key (r, s) must hold the condition of being a unique and nonzero pair [14]. Figure 9.5 shows point doubling and point addition for generator point G, for key generation of (r, s).

The procedure to generate the message signature can be described as follows:

1. Selection of an arbitrary/pseudo random natural number k between 0 and n (not inclusive).
2. Computation of random point on curve $(x1, y1) = kG$ in finite field
3. Calculate $r = x1 \pmod{n}$. If we get r as 0, find another point of the curve and repeat from step 1.
4. Find k inverse $k^{-1} \bmod n$.
5. Find the SHA hash of message m, SHA-1(m) and change the corresponding binary string into an integer e.
6. Compute $s = k^{-1}(e + dr) \bmod n$, where d is the private key.
7. If s is zero, thus not fulfilling the primary condition, revert back to step 1.
8. Thus generated values are (r, s), which form the signature pair for A.

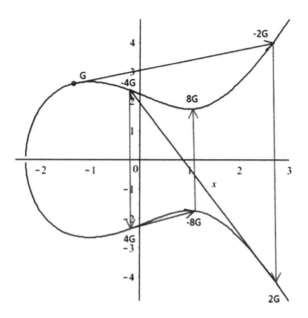

FIGURE 9.5 Pictorial plotting of the multiplication of the generator point G by private key K_{priv} [15].

9.3.3.3 Signature Verification

Since the receiver has a legitimate copy of all the sender's domain parameters and the PubKey Q, we can verify a's message signature (r, s) by doing the following:

1. Assert that received integers (r, s) belong to the interval $I, I \epsilon Z$ and $I > 0$ and $I < n$.
2. Compute the hash of message signal SHA-1(M) and convert this hashed binary string to integer e.
3. Find the modular inverse of s, represented by $w = s^{-1}(\text{mod } n)$.
4. Find $u1 = e * w(\text{mod } n)$ and $u2 = r * w(\text{mod } n)$.
5. Find $X = u1 * G + u2 * G$.
6. The received signal is rejected if $X = 0$.
7. Otherwise convert $x1$, the x-coordinate of X, to integer $x1'$, and compute $v = x1' \ (\text{mod } n)$.
8. If $= r$, the signal is accepted, or else the signal is not verified by the system.

9.3.3.3.1 Proof of the Signal Verification

If message M is derived from entity a, then we infer that

$$s = k^{-1}(e + dr) \text{mod } n$$

Multiplying both sides with s^{-1} and k

$$k = s^{-1}(e + dr)^1 (\text{mod } n)$$

Internet of Things and Blockchain

Using the distributive property over multiplication and multiplying s^{-1} with both terms

$k = s^{-1}e + s^{-1}dr \ (mod \ n)$; since $w = s^{-1}$

Substituting the value of $s^{-1}e \ (mod \ n) \ and \ s^{-1}dr \ (mod \ n)$ calculated in step 3

$$= we + wdr \left(\mod n \right)$$

$$= u1 + u2d \left(\mod n \right)$$

$$u1G + u2Q = \left(u1 + u2 \right) G = kG \tag{9.6}$$

hence $v = r$.

Hence, if the receiver marks the incoming message as being authentic, the precondition $v = r$ is fulfilled. If an external attacker tries to impersonate the sender and generates a message m', such that $m \ ! = m'$, and the impersonator finds the domain parameters a, b, G, h, n, p, and q and computes $Q = (x1, y1)$ points on the curve Choose $r = x1(mod \ n)$, and thus attempts to compute the hash of message m' as $H(m')$ from domain parameters and finds k^{-1}. The effort culminates in proving $v = r$. But since s is dependant on the sender's secret key A, which is only known to the sender, an attacker would require a technique to find the correct secret key without entity A, in other words, solving the ECDSA discrete logarithm problem (i.e., the conceptual fulcrum of blockchain's security) [14].

9.3.4 PROVING THE SPACE AND TIME COMPLEXITIES

9.3.4.1 Hash Function

Most hash functions have been designed with

- Initialization stage (with a fixed performance overhead) $O(1)$
- Compression function
- State update function
- Finalization state (with a fixed performance overhead) $O(1)$

The initialization and finalization of hash functions have fixed overhead time usage, hence the time complexity observed is $O(1)$ for both of these steps to evaluate the computational complexity of the hash function. Compression and state update are considered per block and can be considered together. Figure 9.6 shows a general model of the hash function and time complexity of various steps. Considering that there are total n blocks, and fixing the input by padding of numbers for computation of hash, where there is constant k, per-block overhead padding, and c is the constant time taken to initialize the hash. Then, the total time is calculated as $O(c + kn)$. SHA-256 is one such example of a hash function used in cryptographic encryption techniques like blockchain.

WSN-BASED IOT SYSTEM

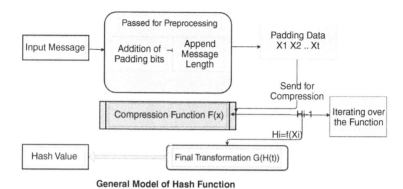

General Model of Hash Function

FIGURE 9.6 Pictorial representation of the general model of a hash function.

9.3.4.2 Traversal Time Complexity

In practice, Merkle trees have been criticized due to their high computational requirement and space and storage costs. Figure 9.7 shows a comparison of two standard approaches of Merkle tree hashing, with a comparison of the space and time trade-off.

Hashing Approach in Merkle Tree	Maximum Space Requirement	Maximum hash Evaluations per Round
Traditional Approach as per Markus Jacobsson [16]	$0.5 \log_2(N)$ units	$2(\log(N))^2$
Time–Space Trade-off Approach [17]	$\frac{3}{2}\log_2(N)/\log(\log(N))$	$2 \log(N)/\log(\log(N))$

Hashing Approach in Merkle Tree	Maximum Space Requirement	Maximum hash Evaluations per round
Traditional Approach As per M.Jacobsson[18]	$0.5\ log_2(N)$ units	$2(log(N))^2$
Time - Space Tradeoff Approach[19]	$\frac{3}{2}log_2(N)/log(log(N))$	$2log(N)/log(log(N))$

FIGURE 9.7 Table comparing space complexity and hash evaluations between hashing approaches in a Merkle tree. This can be further improved by scheduling methods with a logarithmic Merkle tree traversal. It uses scheduling to compute a specific number of nodes at once. This budget is used with 1 for the computation of left nodes and the rest for building the node values with stacks [18].

Internet of Things and Blockchain

9.3.4.3 Elliptic Curve Discrete Logarithm Problem (Ecdlp)

Let there be an elliptic curve E over a finite field F_q where $q = p^n$ and the computational problem is to find two points P, Q such that $P, Q \in E(Fq)$ such that $Q = aP$ [19].

This is the basis of pairing-based cryptography and elliptic curve cryptography. A few algorithms have been proposed for this problem:

- The "Meet-in-the-Middle" algorithm works contrary to a brute force attack.
- For the computation of discrete logarithms, Pollard's rho ρ algorithm is used. With an asymptotic time complexity of $O(\sqrt{n})$ and a space complexity of $O(1)$, it has the basic principle of generating a pseudo-random sequence of points $X1$, $X2,\ldots$, where each $X = a_iP + b_iQ$.
- With the capability of a discrete logarithm in polynomial time, a new quantum algorithm has been proposed, Shor's algorithm, with a worst-case time complexity of $O((\log n)^3)$ and a space complexity of $O(\log n)$.

9.3.5 DISTRIBUTED CONSENSUS MECHANISM BASED ON PROOF OF WORK

Proof of Work (PoW) is used as one of the mechanisms for consensus building and achieving agreement on the distributed network of blockchain to confirm transactions and produce new blocks on the blockchain. With PoW, prospective miners can compete against each other in an effort to validate the various transactions for a reward. We find the probability of being selected for building the next successive block, which is linked to the computation power of the system.

The underlying consensus-building model for blockchain lies on the PoW concept, which relies on a driving and incentive structure that provides a byzantine-fault-tolerant (BFT) distributed network. It is dependent on solving a mathematical puzzle to find a value below a threshold (nonce), which is used for the production of a new block broadcast to the network. Satoshi Nakamoto's protocol design for Bitcoin aims to reach a common coordinated consensus algorithm to authenticate the legitimacy of each transaction. The PoW can be characterized in two properties:

1. It must be computationally intensive and consume greater time for any miner to generate a proof that meets the requirements.
2. The verification of such a proof must not be time-consuming, and the proof's correctness should be easy to verify.

9.4 RISK ANALYSIS AND MATHEMATICAL UNDERSTANDING

9.4.1 SATOSHI NAKAMOTO'S ANALYSIS

The probability that an attacker catches up in authentication when the attacker is lagging behind honest miners by n blocks is:

$$q_n = \left(\frac{q}{p}\right)^n \tag{9.7}$$

200 Blockchain Technology for Data Privacy Management

According to Nakamoto's analysis, the probability that a double-spending attack by a fraudulent miner will be successful is denoted by P_z. When $t = S_z$, there are $N'(S_z)$ mined blocks. The value of the probability of successful attack P_z is

$$P_z = 1 - \sum_{k=0}^{z-1} P\left[N''(S_z) = k\right]\left(1 - q_{z-k}\right) \tag{9.8}$$

9.4.1.1 Satoshi's False Approximation

For $z\epsilon N$, the formula in Satoshi's Bitcoin white paper is:

$$P_{SN}(z) = 1 - \sum_{k=0}^{z-1} \left(\lambda^k e^{-\lambda} / k!\right)\left(1 - (q/p)^{z-k}\right) \tag{9.9}$$

$P_{SN}(z) = P(z,t)$, which is the probability that a double-spending attack will be successful after a total of z, blocks has been validated before a time interval of t.

Python3 code for Satoshi's approximation

```
#function: ProbAttacksuccess
#l: value of lambda
#p: probability
#poiss: poisson distribution value
From math import exp
def ProbAttacksuccess(q,z):
    p = 1 - float(q)
    l = float(z) * (q / p)
    ps =1.0
    for j in range(z+1):
        poiss = exp(-l)
            for (y in range(j+1):
                poiss *= l/y
        ps =ps - poiss * (1.0 - ((q / p)**(z - k))
    return ps
```

This approach was rejected later, due to an underestimation of $P(z)$ with Satoshi's approach.

9.4.1.2 Meni Rosenfeld's Correction

After the representation of a double-spending attack in Satoshi's analysis, the Meni Rosenfeld correction has proven to be the correct analysis of a double-spending attack. The probability of a successful double-spending attack is given by:

$$P(z) = 1 - \sum_{k=0}^{z-1} \left(p^z q^k - q^z p^k\right)\frac{k+z-1}{k} \tag{9.10}$$

Here, $\binom{n}{r}$ denotes ^nC_r (Combination): $^nC_r = (n!)/(((n-r)!)(r!))$
Pseudo code for Meni Rosenfeld's correction

Internet of Things and Blockchain

Q	z	P(z)	$P_{SN}(z)$	q	z	P(z)	$P_{SN}(z)$
0.1	0	1	1	0.3	0	1	1
0.1	1	0.2	0.2045873	0.3	5	0.1976173	0.1773523
0.1	2	0.0560000	0.0509779	0.3	10	0.0651067	0.0416605
0.1	3	0.0171200	0.0131722	0.3	15	0.0233077	0.0101008
0.1	4	0.0054560	0.0034552	0.3	20	0.0086739	0.0024804
0.1	5	0.0017818	0.0009137	0.3	25	0.0033027	0.0006132
0.1	6	0.0005914	0.0002428	0.3	30	0.0012769	0.0001522
0.1	7	0.0001986	0.0000647	0.3	35	0.0004991	0.0000379
0.1	8	0.000063	0.0000173	0.3	40	0.0001967	0.0000095
0.1	9	0.0000229	0.0000046	0.3	45	0.0000780	0.0000024
0.1	10	0.0000079	0.0000012	0.3	50	0.0000311	0.0000006

FIGURE 9.8 A comparison between the probability of success between Meni Rosenfelds's approach and Nakamoto's approach. $P(z)$ are values through Meni Rosenfeld approach and $P_{SN}(z)$ are values in accordance with Nakamoto's approximation.

```
#include<iostream>
#define db double
db probAttac(db α, dbβ){
db probability = 1−β;
db add = 1;
for( int k = 0; k < α; k++){
add-=(pow(probability, α) * pow(β, k)-pow(β,α) * pow( probability, k) )*
choose(k + α-1, k);}
return add;}
```

Its numerical application with plotting z values, according to Figure 9.8, shows the underestimation of probability by Satoshi Nakamoto's analysis.

9.4.1.3 Closed-Form Approach

Using an incomplete beta function, we find the probability of success $P(z)$, after there are total z blocks that have been verified

$$P(z) = I_{4pq}(z, 1/2) \tag{9.11}$$

9.4.1.4 Finer Risk Analysis

Let $T1$ be the time taken for an honest miner to get z blocks. Let the expected time it takes be equal to:

$$E[zT] = \frac{zT_0}{p}, \tag{9.12}$$

and let there be a variable

$$k = \frac{pT_1}{zT_0} \qquad (9.13)$$

In general, in Nakamoto's analysis, this k is assumed to be unity. Instead of finding $P(z)$, let us find $P(z, k)$, keeping k as a contributing factor. Assuming that a total of z blocks has been mined by legitimate miners as $S_z = T_1$,

$$P(z,k) = 1 - Q\left(z, \frac{kzq}{p}\right) + \left(\frac{q}{p}\right)^z * e^{kz\frac{p-q}{p}} Q(z,kz) \qquad (9.14)$$

9.5 BLOCKCHAIN LIMITATIONS

Blockchain has a lack of protocols that are in synchronization with existing HTTP/TCP-IP protocols. Furthermore, its idealization of a decentralized transaction based system with virtually no singular oversight in terms of an intermediary authority/controlling structure limits its credibility in the eyes of the market and big corporations, which have huge stakes in every venture [20,21].

Apart from that intuitive and hesitation-based reason for nonacceptance, the working of blockchain also poses some hurdles:

9.5.1 EXCESSIVE ENERGY REQUIREMENTS

The inclusion of transactions is based on miners' competing in common pools to create the next block to be signed, stamped, authenticated, and verified before it is added. This whole process requires a lot of computational power and energy, to the point that it has become cost-ineffective to be a miner [22,23].

9.5.2 DISTRIBUTION AND DUPLICATION

Even though blockchain works on a network of connections of nodes, there is no distribution of flow or cooperation of computation. Every single node in the distributed system is doing the same work. Hence, a single task is duplicated to millions of nodes, and there is no concept of paralleling or mutual exclusion.

9.5.3 INABILITY TO ADAPT TO USER BASE BURST

Bitcoin is the most common usage of blockchain, and it is not even that popular yet. Even with a very small user base, the speed of authentication and vast instantaneous changes in storage in the whole network for the execution of a single transaction request makes the overall network very slow and sluggish. This leads to the conclusion that when experiencing a burst in user size, blockchain is inefficient.

Internet of Things and Blockchain

9.5.4 LACK OF OVERSIGHT AND CORRESPONDING MANIPULATIONS

Blockchains are assumed to be indestructible, immutable, and uncontrollable by the policies of a country. However, this is not necessarily true, as the location concentration of a blockchain group often yields big chunks (sometimes more than 51 percent) of miners from the same country. This way, due to the equal importance/priority of every node in the distributed network, the policies implemented in individual big nations can manipulate the immutable blockchains [24].

9.5.5 LACK OF ABSTRACTION/PRIVACY

Blockchain was invented with the idea of complete transparency, i.e., any authenticated user has access to the information of every other user. In the real world, this information can often be confidential/not subject to disclosure, which hence can have a negative impact in the field of corporate secrecy.

9.5.6 PROOF OF WORK

To limit the addition of new transactions to a linear blockchain, Proof of Work requires that miners utilize a lot of processing power, in order to slow them down. This is counterproductive, even without the power-consumption consideration. Consequently, this invites a lot of research to discover ways to bypass the PoW, thereby undermining the concept of blockchain.

9.5.7 HYBRIDIZATION NEED AND COMPLEXITY

Blockchains were reputedly not introduced as a perfect implementation, but are the sparks for the idea of a network-wise distributed ledger. To accept this new idea with an open mind is to find a way to incorporate blockchain networks into existing setups. However, this is already complex enough in the existing setups (with different versions of the same variety of centralized networking models trying to come up with protocols for communication and combination). As a result, throwing blockchains into the mix would exponentially increase the complexity of very simple tasks, with a need to create bridges between the distributed and centralized models.

9.5.8 STORAGE

In effect, blockchain is very inefficient. The idea works fine in theory, but in application, maintaining the transaction logs for every node uses a lot of storage. An average estimate is that 200+Gb of data is stored in every node in a blockchain, for a duration of 30 months. Furthermore, adding a new miner in the network in such a situation is complete overkill. In an implemented blockchain having, say, 100 Gb of data, the introduction, downloading, authentication, etc. of a new miner can take up to three to four days.

9.5.9 NETWORK SECURITY RISK

The pride of blockchain is its distributed nature. In an ideal case of a randomized collection of miners in a group, network security is not compromised. But if they combine and assemble in a coven, they can cross the threshold of 51 percent and rewrite or alter the complete network's ledger, thereby compromising network and data security. This is the 51 percent attack. Even though this is very difficult to perform, it can have its own set of profits, ranging from a complex linkage of undisclosed transactions to hide profits, to political manipulation in the network. Apart from such a large-scale attack, individual loophole-based attacks include [25,26]:

9.5.9.1 Mitigating Attacks

Blockchain's peer-to-peer architecture is the major class of blockchain attack surface. The topological asymmetry of networks has been exploited. Blockchain is a permissionless public key cryptosystem. The hash rate can be affected, thus slowing down the efficiency of the system with network layer attacks, leading to transaction stall [27].

9.5.9.2 The Sybil Attack

This attack has a core motive of trying repeatedly to create multiple identities, in order to create a controlling share in the blockchain group. The attacker keeps trying new addresses, to increase its chances of being paired up with the target [28].

9.5.9.3 Race Attack

As the name suggests, the race attack is a very rapidly performed action. The attacker simultaneously creates a lot of unconfirmed transactions on the same bitcoin to a lot of receivers, who then confirm the transaction without waiting for a block confirmation. Later on, that is realized to be a scam; but the repayment transactions have often been completed by that time.

9.5.9.4 Finney Attack

Developed by Hal Finney, this is categorized as a double-spending problem. In this case, a miner who mines blocks normally includes an unreported transaction in the block he/she mines, sending some of the credits/coins back to himself/herself. On finding a pre-mined block, he/she sends the same coins in a second transaction, to be rejected by the other miners, but not before causing a sufficient verification time.

9.5.9.5 Vector76 Attack

The Vector76 attack is a combination of the Finney attack and the race attack. The basis of this attack is that a transaction that has one confirmation can still be rejected and reversed. In this attack, the attacker/miner creates two nodes, one having a well-established connection to peers in the blockchain and the other connected to an exchange node, i.e., one responsible to update values in the blockchain. The miner then creates two vastly different transactions, one having a very high value and the other having a moderately low value, and this difference makes up for the computational and manipulation cost. The miner then pre-mines the high value transaction to an exchange

Internet of Things and Blockchain   **205**

service and quickly pushes the pre-mined transaction to the exchange service receiving the exchange credits. Before the exchange service has been fully executed, the miner sends the low-value transaction to the blockchain. This creates a discrepancy in the transaction record, and the high value transaction is rejected by the verification mechanism, thereby allowing the miner to keep his/her credits.

9.6 CONCLUSION AND FUTURE PROSPECTS

Blockchain has already found implementation in several fields of information security. The IoT has been playing an increasingly important role in our general lifestyle. In both adversarial and civilian contexts, the IoT stands to be a topic of ongoing research interest. Furthermore, by understanding the inner workings of blockchain, which guides and creates a requirement of research in the field of public key cryptographic systems, an analysis of the time-complexity of the proposed methods and a comparison between the present methods and probabilistic proof of security, an implementation of blockchain into the IoT seems pragmatic and beneficial. The IoT has found implementation in civilian and scientific fields alike, hence, the development of protocols and safer standards has found implementation in the field of blockchain in the IoT. Bitcoin is one major example of the implementation of blockchain in the field of networking, as a form of distributed systems and decentralized entities for the management of data and transactions. This makes it very convenient to understand, by creating a bridge between the implemented Bitcoin and the upcoming concept of blockchain in the IoT. As always, the technology is not without its flaws. But we can understand that by tweaking the in-use protocols to merge with the current setup, the pros outweigh the cons. The inspiration was taken from problems in traditional telecommunications networks, where every new technology gets introduced as a seed by eliminating/solving the problems of an existing one. With new research being carried out every day, the practical application of blockchains is being coupled with the exponential expansion of the Iot, which in turn suggests that blockchain will be a new information security standard in a wide range of fields, varying from the setting of alarms for daily life to corporate management. This chapter could serve as a guideline to facilitate understanding about blockchain mechanisms, further realizing the potential research directions in the field of big data and network security for the implementation of data management for data security and concurrency.

REFERENCES

1. Geeks for Geeks, 2020, February 6. What is information security? https://www.geeksforgeeks.org/what-is-information-security/. Accessed on March 14, 2020.
2. Merkle, R. M., 1979. Secrecy, authentication, and public key systems. Ph.D. Dissertation. Stanford University, Stanford, CA, USA. Order Number: AAI8001972.
3. Sharma, D.K., Pant, S., Sharma, M., and Brahmachari, S., 2020. Cryptocurrency mechanisms for blockchains: Models, characteristics, challenges, and applications. In Krishnan, S., Balas, V., Golden, J., Robinson, Y., Balaji, S., Kumar, R. (eds.) *Handbook of Research on Blockchain Technology*. London: Academic Press, Elsevier, pp. 323–348.

4. Nadeem, A., and Younus Javed, M., 2005. *A performance comparison of data encryption algorithms.* In *2005 International Conference on Information and Communication Technologies* (pp. 84–89). Karachi, Pakistan: IEEE. doi:10.1109/ICICT.2005.1598556.
5. Salama, D., et al., 2008. Performance evaluation of symmetric encryption algorithms. *International Journal of Computer Science and Network Security*, 8, pp. 280–286.
6. Hercigonja, Z., 2016. Comparative analysis of cryptographic algorithms. *International Journal of Digital Technology & Economy*, 1(2), 127–134.
7. Zheng, Z., Xie, S., Dai, H.-N., Chen, X., and Wang, H., 2017. *An overview of blockchain technology: Architecture, consensus, and future trends.* In *2017 IEEE International Congress on Big Data (BigData Congress)* (pp. 557–564). Honolulu, HI: IEEE. doi:10.1109/BigDataCongress.2017.85.
8. Merkle, R. C., 1988. A digital signature based on a conventional encryption function. In Pomerance, C. (ed.) *Advances in Cryptology: CRYPTO '87. CRYPTO 1987.* Lecture Notes in Computer Science, vol. 293. Berlin, Heidelberg: Springer.
9. JavaTPoint, 2018. Limitation of blockchain technology. https://www.javatpoint.com/limitation-of-blockchain-technology. Accessed on March 14, 2020.
10. Bayer, D., Haber, S., and Stornetta, W.S., 1993. Improving the efficiency and reliability of digital time-stamping. In Capocelli, R., De Santis, A., Vaccaro, U. (eds.) *Sequences II.* New York: Springer, pp. 329–334. doi:10.1007/978-1-4613-9323-8_24.
11. Rykwalder, E., 2014. The math behind Bitcoin. https://www.coindesk.com/math-behind-bitcoin. Accessed on March 14, 2020.
12. Blackberry Certicom, 2018. Industry solutions and applications. https://www.certicom.com/10-introduction. Accessed on March 14, 2020.
13. Avanzi, R., Cohen, H., Doche, C., Frey, G., Lange, T., Nguyen, K., and Vercauteren, F., 2006. *Handbook of Elliptic and Hyperelliptic Cryptography.* New York: Chapman and Hall/CRC.
14. Johnson, D., Menezes, A., Vanstone, S., 2001. *The elliptic curve digital signature algorithm (ECDSA). International Journal of Information Security*, 1, pp. 36–63.
15. Antonopoulos, A. M., 2014. *Mastering Bitcoin: Unlocking Digital Crypto-Currencies* (1st. ed.). Sebastopol, CA: O'Reilly Media, Inc.
16. Jakobsson, M., n.d. *FractalHashSequenceRepresentationandTraversal. ISIT'02*, p. 437. www.markus-jakobsson.com. Accessed on March 14, 2020.
17. Lipmaa, H., 2002. On optimal hash tree traversal for interval time-stamping. In *Proceedings of Information Security Conference.* Lecture Notes in Computer Science, vol. 24, no. 33. Berlin, Heidelberg: Springer, pp. 357–371.
18. Szydlo M., 2004. Merkle tree traversal in log space and time. In Cachin, C., Camenisch, J. L. (eds.) *Advances in Cryptology: EUROCRYPT 2004.* Lecture Notes in Computer Science, vol. 3027., Berlin, Heidelberg: Springer, pp. 359–366.
19. Galbraith, S. D., and Gaudry, P., 2016. Recent progress on the elliptic curve discrete logarithm problem. *Designs, Codes and Cryptography*, 78, pp. 51–72.
20. Washington, L., (2008). *Elliptic Curves.* New York: Chapman and Hall/CRC. doi:10.1201/9781420071474.
21. Brett, C., 2018, October 15. Blockchain disadvantages: 10 possible reasons not to enthuse. https://www.enterprisetimes.co.uk/2018/10/15/blockchain-disadvantages-10-possible-reasons-not-to-enthuse/. Accessed on March 14, 2020.
22. Khanna, A., Arora, S., Chhabra, A., Bhardwaj, K. K., and Sharma, D. K., 2019. IoT architecture for preventive energy conservation of smart buildings. In Mittal, M., Tanwar, S., Agarwal, B., Goyal, L. (eds.) *Energy Conservation for IoT Devices: Studies in Systems, Decision and Control*, vol. 206. Singapore: Springer, pp. 179–208.

23. Bhardwaj, K. K., Khanna, A., Sharma, D. K., and Chhabra, A., 2019. Designing energy-efficient IoT-based intelligent transport system: Need, architecture, characteristics, challenges, and applications. In Mittal, M., Tanwar, S., Agarwal, B., Goyal, L. (eds.) *Energy Conservation for IoT Devices: Studies in Systems, Decision and Control*, vol. 206. Singapore: Springer, pp. 209–233.

24. Bhagat, A., Mittal, S., and Faiz, U., Sharma D. K., (2020). Data security and privacy functions in fog data analytics. In Tanwar, S. (ed.) *Fog Data Analytics for IoT Applications*. Studies in Big Data, vol. 76. Singapore: Springer. http://doi-org-443. webvpn.fjmu.edu.cn/10.1007/978-981-15-6044-6_15.

25. Chhabra, A., Vashishth, V., and Sharma, D. K., 2018. A fuzzy logic and game theory based adaptive approach for securing opportunistic networks against black hole attacks. *International Journal of Communication Systems*, 31(4). doi:10.1002/dac.3487.

26. Chhabra, A., Vashishth, V., and Sharma, D. K., 2017. *A game theory based secure model against Black hole attacks in opportunistic networks*. In *Proceedings of 51st Annual Conference on Information Sciences and Systems (CISS)*, 22–24 March 2017 (pp. 1–6). Baltimore: IEEE. doi: 10.1109/CISS.2017.7926114.

27. Saad, M., Spaulding, J., Njilla, L., Kamhoua, C., Shetty, S., Nyang, D. H., and Mohaisen, A., 2019. Exploring the attack surface of blockchain: A systematic overview. arXiv:1904.03487.

28. Mikerah, Q.-C., 2019, September 29. Short paper: Towards characterizing Sybil attacks in cryptocurrency mixers. *Cryptology*. ePrint Archive, Report 2019/1111. https://eprint. iacr.org/2019/1111/. Accessed on November 28, 2020.

29. Houtven, L.V., n.d. Crypto 101, Creative Commons Attribution-NonCommercial 4.0 International (CC BY-NC 4.0)

10 Edge-Based Blockchain Design for IoT Security

Pao Ann Hsiung, Wei-Shan Lee, Thi Thanh Dao, I. Chien, and Yong-Hong Liu
National Chung Cheng University, Taiwan

CONTENTS

10.1 Introduction 210
    10.1.1 Overall 210
    10.1.2 Chapter Contributions..212
    10.1.3 Chapter Organization 213
10.2 Related Works..213
10.3 Challenges 216
    10.3.1 Scalability and Interoperability..218
    10.3.2 Limitation with Storage Facility 218
    10.3.3 Data Privacy and Confidentiality 218
    10.3.4 Authentication and Authorization.....................................218
10.4 Edge-Based Blockchain Architecture for IoT Security 218
    10.4.1 Blockchain System Architecture.......................................219
    10.4.2 eBC System Architecture..222
    10.4.3 IoT Application Example Scenarios 226
        10.4.3.1 Data Loss Abnormal (DLA)..............................227
        10.4.3.2 Long-Term Consumption Abnormal (LCA) 228
        10.4.3.3 Short-Term Consumption Abnormal (SCA) 228
10.5 Experiments and Discussions 230
    10.5.1 Performance Evaluation..230
    10.5.2 Information Security 231
        10.5.2.1 Confidentiality...233
        10.5.2.2 Integrity 233
        10.5.2.3 Availability 234
    10.5.3 Security with Virtualization 234
10.6 Conclusions 234
References...235

10.1 INTRODUCTION

10.1.1 OVERALL

Recently, the Internet of Things (IoT) has grown more and more popular and is being deployed in many different applications. IoT devices can connect with each other through wireless or wired technologies and can make smart decisions. Normally, the IoT refers to a system of homogeneous devices having the abilities of sensing, processing, and network connecting, and it relies on a cloud server, where the data from IoT devices can be processed. Since it is a cloud-based system, every device needs to connect to the cloud server to be identified and authenticated, which makes IoT systems more centralized. Since the size and complexity of IoT networks grows every year, cloud servers need to be more and more powerful, which leads to an expensive deployment and a high maintenance cost.

IoT devices usually do not have a powerful ability to calculate, since they are designed just to sense and transmit data to the cloud. In some IoT architecture, there are gateways between the cloud and IoT devices, called fog computing layers, which provide local services and help connect IoT devices and the cloud. To reduce the latency of connection, these fog computing devices are usually deployed near the IoT devices. Thus, fog computing layers are geographically decentralized. Along with the devices' advances in computing ability,, fog computing layers can process more data than before and thus can also handle the calculations required by blockchain technology. The feature of decentralized architecture is also suitable for blockchain. Though IoT devices bring lots of convenience to human life, some issues also arise. The IoT's problems with privacy, security, and users' confidence still do not have effective solutions.

Blockchain is a distributed ledger technology without third-party verification, which is suitable for fog computing. A blockchain system consists of four parts. First, *consensus*, which ensures that the ledger information distributed across all the nodes is of the same version. Second, the *ledger*, which records every operation and every piece of data in a blockchain. It is deployed to every node, and the same version is maintained. Third, *cryptography*, which ensures that data in the ledger and their transmissions among nodes are all encrypted. Only a user with authorized access can decrypt them. Fourth, a *smart contract*, seen as a contract that can be executed automatically when the corresponding condition is satisfied. Since they are self-executable, smart contracts make a blockchain system provide more diversified services than before.

Along with the rapid growth in blockchain applications, several consensus methods have been proposed. The *Proof of Work* (PoW) consensus was first published by Cynthia Dwork and later applied in Bitcoin. Blockchain needs to add a new block behind the old one. In order to add a new block, the PoW will choose the first node that solves a complex math problem, regarded as *work,* and this node will have the right to add a new block. Though it is a well-known consensus method, it consumes extensive energy and time to mine a new block, thus leading to a very high cost. *Proof of Stack* (PoS) is a consensus method that can provide an alternative to PoW. PoS attributes mining power through the amount of coins that a miner holds, instead of through the mining ability that a miner has. Through PoS, the cost

Edge-Based Blockchain Design for IoT Security | **211**

of maintaining a blockchain system is reduced. The *practical byzantine fault tolerance* (PBFT) consensus is an effective consensus that is used in many private blockchain systems. The PBFT consensus selects a node in a blockchain to become a leader. The other nodes need to verify the leader, to make sure the leader is working and is safe. If the leader does not respond to the other nodes, PBFT will select another node to become a new leader. Through the transfer of the leader, the security of the system is enhanced.

In IoT gateways or fog computing layers, data security problems possibly lead to data tampering or data loss. A comprehensive survey on IoT security issues is given by Alaba Ayotunde Fadele et al. [1]. As Sachin Goswami et al. [2] discuss, fog computing can improve the computation and storage capability in centralized architecture, however, it performs very poorly in rising edge processing and protecting data privacy. A large amount of the data collected from a traditional centralized system is unreliable and not to be trusted. These are important issues to consider, especially the contradiction between secured sharing and performance isolation [3]. Many works in IoT application is proposed, and the future challenges associated with IoT big data research are identified by Hang Liu et al. [4]. Traditionally, the IoT applies fog computing to handle the big heterogeneous data stream generated by various terminals [5]. IoT applications are increasingly popular and play an indispensable role in the rapid development of society. Along with that is the integration of new technologies, including various communication devices. Nevertheless, a lack of resources and mechanisms restrict the full implementation of the IoT and limits the practical application of IoT features. But in future, the adoption of blockchain technology will overcome the limitations that the IoT is facing and improve the performance of IoT processing.

Blockchain technologies can provide a permanent forensic record of transactions and a single version of the truth: a network state that is fully transparent and ensures reliability and security for the benefit of all participants. Shaoyong Guo et al. [6] have proposed a distributed and trusted authentication system based on blockchain and edge computing. The main goal of their work is to improve authentication efficiency by an optimized practical Byzantine fault tolerance consensus algorithm. This consensus is designed to construct a consortium blockchain to store authentication data and logs. It guarantees trusted authentication and achieves transaction traceability. Haya Hasan et al. [7] have proposed the blockchain-based creation process of digital twins. Digital twins are necessary for rapid developments in the computing, storage, communication, and networking technologies of the IoT. This proposal was made to ensure secure and trusted traceability, accessibility, and provide a ledger to remain a permanent, indelible, and unalterable history of transactions. In addition, blockchain–based edge computing also has promise for future 5G networks. One important component in the 5G architecture is mobile edge computing. Xumin Huang et al. [8] have proposed exploring mobile edge computing for 5G, which can support many applications and services with low latency. Nasir Abbas et al. [9] have provided a comprehensive survey of the relevant research and developments in the mobile edge computing area, such as its advantages, architectures, and applications. However, the problem of security and data privacy remains. To enhance security in a mobile edge computing system, blockchain technology has been applied in some works. XiaoDong Zhang et al. [10] proposed a security architecture of VANET based on blockchain and mobile edge computing. Xiaoyu Qiu et al. [11] propose a new model-free deep reinforcement learning-based online computation offloading approach for blockchain-empowered mobile edge computing in mining and data processing tasks. Yaodong Huang et al. [12] propose a blockchain system that adapts to the limitations of edge devices.

As summarized by Abid Sultan et al. [13], several blockchain characteristics can be used to address security issues. For example, blockchain's decentralization addresses the single-point-of-failure security issue. Similarly, data privacy can be addressed by blockchain's anonymity. Moreover, blockchain technology meets transparency and scalability challenges [14], which has the ability to make process transactions visible and nondiscriminatory participation. Ruizhe Yang et al. [15] have identified several vital challenges of integrating blockchain and edge computing. Blockchain-based edge computing can enable reliable network access and control. However, self-organization, functions integration, resource management, and new security issues need to be considered before the application of blockchain-based edge computing can be deployed. Four crucial technical challenges in terms of network and storage scalability, throughput, access control, and data retrieval in the practical application of supply chains are summarized and discussed by Hanqing Wu et al. [16].

To overcome the problems mentioned here, we provides a blockchain-based gateway management mechanism with smart contracts in order to increase reliability. Our blockchain was built on the Hyperledger Fabric platform. The effectiveness of the proposed smart contracts is demonstrated by addressing data anomaly problems in smart grids via two application scenarios, namely *abnormal data loss* and *abnormal long/short term consumption*. Based on the results of anomaly detection, smart contracts are triggered, such that a transaction is performed to record the corresponding data. Depending on the specific scenarios, which have different levels of security breaches, the corresponding smart contract notifies the smart grid owners about the current security situation. Thus, our proposed blockchain-based security solution provides not only security, but also fault tolerance at the IoT gateway (fog layer).

10.1.2 CHAPTER CONTRIBUTIONS

In this chapter, we first present related work on blockchain technology and the IoT. Then we discuss the current challenges that exist for blockchain-enabled IoT systems. Afterward, we present the proposed edge-based blockchain architecture for IoT security. In our architecture, we describe how to design and implement each blockchain component. In addition, we describe the main role of edge-based blockchain in IoT security. With different IoT security scenarios, we demonstrate that our proposed architecture can address privacy and security issues. Performance on a Raspberry Pi 3 embedded platform shows that approximately 10 transactions-per-second (TPS) performance can be achieved with query transactions, and around 7 TPS for invoke transactions. The main contributions of this chapter can be summarized as follows.

Edge-Based Blockchain Design for IoT Security

- We propose a blockchain network architecture for the IoT, where we address several typical issues mentioned in previous work, including the consensus mechanism, the privacy mechanism, and smart contract design. Our architecture uses a permissioned blockchain system, Raft consensus, private data collection, and smart contracts to detect attacks and data anomalies. Details can be found in Section 10.4.1.
- Building upon this blockchain network architecture, we further propose an edge-based version called eBC system architecture for edge-based IoT systems. Two connectors were designed into eBC, namely the Blockchain Connector and the Edge Connector. The Blockchain Connector is responsible for permission management, transaction validation, access control, digital signatures, and invoke/query verification. The Edge Connector is responsible for synchronization, processing requests, and historical record logging. Details can be found in Section 10.4.2.
- The proposed eBC IoT architecture was implemented and deployed in gateways of the *advanced metering infrastructure* (AMI), to collect data from smart meters. Three different smart meter security scenarios were considered, including data loss abnormal, long-term consumption abnormal, and short-term consumption abnormal. Details can be found in Section 10.4.3. Corresponding smart contracts were implemented as chaincode in the Hyperledger Fabric v1.4 and verified in our proposed eBC architecture. Further, deep neural network models were also employed for data anomaly detection and prediction. Our proposed eBC architecture for IoT security was able to protect the AMI smart meters from power theft, detect smart meter tampering, and detect smart meter malfunctioning.
- The performance of the proposed system was analyzed in terms of the execution time and the throughput of transactions. Solutions for addressing confidentiality, integrity, availability, and security with virtualization are presented to show the effectiveness of our architecture. Simulation results show that under these different attack scenarios, the proposed system is still reliable.

10.1.3 Chapter Organization

The remainder of this chapter is structured as follows. A discussion of works about blockchain technology and the IoT is presented in Section 10.2. Section 10.3 gives the main shortcomings of current IoT-based blockchain applications and outlines the primary technical challenges. In Section 10.4, our edge-based blockchain architecture and the methodology for evaluating blockchain implementations are presented. Then, a discussion of our performance results and implications are covered in Section 10.5. Finally, Section 10.6 concludes the chapter.

10.2 RELATED WORKS

The *Internet of Things* (IoT) refers to the concept of a set of connected objects, often called *nodes*, that can communicate with each other for a specific purpose, for example, weather monitoring or landslide detection. The nodes are often integrated with ubiquitous equipment and facilities, to realize communication and dialogue through various wireless and wired communication network links (object-to-object message, object-to-person dialogue, person-to-person dialogue), to provide management and service functions.

With the evolution of time, in recent years, the IoT has become an important technology, improving the quality of life, providing convenience, and generating many IoT applications (such as in medical care, smart transportation, smart homes, and smart grids). These applications use many IoT devices, the number of which will reach 5 trillion in 2021 [17]. This also indicates that the market for the IoT is considerable.

Nataliia Neshenko et al. [18] made an exhaustive survey on IoT vulnerabilities by providing a categorization and taxonomy, including layers, security impact, attacks, countermeasures, and situational awareness capabilities. Some common vulnerabilities, such as deficient physical security, insufficient energy harvesting, inadequate authentication, improper encryption, etc., are all considered in the taxonomy. Different kinds of attacks, and remediation strategies, such as access and authorization controls, software assurance, security protocols, situational awareness, and intrusion detection, are all considered. Further, a data-based approach is adopted by scrutinizing 1.2 GB of darknet data collected from an a/8 network telescope and correlating it with Shodan's API, resulting in the discovery of 19,629 unique IoT devices in 169 countries. Tejsi Sharma et al. [19] have given an overview on the basics of blockchain technology and the IoT. Some challenges to blockchain in the IoT are described including scalability, malware detection, a lack of IoT-centric consensus protocol, and privacy. However, the authors did not propose solutions. Ayasha Malik et al. [20] have described several issues that are present in blockchain-based IoT security, such as attacks on hot/cold blocks, phishing, vulnerable signatures, etc. Vikas Hassija et al. [21] have presented a survey of IoT security, especially at the different layers of IoT architecture, including the sensing, network, middleware, gateway, and application layers. The authors discuss both blockchain technology for IoT security and how fog computing architecture can be leveraged for security. Daemin Shin et al. [22] have proposed a security protocol for DMM-based smart home IoT networks, which goes to show that IoT security is now a major concern. Tanishq Varshney et al. [23] discuss the various issues of IoT security and how it differs from traditional IoT security. They also cover the model for IoT threats and state-of-the-art security measures.

Most companies use centralized storage to manage and analyze data. However, traditional infrastructure and cloud computing can no longer meet the requirements of many practical applications. As mentioned earlier, because the number of IoT devices has increased significantly, the amount of data has also increased. If a large amount of data needs to be transmitted at the same time, problems such as increased latency or network congestion occur. As a result, the overall system performance declines. High-availability networks can deal with large amounts of data in real time, in order to solve this problem; but this is not possible on a traditional IoT infrastructure. Nevertheless, *fog* or *edge computing* is a promising design paradigm that solves part of the problem by performing local analysis and the distributed processing of data [24].

Edge-Based Blockchain Design for IoT Security

In edge computing, data is processed near the data collection source, so there is no longer a need to transfer the data to the cloud or to a local data center for processing and analysis. This approach reduces the load on both network and cloud. Due to its ability to process data in real time, and its faster response time, edge computing has a high applicability in the IoT field, especially in the *Industrial Internet of Things* (IIoT). In addition to accelerating the digital transformation of industrial and manufacturing companies, edge computing technologies can also enable innovations including artificial intelligence and machine learning. However, edge computing also faces the problem of deployment, specifically, how to effectively deploy the subordinates at various nodes. In 2017, Rakesh Jain and Samir Tata [25] proposed a deployment method using RED-Node. In this work, a dynamically reconfigurable edge computing architecture is proposed for the IoT based on Docker containers that are automatically orchestrated using Kubernetes.

Docker is an open source project that mainly provides the deployment and automated management of containerized applications. Deploying the Docker engine on an operating system provides a software abstraction layer, such that applications can be automatically deployed in containers through Docker images. It is very lightweight. Compared to traditional virtual machine technology, a Docker container has the following advantages:

- A high-performance virtualized environment,
- Easy migration and extension services,
- Simplified management, and
- A more efficient use of physical host resources

Figure 10.1 compares the differences between Docker and traditional virtualization methods. It can be seen that containers are implemented at the operating system level, using the local host operating system directly, while the traditional method is implemented at the hardware level.

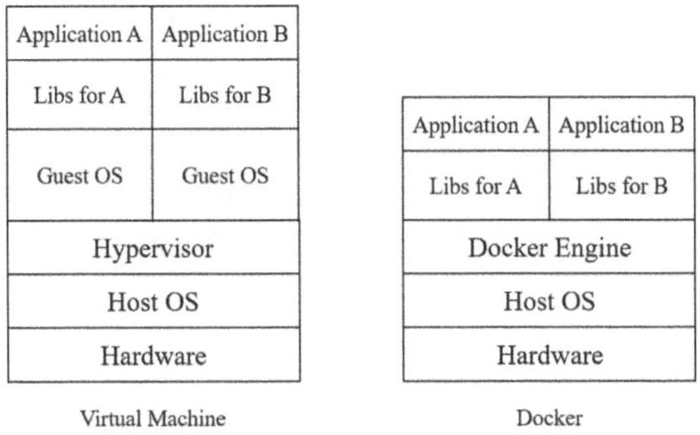

FIGURE 10.1 Virtual Machine vs. Docker Container [The traditional Virtual Machine needs extra Guest OS to support Libs and Applocation. On the contrary, Docker provides Docker Engine.]

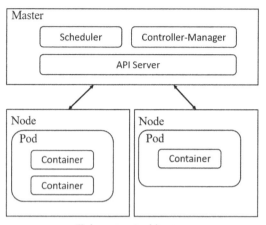

Kubernetes Architecture

FIGURE 10.2 Kubernetes-Based Edge Computing Architecture for IoT Security [There are two Nodes and a Master; the Master can control the Pod of Nodes, which includes Containers.]

Kubernetes is a system developed and open sourced by Google. It is a system that can help us manage microservices. It can automatically deploy and manage multiple containers on multiple machines. The four basic components of Kubernetes are pod, node, master node, and cluster.

First, the pod is the smallest unit of the Kubernetes operation, corresponding to an application service. Pods can have one or more containers. Second, the node is the smallest hardware unit on which Kubernetes operates, corresponding to a machine, whether a physical machine, such as a laptop, or a virtual machine. Third, a master node is the command center operated by Kubernetes and is responsible for managing all the other nodes, acting as a communication bridge between them. A node cannot communicate directly with other nodes; all communications must be through a master node. Finally, a cluster is a collection of multiple nodes and master nodes in Kubernetes, which can be thought of as a unit where all the nodes are grouped together in the same environment.

As shown in Figure 10.2, Kubernetes is also implemented in our proposed edge computing architecture for IoT security. A Kubernetes node corresponds to either a Raspberry Pi 3 platform or a Raspberry Pi 4 platform. In the figure, there are two nodes. A pod is deployed to each node. The left pod has two Docker containers, while the right one has one Docker container. All the nodes are managed by the master node, which is in the cloud, and which consists of a scheduler, a controller-manager, and an API server.

10.3 CHALLENGES

Current IoT ecosystems rely heavily on centralized communication models, such as pure cloud-based architectures, which excel in providing huge data storage and high levels of processing power. However, there are issues in the cloud-based IoT, including extensive security risks, poor scalability, privacy concerns, large data storage, high costs, and inadequate infrastructures. With an ever-growing need for more and more IoT nodes, scalability becomes a major concern in such centralized approaches for constructing an IoT application. Extensive security risks have become a major concern in recent years, due to poor security safeguards in network edge devices such as printers, smoke detectors, etc. Poor scalability results in communication jitters and timeliness problems in real-time applications, including critical safety systems. Sending all the data to the cloud also results in privacy concerns, for example, face recognition systems that send all detected facial images to the cloud servers. In real-world applications, scalability, security, and privacy concerns eventually hinder the adoption of the IoT. A very promising solution is the *edge-based blockchain* (eBC) technology, as proposed in this chapter. This proposal has a distributed architecture with a tamper-proof ledger mechanism. Thus, it can address the three problems, including security (via a tamper-proof ledger), scalability (via distributed processing), and privacy (via edge-based processing).

Blockchain is an emerging technology that started from cryptocurrencies, such as Bitcoin, but has now been widely applied to various different fields, with the IoT being one of the more prominent applications. For IoT nodes to be authenticated when connecting to an IoT infrastructure, a safe decentralized mechanism such as blockchain makes the whole process seamless, automatic, and safe. Thus, the use of blockchain in the IoT has great potential and is very much required. Besides authentication, all interactions among IoT nodes can be validated easily by the blockchain mechanism, via hashing and storage in a tamper-proof ledger. The conventional issue of extensive security risks in a growing complex multi-node architecture also disappears when Blockchain is used, because the greater the number of IoT nodes authenticated, the safer all the transactions will be, because the transactions must be validated by a majority of the existing authenticated nodes. Thus, an IoT platform operated by a decentralized blockchain setup is so robust that it is almost impossible for a hacker to successfully compromise the system, since that then requires knocking down a majority of the nodes. Consensus-based control distributes security responsibilities among the nodes in the blockchain network, preventing hackers from tampering with the network and also protecting the IoT network from being destroyed by DDoS attacks. In this work, we show how edge-based blockchain technology applied to a Docker/Kubernetes-based containerized IoT architecture makes it secure and robust against attacks.

Decentralization also makes such a solution more scalable. Scalability is one of the biggest concerns in the deployment of network security systems on an ever-growing network, with more and more nodes being connected in real-time. In our proposed edge-based IoT security architecture, notification about the addition of new device nodes, the deletion of existing ones, and functional changes in any node goes out to all IoT nodes in real-time, through blockchain smart contracts. This reactive approach of our proposed edge-based blockchain system in our IoT architecture allows the system to be adaptable and flexible enough to expand and evolve without upgrading the entire platform.

Besides security and scalability, privacy is also enhanced in our proposed edge computing IoT architecture, through the containerized technology, where privacy across applications is strictly enforced. For example, access to the image data captured by a camera is only provided to the containers that have been authenticated and authorized. Access to object features extracted by the application in a container is restricted within the container. All access to ledger information in the distributed blockchain architecture is logged within the ledger; thus, privacy is ensured through security.

10.3.1 SCALABILITY AND INTEROPERABILITY

Due to the heavy computations and communications required, there are often performance bottlenecks in a system that uses blockchain technology. With increasing network size, this issue becomes even worse. As a result, the efficiency of the consensus algorithm and the speed of transaction response both decrease. Therefore, blockchain transaction records and related data information require some kind of management, which can cope with better scalability.

Interoperability is the ability to connect and transfer information across different blockchain systems, for the smooth sharing of information. The performance of blockchain technology is reduced by the overhead of interacting with different blockchain networks, for example, a Bitcoin system with an Etherum system.

10.3.2 LIMITATION WITH STORAGE FACILITY

In a blockchain-based IoT system, the ledger records of the blockchain that are stored on each node increase in size, as time passes. Due to the limited storage capacities in IoT nodes, the required storage size will exceed the given capacities of memory storage in the IoT nodes.

10.3.3 DATA PRIVACY AND CONFIDENTIALITY

Since an IoT is often an integration of various devices, services, and networks, the data stored or transferred on a device is vulnerable to privacy violation by compromising nodes in an IoT network. An attacker can attempt to impact data integrity by modifying the stored data or injecting false data into it for malicious purposes.

10.3.4 AUTHENTICATION AND AUTHORIZATION

All devices in an IoT network must be authenticated for privileged access to services. Diversity and heterogeneity in the underlying architectures and environments supporting IoT devices lead to the existence of different authentication mechanisms for the IoT. As a result, there is no global standard protocol for authentication in the IoT.

10.4 EDGE-BASED BLOCKCHAIN ARCHITECTURE FOR IOT SECURITY

In this section, we describe the proposed *edge-based blockchain architecture* for IoT security. First, we describe the whole architecture, along with the details of each component, including the consensus mechanism, the privacy mechanism, and smart contract designs. Then, we discuss the role of edge-based blockchain in IoT security. Finally, we demonstrate the effectiveness of the proposed architecture in different IoT security scenarios.

10.4.1 Blockchain System Architecture

The blockchain system design proposed in this work is shown in Figure 10.3. It consists of three parts, including edge devices with authorization, the blockchain network, and the cloud. The blockchain network is built on the open source Hyperledger Fabric, which is a permissioned consortium blockchain platform. Fabric has several features [26]. In terms of the edge devices, only the nodes (devices) that have the enrollment certificates (ECerts) issued by the Fabric certificate authority (CA) can participate in the network. In other words, the system guarantees information confidentiality, since data cannot be accessed by unauthorized nodes. Furthermore, once a node suffers from malicious attacks or is no longer needed, the authority can revoke its ECerts at any time.

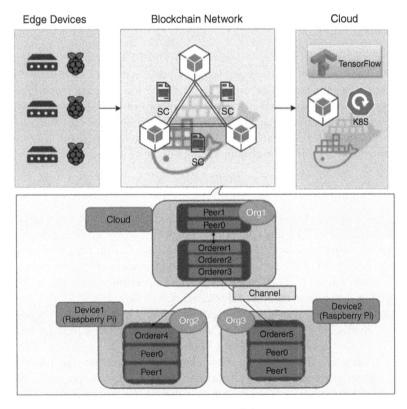

FIGURE 10.3 Blockchain System Architecture [Edge Devices send data to Blockchain Network, then synchronize it to Cloud. Blockchain Network includes three organizations, five orderers, and six peers.]

Inside the Blockchain network depicted in Figure 10.3, we use the Docker Compose tool to deploy the system with three organizations, where an organization is a managed group of device members classified into *peers* and *orderers*. Peers are in charge of the ledgers and invocation of smart contracts, whereas orderers are responsible for packaging transactions into blocks and distributing them to anchor peers. Organization Org1 is deployed in the cloud, while the other two organizations, Org2 and Org3, are deployed on two different nodes, corresponding to two Raspberry Pi 3 platforms: Device1 and Device2. Note that each organization has two peers, namely Peer0 and Peer1. Org1 in the cloud has 3 orderers (Orderer, Orderer2, Orderer3), while the edge nodes have only one orderer each, including Orderer4 on Device1 and Orderer5 on Device2. Thus, in this setup, the cloud organization has three times the chance to be the leader (as described in the Raft consensus later in this chapter), and thus can make the system more robust with a higher fault tolerance, due to the powerful computation ability of the cloud. The leader orderer packages transactions generated by smart contracts into blocks, and distributes them to the anchor peers of other organizations across the network. More important, the cloud administrator can access the data and rapidly be notified as soon as the data of edge devices are abnormal.

The mechanisms used to protect the information security of edge devices, including the consensus mechanism, private data, and smart contracts, are as follows.

- *Practical Byzantine Fault Tolerance* (PBFT): PBFT is a replication algorithm designed by Barbara Liskov and Miguel Castro, in order to solve the Byzantine General's Problem. It solves the defect of the low efficiency of the original Byzantine fault tolerance algorithm (e.g., Proof of Work or Proof of Stake in permissionless blockchains). A PBFT system can tolerate less than one-third malicious nodes in the network, as shown in Equation 1, where R is the number of replicas and f is the maximum number of faulty replicas [27].

$$|R| = 3f + 1 \tag{10.1}$$

- *Raft Consensus*: Despite improving on the shortcomings of the original BFT algorithm, PBFT requires high time-complexity, due to its three stages and the broadcasting to all the other nodes in the network. Raft tries to resolve fault tolerance with $2f + 1$ nodes only, which exhibits more efficiency and supports a larger number of faulty nodes. A brief introduction to Raft and its practical design are given in the rest of this subsection.
  1. *Raft*: Proposed by Diego Ongaro and John Ousterhout, Raft promises a distributed consensus with enhanced understandability from an intuition-focused approach and proof of safety [28]. Nodes are divided into three characters (i.e., Follower, Candidate, and Leader) and go through two main procedures, namely Leader Election and Log Replication. When a leader crashes, a new election is triggered by the heartbeat mechanism. The newly elected leader appends all commands to its log, while issuing AppendEntries RPCs in parallel to each of the other nodes. There is only one leader at any time.

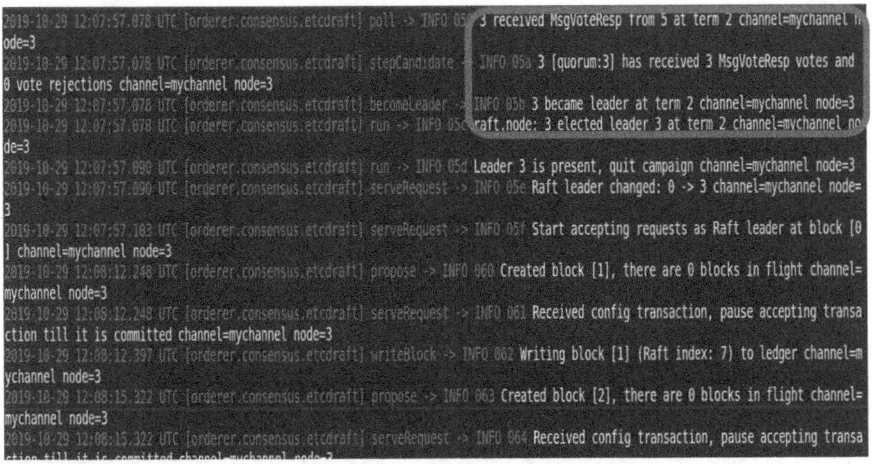

Edge-Based Blockchain Design for IoT Security

FIGURE 10.4 Orderer 3 Elected as Leader in Raft Consensus [Orderer 3 received 3 MsgVoteResp votes from other Orderer and became raft leader at term2.]

2. *Practical Design*: A flexible and practical implementation of the Raft consensus is Etcd/raft, which is a library written in the Golang language and is new, as of Fabric v1.4.1. As shown in Figure 10.4, Orderer3 was the elected leader among Orderers by receiving MsgVoteResp (i.e., its decisions would be replicated by the followers). Now, we bring down the leader Orderer3 as in Figure 10.5. Then, Orderer2 is elected as the new leader, which is called 1st-Fault tolerance; if Orderer2 is also shutdown, another new leader is elected, which is 2nd-Fault tolerance, and so on. The network is more secure and fault tolerant, when there are more and more nodes in it.

- *Private Data*: In some applications, the data information is private in the sense that it should not be visible across specific organizations. An example could be the competitive relationship of data among organizations. In our system, we adopt private data collection, which was newly introduced in Fabric v1.2 to address the privacy issue. Without creating a separate channel, private data stored in an individual private state database [29] can be accessed only by authorized organizers, thus ensuring the privacy of IoT devices, as well as data confidentiality.

- *Smart Contracts*: In contrast to the first generation blockchains (i.e., Bitcoin), the advent of smart contracts in the second generation blockchains gave the ability to implement transaction logic that can be executed automatically under specific circumstances. Predefined conditions and logic instructions are written in smart contracts, and then the participating nodes can invoke them to make decisions or access data, among other functionalities. This can be useful when it comes to IoT security, since all the interactions with smart contracts are tamper-resistant and stored in the blockchain. Additionally, the data information of edge devices can be monitored by the proposed smart contracts in our edge-based blockchain system. More implementation details are described in the next subsection.

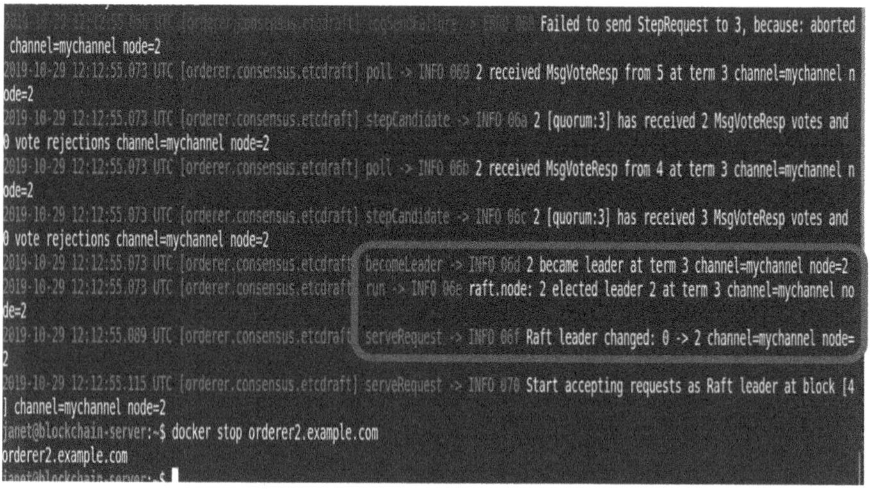

FIGURE 10.5 Orderer 2 Elected as Leader in Raft Consensus [Orderer 2 received 3 MsgVoteResp votes from other Orderer and became raft leader at term3.]

10.4.2 ᴇBC Sʏsᴛᴇᴍ Aʀᴄʜɪᴛᴇᴄᴛᴜʀᴇ

In real-world application domains, the IoT design architecture is often distributed. However, a conventional distributed IoT architecture lacks a distributed management framework that is both secure and scalable. If carefully designed, blockchain network architecture exhibits both of these features. In this chapter, we propose such a solution in the form of an *edge-based blockchain* (eBC) system architecture. Using smart contracts, the proposed architecture also possesses features of synchronization, fault-tolerance, transparency, and enforcement agreements. As shown in Figure 10.6, our proposed eBC system consists of the layers, including a local IoT device layer, an edge computing layer, and a cloud layer, which are described as follows.

- *Local IoT Device Layer*: This layer consists of all the IoT devices, also called nodes, which could be as simple as sensor devices for weather, smart metering, etc., or more complex, as video cameras, etc.
- *Edge Computing Layer*: This layer could include the network gateways that are responsible for collecting data from IoT devices and passing them on to the cloud, either directly or after some basic processing, such as information extraction or edge-based machine learning. The network gateway architecture is designed using Docker containers, such that DNN (deep neural network) models, BC (blockchain) modules, and manager modules are containerized and deployed into different containers. This architecture allows for dynamic orchestration by Kubernetes (in the cloud layer).
- *Cloud Layer*: This layer is responsible for the overall management of the containers in the edge computing layer, the deployment of containerized applications into the containers, including deep neural network and blockchain modules, and the training environment for DNN models.

Edge-Based Blockchain Design for IoT Security

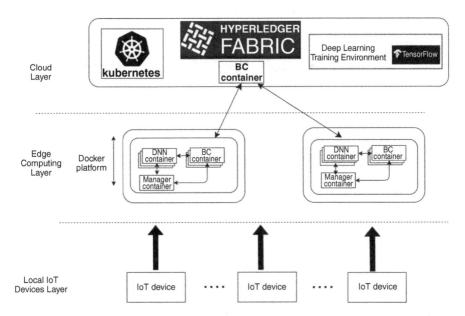

FIGURE 10.6 Overview of the Proposed Edge-Based Blockchain Network [There are three layers, where the Local IoT devices Layer sends data to the Edge Computing Layer then synchronizes it to the BC container on the Cloud Layer.]

The proposed edge-based blockchain network is deployed in one or more containers in the edge computing layer of the overall architecture. Figure 10.7 depicts the role of our proposed edge-based blockchain network. Within the network, there are two connectors, namely, the BC connector and the edge connector. The BC connector is responsible for managing local IoT devices, while the edge connector is responsible for providing services for edge computing. In the following, we describe the functions of the two connectors.

1. Role of Blockchain Connector: The blockchain connector is mainly responsible for the services that are required to maintain a blockchain system running in an IoT network, including permission management, transaction validation, access control, digital signature, and invoke/query verification, as detailed in the following.
    - *Permission Management*: Conventional blockchain systems such as Ethereum and Bitcoin are public, which means that any node can participate in the consensus process to order the transactions and to bundle data into blocks. However, our proposed eBC system is *permissioned*, that is, only specific nodes with permission can participate in the consensus. We use the Hyperledger Fabric, which is permissioned. In our permissioned blockchain system, permission management is used to manage the identity of its participants (IoT devices or Blockchain nodes). In our system, only the Orderers in each organization are permissioned, so that they can participate in the Raft consensus (leader elections). Because IoT systems require high

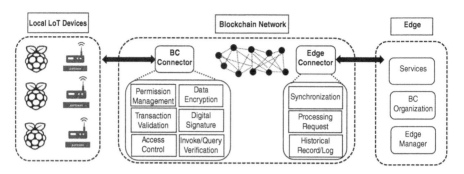

FIGURE 10.7 Overview of Role of Blockchain Network in Edge-based IoT Security [Blockchain Network contains the BC Connector and Edge Connector being responsible for Local IoT devices and Edge computing services.]

scalability and performance, permissioned blockchain is the correct choice, as these kinds of blockchain systems are more efficient in terms of consensus, more cost-effective, and provides private membership. The negative aspects of permissioned blockchains, such as partial decentralization (as opposed to full decentralization) and lower security, can be accommodated by careful design, as in our proposed system.

- *Transaction Validation*: For successful deployment of blockchain in the IoT, instead of the complex transaction validation process in Bitcoin, our edge-based blockchain system adopts a more flexible *execute-order-validate* approach. Before a transaction is committed to the ledger in the blockchain, it needs to be executed by running a smart code (chaincode in Fabric), ordered through a consensus protocol (Raft here), and validated (endorsed by some predefined peers). Some features that make this deployment suitable for the IoT include (a) efficient endorsing. where a transaction needs to be endorsed by only a subset of peers, and (b) parallel execution, where transactions can be endorsed in parallel by different peers.
- *Access Control*: In a permissioned blockchain, access control is one of the critical mechanisms used to deal with the protection and privacy issues in the IoT. With access control, the users of IoT applications have full access to their data and can control how the information is shared. Our method uses *access control lists* (ACLs) to manage access to resources by associating a *policy* with the resource. A policy specifies a rule that evaluates to true or false, given a group of identities and the target resource. Hyperledger Fabric includes a variety of default access control lists. To access a resource with an associated policy, the accessing identity (or set of identities) needs to be checked against the policy. There are two types of default policies, namely *Signature* and *ImplicitMeta*. Signature policies specify the combination of endorsing peers needed for signing a resource requesting transaction before the transaction is allowed access to the resource. ImplicitMeta policies aggregate the results of policies, such as Signature policies.

Edge-Based Blockchain Design for IoT Security

- *Digital Signature*: The confidentiality, integrity, and authenticity (CIA) of the data in an IoT needs to be ensured. Confidentiality can be ensured by access control. Integrity is ensured by the hashing mechanism in blockchain. Authenticity is ensured in our system through the notarization of documents using digital signatures. Each IoT node (blockchain user) enrolls with the Fabric CA and collects the signing key and an authentication certificate (e.g., X.509), stored in its own wallet. For each document (or data) to be signed, a hash (e.g., SHA-256) is computed from the document and a signature of the hash using the signing key from the user's wallet. For authentication, the document to be verified is obtained from the ledger. A *hash* is computed from the document and compared to the *computed hash*, which is obtained by decrypting the signature using the public key in the authentication certificate. The blockchain connector thus implements the authentication procedures for the local IoT edge devices.
- *Invoke/ Query Verification*: The blockchain connector also provides the chaincode services for IoT nodes, including the invoke and query functions for application logic. For example, in the smart meter application, meter data to be logged into the ledger consists of MeterId, Timestamp, Status (normal/ abnormal), Consumption, and GatewayId, which are explained in detail later in the chapter. Our chaincodes are implemented in the GoLang language. Create, update, and delete operations are all encapsulated within the invoke/ query method. IoT security policies are implemented as TLS certificates in the X.509 public key certificate format. Different channels are used for different purposes or organizations, depending on the IoT application.

2. Role of Edge Connector: The edge connector is mainly responsible for managing the application logic in an IoT network, including mainly the edge functions, such as synchronization, processing request, and historical record/log, as detailed in the following.

- *Sychronization*: An inherent characteristic of a blockchain network is state replication, where the participating nodes can always update their ledger states to the latest committed version. This characteristic is achieved via a blockchain's sync mechanism, where a node in the network communicates with other nodes to request and import the full history of blocks on the ledger. It involves multiple components of the kernel, notable among them being: the peer-to-peer protocol, which enables communication, and the consensus algorithm, which ensures that all nodes agree on how the information in each block is interpreted. For IoT deployment, the synchronization service helps maintain consistency across all the nodes at the edge layer. In other words, the conventional method of using cloud-based MQTT servers for schronization (publish/subscribe topics) can be fully replaced with the more secure and scalable, distributed blockchain sychronization mechanism. This is also one of the innovations of this work.
- *Processing request*: The edge connector provides service for processing requests from the application logic, in the form of smart contracts or chaincodes. Security is ensured in the processing request, because each

communicated message is signed by its sender IoT node and validated by the endorsing peers, before being committed by the leader node. Thus, even if there exists a malicious leader, it cannot compromise the network, because synchronization from other nodes will eventually lead to the eradication of the effects of the malicious leader's attack, such as man-in-the-middle, impersonification, etc.

- *Historical Record/Log*: To support client application requirements, the proposed edge-based blockchain system design provides services to access and query transaction history. This service focuses on the precision and reliability with which the system can record, prove, and audit events related to data in the ledger.

10.4.3 IoT Application Example Scenarios

Based on the previous section, we actually implemented an edge-based IoT system architecture and ensured its security using deep neural network models and an edge-based blockchain network. The implemented system architecture was as shown in Figure 10.6, where the manager container was responsible for the dynamic configuration of the containers, the DNN container was deployed with a long-short-term memory (LSTM) model, and the BC container was deployed with the proposed edge-based blockchain system module. The application was smart meters, and the edge-based design was implemented in a smart IoT gateway that collected electricity consumption data from 200 smart meters. The main goal of the IoT security for this smart meter application was to detect power theft, smart meter tampering, and smart meter malfunctioning.

The corresponding smart grid architecture is shown in Figure 10.8, where abnormal data can be detected using our IoT security scheme, recorded in the ledger, and

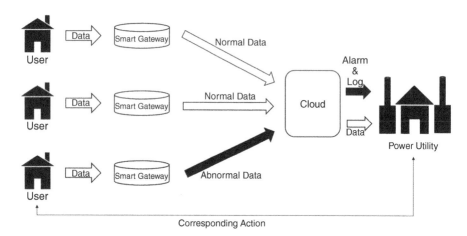

FIGURE 10.8 Smart Grid Architecture for IoT Security [User Data is sent to the Smart Gateway, once there is Abnormal Data, the Power Utility will receive an Alarm, as well as log and take the Corresponding Action.]

Edge-Based Blockchain Design for IoT Security

also identified to the power utility manager through alarms. In Figure 10.8, User represents the residence or the micro-grid where a smart meter is installed. Data represents the smart meter data sent from the User to the smart gateway. The results of anomaly detection by the LSTM model deployed in the smart gateway, namely either Normal Data or Abnormal Data, are both stored locally at the gateway and also sent to the cloud, via our proposed edge-based blockchain network. The blockchain organization deployed in the cloud is responsible for notifying the power utility manager, via smart contract, in case of data anomaly. The smart meter data contains the *MeterID, Timestamp, Consumption*, and *GatewayId*. One more piece of information is added to the smart meter data record, via another smart contract, after the meter data is verified via the LSTM model, that is, whether the data has a normal or abnormal status: *Status (normal/abnormal)*. Smart meter data stored in the blockchain nodes in the cloud can also be queried using smart contracts. For example, we can query how many smart meters were compromised and how severe (frequent) were the smart meter attacks.

As shown in Figure 10.9, IoT security for the smart meter application [30] was mainly targeted at anomaly detection in the different scenarios, including *Data Loss Abnormal* (DLA), *Long-Term Consumption Abnormal* (LCA), and *Short-Term Consumption Abnormal* (SCA), which are described as follows.

10.4.3.1 Data Loss Abnormal (DLA)

When a smart meter is unintentionally broken or physically tampered with, or its communication is disconnected, the IoT gateway will not receive any data from it. This is called *Data Loss Abnormal* (DLA). In our proposed edge-based blockchain system, we have designed a smart contract that can detect such data loss abormal events and notify the power utility about the defective smart meter. As a result, it prevents a power utility from having losses because of missing user power consumption data. The power utility can thus send manpower to check and fix the problematic meter.

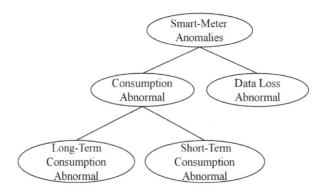

FIGURE 10.9 Smart Gateway Abnormal Classification Tree [Smart Meter Anomalies can be divided into Consumption Abnormal and Data Loss Abnormal, where Consumption Abnormal includes Long-Term and Short-Term.]

10.4.3.2 Long-Term Consumption Abnormal (LCA)

Due to permanent power leakages in electrical circuits or energy theft, there could be long-term anomalies in the load data sensed by smart meters, which we call *Long-term Consumption Abnormal* (LCA) events. In our proposed edge-based IoT security system, we use the LSTM model to detect LCA by first collecting enough data, for example, over weeks, months, or even years. As a result, we can help users and power utilities to detect problems earlier and thus save cost and electricity. Kubernetes orchestration aids in deploying different Docker containers with LSTM models to cater to different time windows.

The blockchain peers in the cloud organization can execute chaincode queries to retrieve historical consumption data for each smart meter stored in the ledger, over a time span of one week, one month, half a year, one year, or more. The retrieved history data is then provided to the LSTM model in another container that can detect LCA. Upon detection, warnings are then sent to the user and the power utility, via another chaincode. The manager container application helps in configuring and switching the connections among different LSTM containers and blockchain containers. A table of configured sockets for communication among the containers is maintained and replicated in all the containers.

10.4.3.3 Short-Term Consumption Abnormal (SCA)

Since most smart meters provide half-hourly load data, a more real-time analysis and detection can be performed, which is called *Short-Term Consumption Abnormal* (SCA). Similar to LCA, SCA also employs LSTM modules deployed in Docker containers and blockchain modules deployed in Docker containers. For short-term analysis, four levels of anomaly risk are defined, based on which users can be warned or isolated. In order of increasing forecasted risks, the four classes include Warn user, Dangerous user, High-risk user, and Emergency user. For more accurate SCA, *load profiling* [31] is performed for different classes of users, and different thresholds are used to detect anomalies in different classes [32] .

In the Hyperledger Fabric, multiple smart contracts can be defined within the chaincode where the transaction logic was packaged. Considering DLA first, when the smart gateway receives the signal -999 (i.e., the meter is broken) and tries to store it in the blockchain network, invoking the "save_data" function in chaincode, as shown in Figure 10.10, the node in the cloud sends an alarm to the power utility, with the broken meter information. The information sent by a smart meter contains MeterId, Timestamp, Status (normal/abnormal), Consumption, and GatewayId. Before saving the new information, the function checks to determine whether a similar record exists. If a similar record is found, then it is updated with the new information. No duplicate records are stored in the ledger.

When the smart grid manager needs to retrieve the meter data from the ledger to analyze the problem of LCA, the function "query_all," as shown in Figure 10.11, is invoked, which returns all the meter data in the JSON format. Furthermore, as shown in Figure 10.12, the manager can also retrieve meter data from the ledger in a particular range, for example, from January to March of a particular year. In the case

Edge-Based Blockchain Design for IoT Security

FIGURE 10.10 DLA Chaincode [The save_data function saves data from the smart meter and check for any missing data labeled with "–999."]

FIGURE 10.11 Query all Chaincode [The query_all function queries data from ledger and writes them into buffer.]

```
// flag used to distinguish query or compute function by month
func (t *SmartContract) query_by_month(stub shim.ChaincodeStubInterface, args []string, flag int) pb.Response {

    if len(args) != 4 {
        return shim.Error("Incorrect number of arguments. Expecting 4(Id, gatewayId, monthStart, monthEnd)")
    }

    id := args[0]
    gatewayid := args[1]
    monthStart := args[2]
    monthEnd := args[3]

    //Get State of CompositeKey "Id-GIdTimestamp"
    idResultsIterator, err := stub.GetStateByPartialCompositeKey("Id-GIdTimestamp", []string{id})

    if err != nil {
        return shim.Error("Fail to partial Composite key!")
    }
    defer idResultsIterator.Close()

    var buffer bytes.Buffer
    buffer.WriteString("\n[")

    bArrayMemberAlreadyWritten := false
    for idResultsIterator.HasNext() {
        queryResponse, err := idResultsIterator.Next()
        if err != nil {
            return shim.Error(err.Error())
        }

        objectType, compositeKeys, err := stub.SplitCompositeKey(string(queryResponse.Key))
        fmt.Printf("%s", objectType)
        returnKey := compositeKeys[1]

        // Get the attribute of return key
        valAsbytes, err := stub.GetState(returnKey)
        if err != nil {
            return shim.Error("Failed to get smartmeter:" + err.Error())
        } else if valAsbytes == nil {
            return shim.Error("Smartmeter does not exist")
        }
```

FIGURE 10.12   LCA Chaincode [The query_by_month function queries data from ledger during a period of time and writes them into buffer.]

of SCA, the manager can also set the threshold for the frequency of abnormal events within a week (i.e., how many abnormal events have occurred in a certain interval), as a parameter of the chaincode function.

10.5   EXPERIMENTS AND DISCUSSIONS

In this section, we first evaluate the performance of the proposed edge-based blockchain system for IoT security and then give a detailed discussion of its security features, with respect to confidentiality, integrity, and availability. Finally, we touch upon how Docker container-based application virtualization helps in IoT security and privacy.

10.5.1   Performance Evaluation

The edge-based blockchain security system for the IoT was implemented and experiments performed on a Raspberry Pi 3B platform with a Quad Core Broadcom BCM2837 that runs at a frequency of 1.2 GHz. The size of the main memory is 1 GB, and the operating system is 64-bit Ubuntu Ultimate. The blockchain system base used was Hyperledger Fabric v1.4.

We evaluated the performance of the proposed blockchain system by measuring the throughput of the smart contract operations, such as invoke and query transactions. The performance was assessed in terms of execution time and throughput, and estimated by varying the workload (number of transactions and requests (query or invoke) that were requested simultaneously by the peers) on each platform, with a maximum of up to 4,000 transactions.

Edge-Based Blockchain Design for IoT Security

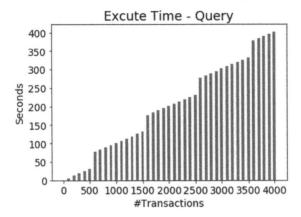

FIGURE 10.13 Execution Time for Query [A bar chart showing Query Seconds on the y-axis increasing as Number of Transactions increases on the x-axis.]

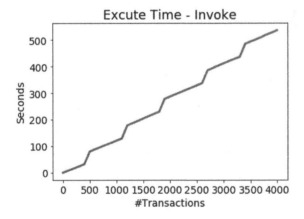

FIGURE 10.14 Execution Time for Invoke Function [A bar chart showing Invoke Seconds on the y-axis increasing as Number of Transactions increases on the x-axis.]

Figures 10.13 and 10.14 show the execution time for query transactions and invoke transactions, respectively. The execution time is measured as the time required for a platform to add and execute a transaction successfully. In general, execution times increase with increases in the number of transactions executed.

As shown in Figures 10.15 and 10.16, we also evaluated the throughput of the system in terms of the number of successful transactions per second (TPS).

For both scenarios, the transaction time was measured from the submission of the transaction for consensus by the peers up to committing the transaction to a block.

10.5.2 Information Security

IoT security considers how to protect networks and connected devices in the IoT. It allows devices connected in the network to avoid serious issues they once faced, such as the problem of information leakage [33].

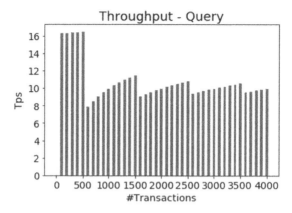

FIGURE 10.15 Throughput for Query Function [A bar chart showing Query tps on the *y*-axis increasing as Number of Transactions increases on the *x*-axis, peaking at sixteen Tps at two hundred and fifty to five hundred transactions.]

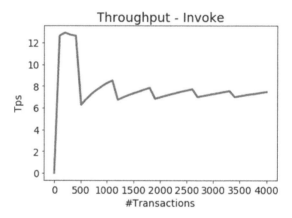

FIGURE 10.16 Throughput for Invoke Function [A bar chart showing Invoke tps on the *y*-axis increasing as Number of Transactions increases on the *x*-axis, peaking at thirteen Tps at two hundred and fifty transactions.]

To ensure information security, we will discuss how our proposed edge-based blockchain network for the IoT can be used to address the CIA triad, namely *confidentiality*, *integrity*, and *availability*. Before discussing how each of the three security features is addressed in our system, we elaborate on the intuition behind the choice of key mechanisms.

- Choice of Hyperledger Fabric: We chose Hyperledger Fabric because it provides permissioned membership as well as pluggable module architecture. What's more, channels and private data in Hyperledger Fabric protect data received from IoT devices securely. All of these features make Hyperledger Fabric quite suitable for IoT deployment, as it is secure, scalable, and high performing.

Edge-Based Blockchain Design for IoT Security 233

- Choice of Consortium Blockchain: Since IoT nodes need privacy, but they also need openness to the inclusion of new nodes, consortium blockchain is the best choice for the IoT. Public blockchains cannot ensure privacy, and private blockchains is so closed that it creates problems for scaling the IoT automatically and easily.
- Choice of Raft Consensus: Availability is an important feature of any IoT application, Raft Consensus is more lightweight than PBFT and can reelect a new leader node efficiently. Thus Raft is a good choice for IoT deployment.
- Choice of Containerized Architecture: Security across different applications and across different users in the same application can be ensured through the use of container technology, where data and communication can be safeguarded within the realms of the container. The consequences of attacks against applications or users can be restrained in the corresponding containers, thus ensuring threat isolation.

As summarized in Table 10.1, now let us discuss how our proposed edge-based blockchain system addresses the CIA triad within each of the three tiers.

10.5.2.1 Confidentiality

Confidentiality is achieved in our eBC system via several means, including the permissioned blockchain system, where all IoT nodes need to enroll (register) themselves by obtaining a digital signature along with a public key certificate, before they can perform any transaction. Further, we also have channel-based access control and private transactions, where a user can always choose the legitimate users that can see the data related to its currently executing transactions. The whole set of mechanisms is implemented into the Fabric CA.

10.5.2.2 Integrity

In our proposed eBC system, data integrity at the edge and the cloud layers is maintained through the *gossip data dissemination protocol* of Hyperledger Fabric, along with dynamic leader elections. At the IoT device layer, data integrity is maintained

TABLE 10.1

Ensuring the CIA Triad in the Proposed eBC System

	Confidentiality	Integrity	Availability
Local IoT Devices	✓ Smart meter enrollment	✓ AES-128 encryption	✓ ACK mechanism
Edge Computing Layer	✓ Fabric CA ✓ Private data ✓ Access control	✓ BC-tamper resistant (hash)	✓ Raft consensus ✓ Docker container ✓ Distributive allocation
Cloud Layer	✓ Fabric CA ✓ Private data ✓ Access control	✓ BC-tamper resistant (hash)	✓ Raft consensus ✓ Docker container ✓ Kubernetes deployment

by the security of the communication protocol. For example, in our smart meter application, AES-128 encryption [34] was used in the LoRaWAN protocol to protect data.

10.5.2.3 Availability

When authorized users need to access the data in the ledger, the system should be available at any time and provide uninterrupted service. Further, the system must be highly fault tolerant, even under the circumstances of broken components, such as sensor devices. In our proposed eBC architecture, when IoT devices fail to transmit data and/or did not get any ACK reply from a network gateway, they would retransmit up to three times. If a blockchain node is broken or misbehaves at the edge computing layer and/or cloud layer, a new leader will be automatically reelected based on the Raft consensus, and as a result, eBC maintains availability. Two main differences in availability maintenance in the two layers are that the edge layer employs *distributive allocation*, while the cloud layer employs Kubernetes deployment; the former can reduce the computational burden on the cloud layer, and the latter can reorchestrate the crashed containers. Both of them ensure high availability.

10.5.3 SECURITY WITH VIRTUALIZATION

In our system architecture, application-level virtualization is realized using Docker containers. Nevertheless, third-party suites might have security concerns, for example, Theo Combe et al. [35] investigated Docker security issues, such as host security attacks via privileged containers. Combe et al. also gave a method for setting file parameters so as to improve Docker security. In our proposed eBC system, the use of containers is made secure by isolating between the container and the host, so that the host is protected from malicious container attacks, such as containers illegally accessing the host's data. Furthermore, the containers are deployed in internal networks, that is, the network is not open to the outside world, thus it is impossible to carry out network attacks on the container, and all operations on the container are performed only through the host, thereby ensuring security.

10.6 CONCLUSIONS

We proposed an edge-based blockchain system for IoT security. We not only employed state-of-the-art technologies, including blockchain, Docker, and Kubernetes, but also implemented the proposed architecture in a smart meter application and addressed the problems of Data Loss Abnormal, Long-Term Consumption Abnormal, and Short-term Consumption Abnormal. Different from a cloud-based blockchain network implementation, the proposed edge-based blockchain system was purposely designed for IoT security. It was implemented on the open source Hyperledger Fabric v1.4, along with the Raft consensus mechanism. Due to the memory limitations of Raspberry Pi 3 devices, our blockchain system was divided into multiple organizations, deployed across different Pi devices and the cloud.

Experiments of the eBC system implemented with the Hyperleger Fabric v1.4 on Raspberry Pi 3 show that 4,000 query transactions can be executed in 400 seconds, which gives approximately 10 transactions per second (TPS) performance. In the case of invoke transactions, the performance is around 7 TPS. Overall, the performance analysis results demonstrate that our proposed blockchain network could be used to enhance security in specific IoT applications, such as smart meters.

In the future, we plan to enhance the performance of the implementation on Raspberry Pi 4 and Nvidia Jetson Nano. Further, we also plan to apply Kubeflow to the dynamic MLOps development of the containerized modules (DNN and BC) in the proposed smart gateway for IoT security.

REFERENCES

1. Alaba, F. A., Othman, M., Targio, I. A. H., and Alotaibi, F., 2017, June. Internet of things security: A survey. *Journal of Network and Computer Applications*, 88, pp. 10–28.
2. Goswami, S. A., Padhya, B. P., and Patel, K. D., 2019. *Internet of Things: Applications, challenges and research issues*. In *2019 Third International Conference on I-SMAC (IoT in Social, Mobile, Analytics and Cloud) (I-SMAC)* (pp. 47–50). Palladam, India: IEEE.
3. Cai, H., Xu, B., Jiang, L., and Vasilakos, A. V., 2017. Iot-based big data storage systems in cloud computing: Perspectives and challenges. *IEEE Internet of Things Journal*, 4(1), pp. 75–87.
4. Liu, H., Eldarrat, F., Alqahtani, H., Reznik, A., de Foy, X., and Zhang, Y., 2017. Mobile edge cloud system: Architectures, challenges, and approaches. *IEEE Systems Journal*, 12(3), pp. 2495–2508.
5. Khan, L. U., Yaqoob, I., Tran, N. H., Kazmi, S. M. A., Dang, T. N., and Hong, C. S., 2020. Edge computing enabled smart cities: A comprehensive survey. *IEEE Internet of Things Journal*, 7(10), pp. 10200–10232.
6. Guo, S., Hu, X., Guo, S., Qiu, X., and Qi, F., 2020. Blockchain meets edge computing: A distributed and trusted authentication system. *IEEE Transactions on Industrial Informatics*, 16(3), pp. 1972–1983.
7. Hasan, H. R., Salah, K., Jayaraman, R., Omar, M., Yaqoob, I., Pesic, S., Taylor, T., and Boscovic, D., 2020. A blockchain-based approach for the creation of digital twins. *IEEE Access*, 8(34), pp. 113–134.
8. Huang, X., Yu, R., Kang, J., He, Y., and Zhang, Y., 2017. Exploring mobile edge computing for 5G enabled software defined vehicular networks. *IEEE Wireless Communications*, 24(6), pp. 55–63.
9. Abbas, N., Zhang, Y., Taherkordi, A., and Skeie, T., 2018. Mobile edge computing: A survey. *IEEE Internet of Things Journal*, 5(1), pp. 450–465.
10. Zhang, X., Li, R., and Cui, B., 2018. *A security architecture of VANET based on blockchain and mobile edge computing*. In *In 2018 1st IEEE International Conference on Hot Information-Centric Networking (HotICN)* (pp. 258–259). Shenzhen: IEEE.
11. Qiu, X., Liu, L., Chen, W., Hong, Z., and Zheng, Z., 2019. Online deep reinforcement learning for computation offloading in blockchain-empowered mobile edge computing. *IEEE Transactions on Vehicular Technology*, 68(8), pp. 8050–8062.
12. Huang, Y., Zhang, J., Duan, J., Xiao, B., Ye, F., and Yang, Y., 2019. *Resource allocation and consensus on edge blockchain in pervasive edge computing environments*. In *2019 IEEE 39th International Conference on Distributed Computing Systems (ICDCS)* (pp. 1476–1486). Dallas: IEEE.

13. Sultan, A., Mushtaq, M., and Abubakar, M., 2019, March. *IoT security issues via Blockchain: A review paper*. In *Proceedings of the International Conference on Blockchain Technology* (pp. 60–65). Honolulu: Assocation for Computing Machinery.

14. Anilkumar, V., Joji, J. A., Afzal A., and Sheik, R., 2019. *Blockchain simulation and development platforms: Survey, issues and challenges*. In *2019 International Conference on Intelligent Computing and Control Systems (ICCS)*. (pp. 935–939). Madurai, India: IEEE

15. Yang, R., Yu, F. R., Si, P., Yang, Z., and Zhang Y., 2019. Integrated blockchain and edge computing systems: A survey, some research issues and challenges. *IEEE Communications Surveys Tutorials*, 21(2), pp. 1508–1532.

16. Wu, H., Cao, J., Yang, Y., Tung, C. L., Jiang, S., Tang, B., Liu, Y., Wang, X., and Deng, Y., 2019. *Data management in supply chain using blockchain: Challenges and a case study*. In *2019 28th International Conference on Computer Communication and Networks (ICCCN)* (pp. 1–8). Valencia, Spain: IEEE.

17. Routh, K., and Pal, T., 2018, February. *A survey on technological, business and societal aspects of Internet of Things by Q3, 2017*. In *Proceedings of the 3rd International Conference On Internet of Things: Smart Innovation and Usages (IoT-SIU)* (pp. 1–4). Bhimtal: IEEE.

18. Neshenko, N., Bou-Harb, E., Crichigno, J., Kaddoum, G., and Ghani, N., 2019. Demystifying IoT security: An exhaustive survey on IoT vulnerabilities and a first empirical look on internetscale IoT exploitations. *IEEE Communications Surveys Tutorials*, 21(3), pp. 2702–2733.

19. Sharma, T., Satija, S., and Bhushan, B., 2019. *Unifying blockchian and iot:security requirements, challenges, applications and future trends*. In *2019 International Conference on Computing, Communication, and Intelligent Systems (ICCCIS)* (pp. 341–346). Greater Noida, India: IEEE.

20. Malik, A., Gautam, S., Abidin, S., and Bhushan, B., 2019. *Blockchain technology-future of IoT: Including structure, limitations and various possible attacks*. In *2019 2nd International Conference on Intelligent Computing, Instrumentation and Control Technologies (ICICICT)* (pp. 1100–1104). Kannur, Kerala, India: IEEE.

21. Hassija, V., Chamola, V., Saxena, V., Jain, D., Goyal, P., and Sikdar, B., 2019. A survey on IoT security: Application areas, security threats, and solution architectures. *IEEE Access*, 7, pp. 82721–82743.

22. Shin, D., Yun, K., Kim, J., Astillo, P. V., Kim, J.-N., and You, I., 2019. A security protocol for route optimization in DMM-based smart home IoT networks. *IEEE Access*, 7, pp. 142531–142550.

23. Varshney, T., Sharma, N., Kaushik, I., and Bhushan, B., 2019. *Architectural model security threats and their countermeasures in IoT*. In *Proceedings of the International Conference on Computing, Communication, and Intelligent Systems (ICCCIS)* (pp. 424–429). Greater Noida, India: IEEE.

24. Ahmed, A., and Ahmed, E., 2016, January. *A survey on mobile edge computing*. In *Proceedings of the 10th International Conference on Intelligent Systems and Control (ISCO)* (pp. 1–8). Coimbatore: IEEE.

25. Jain, R., and Tata, S., 2017, June. *Cloud to edge: Distributed deployment of process-aware IoT applications*. In *Proceedings of the IEEE International Conference on Edge Computing (EDGE)* (pp. 182–189). Honolulu, HI: IEEE.

26. Androulaki, E., Barger, A., et al., 2018, January. *Hyperledger Fabric: A distributed operating system for permissioned blockchain*. In *Proceedings of the Conference on the EuroSys*(pp. 1–15). Porto, Portugal: Assocation for Computing Machinery.

27. Nolan, S., 2018, November. *pBFT: Understanding the consensus algorithm*, https://medium.com/coinmonks/pbft-understanding-the-algorithm-b7a7869650ae. Accessed on November 18, 2020.

28. Howard, H., 2014, July. *ARC: Analysis of raft consensus*, https://www.cl.cam.ac.uk/techreports/UCAM-CL-TR-857.pdf. Accessed on November 18, 2020.

29. Thummavet, P., 2019, May15. *Demystifying Hyperledger Fabric (2/3): Private data collection*,https://www.serial-coder.com/\post/demystifying-hyperledger-fabric-private-data-collection. Accessed on November 18, 2020.

30. Tabrizi, F. M., and Pattabiraman, K., 2019, May. Design-level and code-level security analysis of IoT devices. *ACM Transactions on Embedded Computing Systems*, 18(3), pp. 1–25.

31. Wang, Y., Chen, Q., Hong, T., and Kang, C., 2019, May. Review of smart meter data analytics: Applications, methodologies, and challenges. *IEEE Transactions on Smart Grid*, 10(3), pp. 3125–3148.

32. Farah, E., and Shahrour, I., 2017, September. *Smart water for leakage detection: Feedback about the use of automated meter reading technology*. In *Proceedings of the IEEE International Conference on Sensors Networks Smart and Emerging Technologies (SENSET)* (pp. 1–4) Beirut: IEEE.

33. Arora, A., Kaur, A., Bhushan, B., and Saini, H., 2019. *Security concerns and future trends of internet of things*. In *2019 2nd International Conference on Intelligent Computing, Instrumentation and Control Technologies (ICICICT)* (pp. 891–896). Kannur, Kerala, India: IEEE.

34. Tsai, K., Huang, Y., Leu, F., You, I., Huang, Y., and Tsai, C., 2018. AES-128 based secure low power communication for LoRaWAN IoT environments. *IEEE Access*, 6, pp. 45325–45335.

35. Combe, T., Martin, A., and Di Pietro, R., 2016. To docker or not to docker: A security perspective. *IEEE Cloud Computing*, 3(5), pp. 54–62.

11 Blockchain for the Security and Privacy of IoT-Based Smart Homes

Somya Goyal
Manipal University Jaipur, India and Guru Jambheshwar University of Science and Technology, India

Sudhir Kumar Sharma
Institute of Information Technology and Management, India

Pradeep Kumar Bhatia
Guru Jambheshwar University of Science & Technology, India

CONTENTS

11.1 Introduction 239
11.2 Blockchain Technology and Smart Homes...241
    11.2.1 Security and Privacy Threats to Smart Homes 241
    11.2.2 Blockchain as a Security Solution 241
    11.2.3 Cryptographic Aspects of Blockchain 242
11.3 Case Study 244
    11.3.1 Framework 244
        11.3.1.1 Layer 1: Primary Data Layer..245
        11.3.1.2 Layer 2: BC layer 245
        11.3.1.3 Layer 3: Applications Layer..245
        11.3.1.4 Layer 4: UI Layer..246
    11.3.2 Proposed Model 246
    11.3.3 Smart Home System: An Illustration 247
    11.3.4 Experimental Results 249
11.4 Conclusion...250
References...250

11.1 INTRODUCTION

The Internet of Things (IoT) has given people a gift, in the "smart home," which is a coordinated living arrangement that offers increased comfort and convenience. Smart homes can expand personal satisfaction. Inside a smart home, a system interconnects various devices, such as cell phones, sensors, and actuators. This immense, smart system is based on the IoT. Smart home frameworks find an enormous scope of utilization in our day-to-day lives, which has become a focal point of fascination for both clients and designers. The applications include monitoring, home automation, health care, and so on. The worldwide market for smart homes is constantly developing.

A smart home is an IoT-based system linking heterogenous devices, which can make people's life more pleasant [1]. Blockchain innovation is supplementing the smart home framework, on account of its decentralization, straightforwardness, adaptability, and versatility [2]. The blockchain-based smart home biological system can be seen as a four-layer structure. This structure includes an IoT sensor layer or information source layer, a blockchain layer, a smart home layer, and a user interface layer or customer layer [3]. The IoT sensor layer or information source includes sensors to monitor the state, condition, and occupants of a smart home. As an example, an indoor thermostat can be introduced to this layer, to quantify and manage room temperature. Closed-circuit TV (CCTV) could be another constituent of this layer. Information gathered from these sources is united and stored on a unified server or in some decentralized environment, for example, the blockchain [4]. Blockchain innovation is at the top of the information sources. It has two significant parts: the blockchain information structure and the smart contract. Hash values cryptographically interface blocks. A miner (a home server) is responsible for confirming and adding new transactions to new blocks, while smart contracts adhere to predefined guidelines and encourage decentralized transactions [5]. The smart home layer is made to encourage different smart home applications. These include, for example, information commercial center, get to the executives, homecare and social insurance interoperability, and mechanized utility installment and smart city administrations. At the height of the progression comes the user interface UI customer layer, which permits outside partners to profit from blockchain-based smart home applications, for example, a microgrid, retail shops, specialist co-ops, security devices, and so on. The general engineering is contained, so as to counter the protection and security issues of smart home applications [6]. It is essentially comprised of smart agreements, private blockchains, and open blockchains. Smart agreements are built into the devices of a smart home; at the main layer, this implies specifically IoT information sources. Private blockchains permit safe communication among the devices inside the home [7]. An open blockchain is connected for distributed correspondence between various smart homes. This engineering permits the incorporation of blockchain inside the smart home design, to guarantee the security and protection of the smart home.

As shown, research has contributed to protocol design and standards, payment gateways, and smart city projects [8]. But, the contribution to smart homes and ensuring the security of data and access to devices is an area for future research and experimentation [9].

This chapter makes the following contributions. It gives a thorough review of the emerging technology: blockchain, in simplified terms. It explains the workings of blockchain and information blocks, with their linking techniques. It proposes a case study of smart home security with blockchain, along with implementation and comparison to a traditional system, which could raise the entire digital world into a higher dimension.

Blockchain for the Security and Privacy of IoT-Based Smart Homes 241

The chapter is organized as follows. Section 11.2 presents the central blockchain innovation and smart home application system. In Section 11.3, we provide contextual analysis with a case study of smart home security using blockchain. The subsections involve the model architecture proposed, its implementation, and an evaluation of the results. The work is concluded in Section 11.4, with remarks on the possible future scope of this technology.

11.2 BLOCKCHAIN TECHNOLOGY AND SMART HOMES

This section presents the structure on which blockchain innovation is executed. It is progressive in nature and permits disseminated trust to guarantee security and protection. It makes the blockchain arrangement progressively suitable for use with the IoT in smart homes.

11.2.1 SECURITY AND PRIVACY THREATS TO SMART HOMES

While smart homes possess great advantages, they are still defenseless against the possible dangers of vindictive digital assaults, because of the expanding assortment and sharing of the live data of property holders. Researchers predict that in 2020, malware, viruses, and web-based attacks will be driven into approximately a great many thousands [10]. This could risk the security and protection of clients connected by means of smart home networks. The effort to achieve the security and protection of IoT-based smart homes has motivated IoT-based innovations from its early stages to the present development. When applied to smart homes, conventional security arrangements are considerably vulnerable to attack, because of the centralization of the system. This prompts the appropriate utilization of blockchain, which is a disseminated, decentralized, open record. The three specific mainstays of blockchain are decentralization, transparency, and immutability, making it useful for guaranteeing security and protection to the proprietors of smart homes.

11.2.2 BLOCKCHAIN AS A SECURITY SOLUTION

Blockchain addresses each of the three CIA parameters: confidentiality, integrity, and availability. The widespread acknowledgment that blockchain innovation can protect transactions has opened new ways of implementing protection and security in the smart home. Blockchain has accomplished surprising execution for home access control and information sharing for smart home applications. The security issues of smart home are being resolved utilizing blockchain, specifically, because of the use of miners. The miner is responsible for keeping track of the heterogenous devices associated by means of the IoT. Additionally, blockchain encrypts the correspondence between different devices. The main levels of the structure are:

    I. Smart Home
    II. Cloud Storage
    III. Overlay

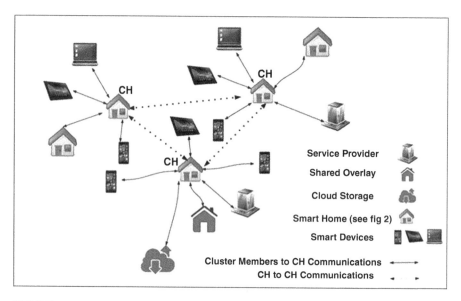

FIGURE 11.1 Overlay Architecture.

Smart Home Level: All the smart devices are introduced inside this level and are controlled through the excavator in a unified manner. At this level, the miner organizes overlay alongside service providers (SP), cloud storage, and clients' devices (cell phones or PCs).

Cloud Storage: Distributed storage is utilized by the smart home devices to store and offer information.

Overlay: At this level, the hubs are isolated into groups, so as to diminish the delay in the system. A cluster head (CH) is chosen by every one of the clusters. The reason for CHs is to keep up an open BC, with two key records: requester key records and requestee key records. This overlay engineering is shown in Figure 11.1 [11].

Further, the components of a smart home are shown in Figure 11.2, including:

i. *Transactions*: the communication occurring between network devices or overlay hubs.
ii. *Local BC*: monitors exchanges. It additionally keeps up an arrangement header which stores the clients' strategy for approaching and active exchanges.
iii. *Home miner*: A smart home miner is a device that centrally processes incoming and outgoing transactions to and from the smart home.
iv. *Local Storage*: a reinforcement drive, to store information locally.

11.2.3 CRYPTOGRAPHIC ASPECTS OF BLOCKCHAIN

In 2008, Satoshi Nakamoto [12] introduced the concept of blockchain. It is the underlying platform of cryptocurrencies (e.g. Bitcoin) that facilitates a P2P transaction system to eliminate third-party and double spent problem. It is not a centralized

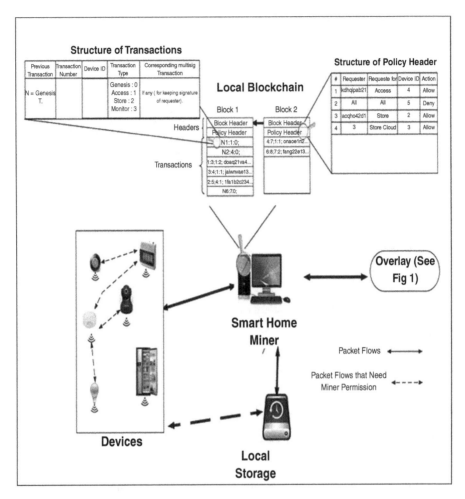

FIGURE 11.2 Smart Home Based on Blockchain Architecture.

information hub. In blockchain, each portion of information is cryptographically associated with past portions, utilizing hashing characteristics, for instance, the secure hash algorithm (SHA-256). Such a portion of information is called a block, with associated properties such as block number, past blocks' hash, timestamp (TS), exchange information, and nonce [12]. Figure 11.3 portrays the blockchain plan, inside structure, and work process. The following steps describe the essential functionalities.

1. Every connected IoT device in a blockchain network is called a node. It stores all current transactions, which are in a waiting queue in a memory pool.
2. All the transactions are verified by a Merkle tree.
3. The validated and verified transactions are added to the block.
4. Miners change the nonce and timestamp to generate a hash of block.
5. The generated hash is compared by the system with the target.

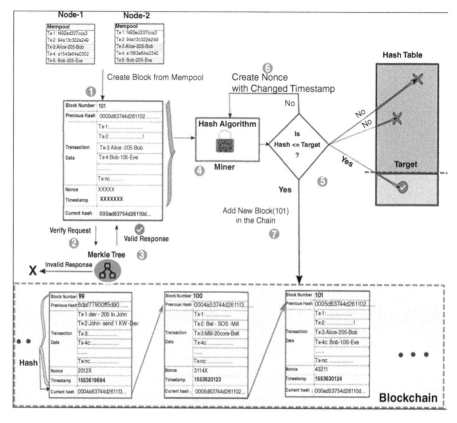

FIGURE 11.3 Workflow of Blockchain.

6. If the hash value is found to be above the target value, then go to step 4.
7. If the hash value is less than the target value, then it is a successful case. It has successful verification of the PoW.

11.3 CASE STUDY

In this section, we discuss a completely implemented case study of blockchain in a smart home application.

11.3.1 Framework

Blockchain technology allows the easy integration of multiple IoT devices from a diverse range within the smart home. The conceptual framework consists of four layers. Figure 11.4 shows a smart home based on the blockchain framework. The *four comprising layers* are:

  i. Primary data layer
  ii. BC layer

Blockchain for the Security and Privacy of IoT-Based Smart Homes 245

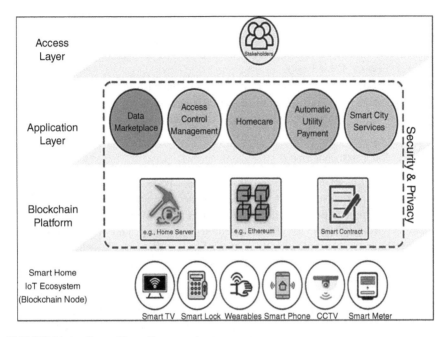

FIGURE 11.4 Smart Home Framework.

iii. Applications layer
iv. UI layer

11.3.1.1 Layer 1: Primary Data Layer

The IoT information source layer creates information from devices that assume an indispensable job in monitoring the state, condition, and inhabitants of a smart home. These devices are generally arranged into three principle classes: sensors, sight and sound, and social insurance. Sensors measure ecological elements. For instance, the indoor regulator is utilized to gauge and control room temperature. Information gathered from all these devices is combined and stored in an incorporated server, installed on the next layer as a blockchain platform.

11.3.1.2 Layer 2: BC layer

Blockchain innovation lives at the head of the IoT biological system and comprises two significant segments: the blockchain information structure, and a smart agreement. Hash values cryptographically associate blocks.

11.3.1.3 Layer 3: Applications Layer

The application layer is made to encourage different smart applications for smart homes, along with their collaboration with blockchain platform. It is the most important layer for incorporating smart applications for smart homes, for example, data marketplace, access management, homecare and healthcare interoperability, and

246   Blockchain Technology for Data Privacy Management

automated utility payment and smart city services. Vast numbers of these developing applications are utilizing blockchain, and some are as yet under examination.

11.3.1.4 Layer 4: UI Layer

Ultimately, at the head of the pecking order comes the customer layer, which permits outsider partners to profit by blockchain-based smart home applications, for example, a microgrid, retail shops, specialist co-ops, caregivers, etc.

11.3.2 PROPOSED MODEL

In this modern digital era, the installation of sensor-based devices incorporated with IoT platforms and automation technology has become attractive and essential in a smart home. Home automation and smart homes set up with distinctive IoT devices depend upon gateways for their assembled working and collaborative task accomplishments. The gateways play a vital and indispensable role in the smart homes [13]. Nevertheless, there are different security vulnerabilities to which the IoT-based smart home is exposed by being connected to the internet. For example, eavesdroppers can access the private cameras of a smart home and breach the privacy of home members. We propose a novel smart home gateway architecture based on blockchain, to address and resolve all such security issues. The proposed gateway architecture counters all vulnerabilities anticipated in smart homes. The proposed framework consists of a layered stack of three levels. Level 1 contains all the devices, Layer 2 carries the gateway, and Layer 3 contains the cloud itself [14]. Layer 2 is proposed to have the most crucial component, i.e., the blockchain. It is responsible for accepting data and exchanging the information blocks. It supports the decentralization of information processing. The blockchain available at Layer 2 provides the authentication and authorization of the bulk data coming from an outside network to the secured inside network of the smart home. It provides reliable communication between household devices and external devices [15]. We advocate the implementation of the proposed framework on the Ethereum blockchain. In this study, the development, implementation, and performance evaluation considering the security, response time, and accuracy is done. From the experimental results, we infer that the proposed security model for a smart home based on blockchain outperforms current state-of-art techniques.

IoT devices [16,17] installed in smart homes are associated with one gateway. From that gateway, a unique identifier is allotted, as a device ID. These communication channels and IoT devices are assigned fixed IDs and have the cryptographical ability to work encryption and decryption computations with PKI and SHA2. The selection of certificates, data representational structures for device and their interconnections are shown in Figure 11.5.

1. Devices confirmed by a gateway should consistently be confirmed intermittently. The Dn in the device layer endeavors to enroll straightforwardly with or naturally interface with the gateway. The gateway requests an ID from the device that is connected or requested, to acquire data about the connected device.

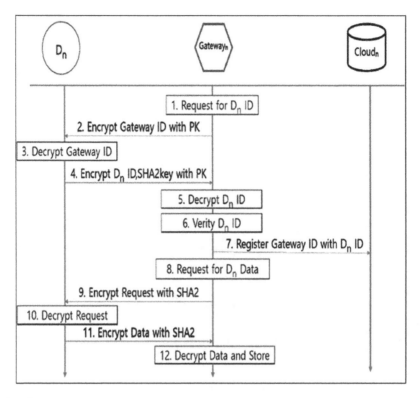

FIGURE 11.5 Proposed Gateways in Smart Home Model.

2. The device's gateway executes a cryptographic calculation to scramble the passage data to the device and sends the message. Devices translate the encrypted messages, with the help of previously shared keys.
3. Encrypted messages containing passage data are unencrypted and sent to the unregistered or unencryptable entryway when they are received.

11.3.3 SMART HOME SYSTEM: AN ILLUSTRATION

This section demonstrates how the proposed model implements security. Let us consider an illustration. A visitor wants to enter the home. A cat (pet) is inside the home, it is dinner time for the pet, and it is alone. The visitor is there to feed the hungry cat. Now, the visitor can be permitted modified granted access to the home. Figure 11.6, shows the steps following which our proposed blockchain-based framework allows secure access:

1. The visitor knocks on the door. There is an access control list, on which levels of access are assigned to different users. Level_1 (highest priority) is assigned to the owner; Level 2 is assigned to the spouse and children, and so on. Now,

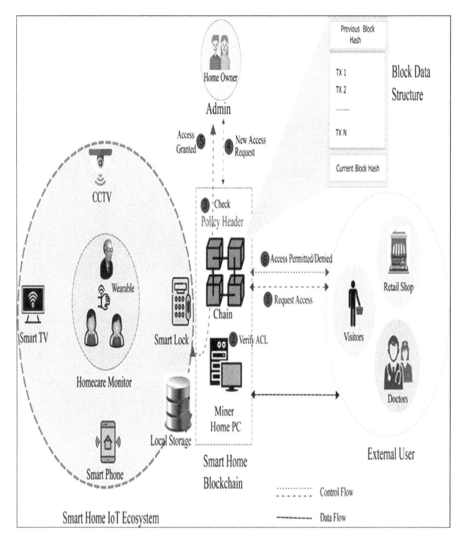

FIGURE 11.6 An Illustration.

this new visitor has Level 0, meaning no access to the home. So, at step (1), he requests permission to let himself in.

2. The request from the visitor ticks the home server. The home server checks the access control list (ACL) and also forwards the details of the request to the blockchain, to verify the policies for that particular type of visitor.
3. Now, the policy header plays its role. This part of a blockchain stores all the access control lists, along with details of the devices attached and of the control policy's implementation.
4. The admin comes into play at this point, when the request ticket moves to admin from the security system. Now, admin can favor or deny the request.

Blockchain for the Security and Privacy of IoT-Based Smart Homes

5. The visitor is allowed to enter the home, after which the blockchain will add all the information in the policy header.
6. The visitor can feed the cat inside the house, with restricted permission, as per the access control policies and permissions.

In this way, controlled access is being implemented to avoid security breaches, using the blockchain system and its characteristic functions.

11.3.4 Experimental Results

This section describes the experimental setup for the execution of the proposed framework. The emulating environment is set up using Mininet [18]. It emulates the open switches and connecting nodes, such as sample IoT devices. The platform is installed as a Linux server, by using 15 desktops. The configuration for each desktop is an i7 processor and a 64 GB DDR3 RAM. The SDN controller is run to configure the gateway. It is run in autonomous VMs. An Amazon EC2 cloud data server [19] is employed for cloud services. The proposed framework is implemented using the Ethereum blockchain [20]. For performance evaluation [21], the metrics of this decentralized system are compared with the traditional centralized one. The competing centralized framework is being implemented using an Amazon EC2 cloud server and its gateway [22,23]. The presentation assessment is plotted in Figures 11.7 and 11.8, wherein standard evaluation estimations of the security, which are response time and precision as accuracy, are assessed with the variation of data traffic quantity.

From Figure 11.7, it is clear that the proposed architecture performs better than the current state-of-art-architectures. The faster response time in our proposed architecture is due to the fact that the gateway lies near the device and hence responds faster [24–27].

Figure 11.8 shows that the proposed model has better performance with increasing loads, compared to the centralized model.

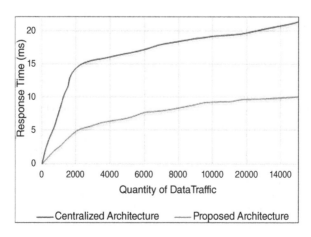

FIGURE 11.7 Response Time.

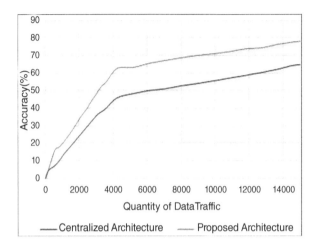

FIGURE 11.8 Accuracy.

11.4 CONCLUSION

It can be concluded that blockchain has the capability to transform an ordinary house into a smart home automatically which is safe and secure. The characteristic on which block chain depends for this magical transformation is its decentralized and autonomous nature. A thorough review of emerging technology—the blockchain—has been given in a simplified manner. A deeper insight into the workings of block chain and information blocks, with their linking techniques is modeled successfully. A case study for smart home security with block chains is proposed and implemented. The comparative study shows that blockchain has the potential to uplift the entire digital world into a higher dimension. It ensures the privacy, confidentiality, authentication, and integrity of data for smart homes. For secure digital smart homes, blockchain technology can be deployed and in future, this work could be carried out at a larger scale, to develop a prototype for research purposes.

REFERENCES

1. PR Wire, 2016. Gartner: Blockchain and connected home are almost at the peak of the hype cycle. https://prwire.com.au/pr/62010/gartner-blockchain-andconnected-home-are-almost-at-the-peak-of-the-hype-cycle. Accessed on December 28, 2019.
2. Sharma, P. K., Moon, S. Y., and Park, J. H. 2017. Block-VN: A distributed blockchain based vehicular network architecture in smart city. *Journal of Information Processing Systems*, 13(1), pp. 184–195.
3. Bharat, B., Aditya, K., Martin Sagayam, K., Sudhir Kumar, S., Mohd Abdul, A., and Debnath, N. C., 2020. Blockchain for smart cities: A review of architectures, integration trends and future research directions. *Sustainable Cities and Society*, 61(102360), pp. 2210–6707. doi:10.1016/j.scs.2020.102360.

Blockchain for the Security and Privacy of IoT-Based Smart Homes **251**

4. Malik, A., Gautam, S., Abidin, S., and Bhushan, B., 2019. *Blockchain technology–future of IoT: Including structure, limitations and various possible attacks.* In *2019 2nd International Conference on Intelligent Computing, Instrumentation and Control Technologies (ICICICT)* (pp. 1100–1104). Kannur, Kerala, India: IEEE. doi:10.1109/icicict46008.2019.8993144.

5. Ahram, T., Sargolzaei, A., Sargolzaei, S., Daniels, J., and Amaba, B., 2017. *Blockchain technology innovations.* In *2017 IEEE Technology & Engineering Management Conference (TEMSCON)* (pp. 137–141). Santa Clara, CA: IEEE. doi:10.1109/TEMSCON.2017.7998367.

6. Lee, Y., Rathore, S., Park, J. H., and Park, J. H., 2020. A blockchain-based smart home gateway architecture for preventing data forgery. *Human-Centric Computing and Information Science*, 10(9). doi:10.1186/s13673-020-0214-5.

7. Arora, A., Kaur, A., Bhushan, B., and Saini, H., 2019. *Security concerns and future trends of Internet of Things.* In *2019 2nd International Conference on Intelligent Computing, Instrumentation and Control Technologies (ICICICT)* (pp. 891–896). Kannur, Kerala, India: IEEE. doi:10.1109/icicict46008.2019.8993222.

8. Bhushan, B., Sahoo, C., Sinha, P., and Khamparia, A., 2020. Unification of blockchain and Internet of Things (BIoT): Requirements, working model, challenges and future directions. *Wireless Networks*. doi:10.1007/s11276-020-02445-6.

9. Khamparia, A., Singh, P. K., Rani, P., Samanta, D., Khanna, A., and Bhushan, B., 2020. An internet of health things-driven deep learning framework for detection and classification of skin cancer using transfer learning. *Transactions on Emerging Telecommunications Technologies*. doi:10.1002/ett.3963.

10. Moniruzzaman, Md., Khezr, S., Yassine, A., and Benlamri, R., 2020. Blockchain for smart homes: Review of current trends and research challenges. *Computers & Electrical Engineering*, 83, p. 106585. doi:10.1016/j.compeleceng.2020.106585.

11. Nakamoto, S., 2008. Bitcoin: A peer-to-peer electronic cash system. https://Bitcoin.org/Bitcoin.pdf. Accessed on December 28, 2019.

12. Goyal, S., Parashar, A., and Shrotriya, A., 2018. Application of big data analytics in cloud computing via machine learning. In *Data Intensive Computing Applications for Big Data*, IOS PRESS-2018. *Advances in Parallel Computing Series*, vol. 29, pp. 236–266. doi:10.3233/978-1-61499-814-3-236.

13. Goyal, S., and Bhatia, P., 2021. *Empirical Software Measurements with Machine Learning. Computational Intelligence Techniques and Their Applications to Software Engineering Problems*, vol. 1 Boca Raton: CRC Press, Taylor & Francis Group, pp. 49–64. doi:10.1201/9781003079996.

14. Parashar, A., Parashar, A., and Goyal, S., 2018. Big data analysis using machine learning approach to compute data. In *Data Intensive Computing Applications for Big Data*, IOS PRESS-2018, *Advances in Parallel Computing Series*, vol. 29, pp. 133–160. doi:10.3233/978-1-61499-814-3-133.

15. Sharma, P. K., and Park, J. H., 2018. Blockchain based hybrid network architecture for the smart city. *Future Generation Computer Systems*, 86, pp. 650–655. doi:10.1016/j.future.2018.04.060.

16. Goyal, S., and Parashar, A., 2017. Selecting the COTS components using Ad-hoc approach. *International Journal of Wireless and Microwave Technologies (IJWMT-2017)*, 7(5), pp. 22–31. doi:10.5815/ijwmt.2017.05.03.

17. Goyal, S., and Bhatia, P. K., 2020. *Cloud assisted IoT enabled smoke monitoring system (e-Nose) using machine learning techniques.* In *Proceeding of SSIC 2019, Smart Systems and IoT: Innovations in Computing* (Edition Number 1, Series Vol. 141), pp.743–754). Singapore: Springer. doi:10.1007/978-981-13-8406-6_70.

18. Xue, J., Xu, C., and Zhang, Y., 2018. Private blockchain-based secure access control for smart home systems. *Ksii Transactions on Internet and Information Systems*, 12(12), p. 6057. doi:10.3837/tiis.2018.12.024.

19. Wang, J., Gao, Y., Liu, W., Sangaiah, A. K., Kim, H. J., 2019. Energy efficient routing algorithm with mobile sink support for wireless sensor networks. *Sensors*, 19(7), pp. 1468–1494.

20. Rathore, S., and Park, J. H., 2018. Semi-supervised learning based distributed attack detection framework for IoT. *Applied Soft Computing*, 72, pp. 79–89. doi:10.1016/j. asoc.2018.05.049.

21. Somya, G., and Pradeep Kumar, B., 2020. Comparison of machine learning techniques for software quality prediction. *International Journal of Knowledge and Systems Science (IJKSS)*, 11(2), pp 21–40. doi:10.4018/IJKSS.2020040102.

22. Huang, X., Yu, R., Kang, J., Xia, Z., and Zhang, Y., 2018. Software defined networking for energy harvesting internet of things. *IEEE Internet of Things Journal*, 5(3), pp. 1389–1399. doi:10.1109/JIOT.2018.2799936.

23. Magurawalage, C. M. S., Yang, K., Hu, L., and Zhang, J., 2014. Energy-efficient and network-aware offloading algorithm for mobile cloud computing. *Computer Networks*, 74, Part B, pp. 22–33. doi:10.1016/j.comnet.2014.06.020.

24. Rathore, S., Kwon, B. W., and Park, J. H., 2019. BlockSecIoTNet: Blockchain-based decentralized security architecture for IoT network. *Journal of Network and Computer Applications*, 143, pp. 167–177. doi:10.1016/j.jnca.2019.06.019.

25. Choi, B., Lee, S., Na, J., and Lee, J., 2016. Secure firmware validation and update for consumer devices in home networking. *IEEE Transactions on Consumer Electronics*, 62(1), pp. 39–44. doi:10.1109/TCE.2016.7448561.

26. Goyal, S., Sharma, N., Kaushik, I., Bhushan, B., and Kumar, A., 2020. *Precedence & issues of IoT based on edge computing.* In *2020 IEEE 9th International Conference on Communication Systems and Network Technologies (CSNT)* (pp. 72–77). Gwalior, India: IEEE. doi:10.1109/csnt48778.2020.9115789.

27. Varshney, T., Sharma, N., Kaushik, I., Bhushan, B., 2019. *Architectural model of security threats & their countermeasures in IoT.* In *2019 International Conference on Computing, Communication, and Intelligent Systems (ICCCIS)* (pp. 424–429). Greater Noida, India: IEEE. doi:10.1109/icccis48478.2019.8974544.

12 A Framework for a Secure e-Health Care System Using IoT-Based Blockchain Technology

T. Sanjana
Department of Electronics and Communication Engineering, B. M. S. College of Engineering, India

B. J. Sowmya, D. Pradeep Kumar
Department of Computer Science and Engineering, M S Ramaiah Institute of Technology, India

K. G. Srinivasa
National Institute of Technical Teachers Training and Research, India

CONTENTS

12.1 Introduction 254
12.2 Related Work 256
12.3 Design 259
    12.3.1 Information Collector Using Sensors (IoT Techniques) 260
    12.3.2 Processing of Health Care Records in Fog Nodes 261
        12.3.2.1 Preprocessing of Electronic Health Records (EHR) 261
        12.3.2.2 Extract Database Entries 261
        12.3.2.3 Define Features 261
        12.3.2.4 Process Data 261
        12.3.2.5 Assess Feature Values 262
        12.3.2.6 Integrate Data Elements 262
    12.3.3 Effective Analysis Using Machine Learning 262
    12.3.4 Securing of Health Records Using Blockchain 262
        12.3.4.1 Initialization 262
        12.3.4.2 Generating a New Block 263
        12.3.4.3 Verifying the New Block 263
        12.3.4.4 Appending the New Block 263
12.4 Implementation and Results 263
    12.4.1 Sensors Layer 263
        12.4.1.1 Fog Computing Layer 264
        12.4.1.2 Cloud Computing Layer 264

	12.4.1.3 Data Preprocessing	264
	12.4.1.4 Effective Analytics Using Machine Learning	264
	12.4.1.5 Building a Blockchain and Cryptocurrency	266
12.5	Conclusion	269
References		270

12.1 INTRODUCTION

A quality health care system is a prime area of concern in every country. A health care system consists of hospitals, clinics, diagnostic labs, medical devices, equipment, remote health care, health insurance, and telemedicine. In the present era, growth in the health care industry has been revolutionary. Factors such as the need for the creation, integration, and maintenance of EHRs, the analysis of huge amounts of data generated from wearable devices, smart devices, and medical equipment, the expectation of a high quality of service from health care providers, and patient-centered health care have influenced the health care system. Recent advances in telemedicine, 3D printing, artificial intelligence, the IoT, cloud computing, fog computing, 5G, cryptography, and virtual reality have taken the health care industry to the next level, called Health 4.0. All of the services and sectors of health care are expected to be impeccable, as this is vital for human well-being and the economic growth of a country. This being the importance of health care, it compels us to focus on the shortcomings of the health care industry. The challenges faced by the current health care system include medical record data management, updating health records, errors in medication, counterfeit drugs, difficulty in accessing electronic health records, the need for the security of medical records, tracking pharmaceuticals, medical credentials, clinical trials, health care transactions, insurance claims, preventing and monitoring diseases, and many more [1]. Additionally, new demands are being made for m-health (mobile-health), medical reconciliation, virtualization, decentralized systems, interoperability, high quality, and real-time service. The urge to meet new demands and address different challenges in health care brings blockchain and IoT technology into the limelight.

Blockchain is a state-of-art technology that is capable of providing transparency, security, privacy, authentication, confidentiality, and the validation of data in a system. It is a distributed, safe, auditable, and immutable ledger [2]. The key features offered by blockchain include consensus algorithms, smart contracts, peer-to-peer networks (decentralized and distributed), hash cryptography, mining, and an open immutable ledger. These overwhelmingly useful features of blockchain have provoked its use in diverse areas. One of the most promising applications of blockchain is in the IoT framework of the health care system.

Blockchain is a linkage of blocks, each of which has a header containing the previous block's hash. The previous block is the parent block to the current block. Only one parent block is possible for a block. The first block in a blockchain, which has no parent block, is called the genesis block. A block consists of a header and a body. The header contains the version, Merkle root hash, timestamp, target, nonce, and parent block hash [3]. The body of the block contains transaction data or any data with value. The size of the block is limited to the maximum size allowed. Each block has

A Framework for a Secure e-Health Care System

a unique digital signature, generated from a cryptographic hash function. The digital signature makes the block immutable, because any minute change in the block changes its signature and as a result, disconnects it from the blockchain, as this will not be accepted by others on the blockchain. The nonce in a block header is a totally arbitrary string of numbers. The procedure of changing the nonce and generating a valid digital signature for a block, in order for the block to be added to the blockchain, is called mining, and this is done by miners. Miners use computational resources to perform mining. The more computational resources they have, the quicker they can hash and the quicker they are to locate a valid signature. It is not easy for any miner or user to corrupt a blockchain. One possible attack is the 51 percent attack, which is practically impossible, because it is difficult for a selfish miner to control 50 percent of a blockchain network's computing power. This is rarely possible, but it could happen when a miner is quicker and has more computational power than the rest of the network, in order to change the digital signatures of existing and newly added blocks [4]. Each blockchain user has private and public keys. A private key is used to encrypt the user's own transaction/data, whereas the public key is visible to all users of the blockchain and is used for validation. The most important characteristics of a blockchain are its decentralized and distributed system, cryptographically sealed and timestamped chronologically, smart contracts, consensus, immutability, security, and validation [5,6]. These characteristics make it favorable to the utmost for deployment in the health care industry.

The IoMT is an interconnection of medical devices (equipped with sensors), software applications, and health care services. The IoMT facilitates the better care of patients in health critical situations, and also of those patients who are remotely placed. It is easily accessible, fast, efficient patient care. Sensor-based devices such as wearables, smart pills, implanted sensors, standalone devices (such as an X-ray machine, CT scanner, or MRI), and smart equipment collect data related to a patient's health and transfer it to the cloud or storage. This information can be accessed and analyzed by doctors and other care providers, to suggest further actions to be taken by the patient. The IoMT thus aids in remote monitoring and preventing frequent visits to the doctor, providing quick diagnoses, and preventing the adverse effects of chronic diseases. It is used to monitor patient activity such as heart rate, blood pressure, glucose levels, body temperature, etc. Artificial intelligence (AI) can be used to process the data stored in the cloud and provide doctors with relevant information. AI, deep learning, seamless wireless connectivity, and cloud and fog computing have enhanced the capabilities of the IoMT.

Among the advantages of the IoMT are remote monitoring and care, saving patients time, greatly aiding elders and patients who need frequent or emergency medical assistance, providing easy accessibility and implementation, providing preventive health care, reducing medication errors, and improving efficiency. However, the system is also vulnerable to security threats and attacks, and lacks regulatory measures and standardization. The architecture of the IoMT is divided into three layers: the layer of things, the intermediate layer, and the computing layer [7]. The things layer includes wearables, smart devices and medical equipment, and all stakeholders, including patients and health care service providers. They communicate through different protocols with the intermediate layer, which consists of a gateway.

The back-end computing layer preprocesses and analyzes data and transfers it to servers. Here, big data analytics and cloud computing play major roles. Complex security protocols are used in this layer. The challenges to the IoMT involve performance, interoperability, device constraints, security, and privacy [8]. The main hindrance to the performance of the IoMT is distributed denial-of-service (DDoS), ransomware, and other attacks. When security and privacy schemes are not robust, patients' sensitive medical information, and the details of hospital staff, may be misused or stolen by unauthorized users, when stored in the cloud.

As patients and hospital/clinical staff use their personal devices for the transfer of or access to health-related information, they are vulnerable to various security attacks. The main security principles that need to be supported in the IoMT are confidentiality, integrity and availability (CIA). Security is essential at three different levels: device, connectivity, and cloud. Private keys or security tokens can be used to authenticate devices to the gateway, which ensures device security. By managing registry and protocols used in the IoT gateway, connectivity and security can be guaranteed. Security issues that can occur at the cloud level, such as a data breach, account hijacking, DoS, and malicious nodes, can be prevented by designing suitable security schemes and network protocols. Privacy is also very important in e-health. Personal identifiable information (PII) may have to be disclosed to third parties for proper functioning of the IoMT system. In order to overcome this and protect privacy, various techniques are used, such as policies where the PII owner decides what amount of information is to be shared to whom; confining storage; deleting old data; avoiding repeated queries, and anonymous privacy schemes. However, these fail to protect privacy during certain circumstances and attacks, and moreover, they do not maintain an immutable ledger of patient health records [9]. The risk-assessment model developed in [10] for existing attacks can help quantify the security and privacy of medical devices used in the IoMT, helping users to choose appropriate devices that satisfy the metric. But every day, there is the chance of new and unknown attacks, which is challenging to handle. Thus, blockchain is a preferred technique to be integrated with the IoMT to ensure security, privacy, and immutability in health care systems.

12.2 RELATED WORK

Sincere attempts have been made to integrate blockchain with the IoT for health care applications. It is worth sharing some of the proposed architectures in the literature, as use cases addressing issues such as scalability, latency, storage, interoperability, computational complexity, throughput, and others.

The architecture proposed in [11] uses blockchain to store a hash of the user's health care data and the policies for user data access. Here, hospitals establish the blockchain by registering patients and hospital staff as users. To overcome network overhead, clustering and the practical byzantine fault tolerance (PBFT) method of consensus are employed. Another architecture consists of IoT devices, fog nodes, and the mobile phones of patient and doctor, which are assigned private and public keys. Signals measured from IoT devices are signed using an elliptic curve signature algorithm (ECDSA) and its private key. A fog node, which is a Raspberry Pi3, stores

A Framework for a Secure e-Health Care System

the data from the IoT devices for few hours and uses a machine learning (ML) algorithm to classify data based on whether a patient needs attention or not [12]. An analytic framework in blockchain is proposed in [13], where patient data collected in smart phones from wearable devices using Bluetooth is analyzed using ML. An analysis based on symptoms, vital parameters, age, and the health trend over a certain period of time helps doctors make better and quicker diagnoses. BiiMED, a blockchain-based framework is proposed in [14]. This system uses a health information system (HIS), in line with ICD-10 (International Classification of Diseases) standards. It has a front-end web portal and a back end with blockchain and the cloud. When evaluated in terms of cost and scalability, it was proven to be good. A consortium blockchain operated by a group of hospitals is implemented in a Hyperledger Fabric framework [15].

A customized blockchain for remote patient monitoring (RPM) is put forth in [16]. The main unit of this model is a patient-centric agent (PCA), which classifies patient data as an eventful or uneventful case. A two-way authentication based on a keyed hash message authentication code (HMAC) and voice-based proximity identification is used, because it is fast. The detection and better diagnosis of dyslexia are proposed in [17]. In this case, the subject takes an online test, using a smartphone or other electronic device. Multimedia data is captured and auto-graded in the mobile edge. Multimedia data is stored in BigchainDB, a decentralized database. The hash of these files and test results is available to the stakeholders, via blockchain. A cost effective blockchain-based model for home therapy monitoring is developed in [18]. Physically challenged patients are served using a mobile edge computing enabled IoT platform and secure communication, using Tor-based blockchain. Gesture tracking sensors and other IoT sensors generate multimedia data which are stored, analyzed, and processed at the mobile edge. Blockchain also stores a pointer to the multimedia files, which are in distributed storage with enhanced security.

The DITrust blockchain model takes care of interoperability and scalability in IoT environments. Public blockchain is used to connect every IoT node. Ripple blockchain is implemented at the edge, which involves itself in verifying and validating transactions, access control, and privacy protection [19]. A better visualization of patient data and tracking patients during emergencies is made possible in [20]. The architecture developed here consists of two blockchains: a medical device blockchain for patients and a consultation blockchain for doctors. A live monitoring system through a graphical user interface (GUI) is used to send alerts to doctors, with the patient's location in case of an emergency. Blockchains are implemented in Hyperledger Fabric.

Blockchain and the IoT framework for patients to manage diabetes care by themselves is described in [21]. Hospitals validate, create, and add blocks, and store the hash of data in off-chain distributed storage. Proof of Authority (POA) is chosen as a consensus algorithm, since it requires less computation and energy compared to other existing consensus algorithms. A secure health care system with a combination of IoT, blockchain, and machine learning is detailed in [22]. Machine learning is used to detect anomalies in patient data and also forecast conditions faced by patients, by utilizing data over a period of time. There are two blockchains used, namely personal health care (PHC) and external record management (ERM). A PHC blockchain holds and regulates data from medical devices stored in external storage. An ERM blockchain stores feeds from hospitals.

An optimized blockchain framework for health care is proposed in [23], consisting of three tiers: IoT devices, an overlay network, and the cloud. Data is encrypted with a digital signature before being sent to the cloud. A root hash of the data generated by the cloud server using SHA-3 is sent to the overlay network, which stores it. Anik Islam and Soo Young Shin have proposed a safe outdoor health care model, BHMUS [24]. It uses an unmanned aerial vehicle (UAV), mobile edge computing (MEC) servers, and blockchain. Medical data from wearable devices are encrypted and carried via UAV to an MEC server. The MEC server validates the data received, using an η-hash bloom filter, and checks for any abnormalities. If an abnormality is found, then the user and the nearest hospital are notified.

Designing smart contracts for a secure remote health monitoring system is provided in [25]. Abnormal data is detected and sent to the blockchain by changing its gas price (i.e., a fee to conduct a transaction). An emergency message, along with health parameters and patient location, are sent to the doctor concerned. The main components of this system include a Raspberry Pi 3, a GSM module, smart devices with sensors, and GPS. A blockchain-based payment verification system is proposed by Partha Pratim Ray, Neeraj Kumar, and Dinesh Dash in [26]. In an IoT e-health care environment, patient data from the sensor is processed and sent to the cloud. Later, a simplified payment verification (SPV) is performed, with the hospital using TestNet bitcoin.

Two important concerns about incorporating blockchain in IoT applications are scalability and throughput. These two issues are addressed by designing an efficient field-programmable gate array (FPGA) caching system on a Virtex 6 board, to reduce the load on the blockchain and a custom SHA-256 hash function, to get a better hash table. A network interface controller (NIC) on an FPGA will check the cache, whenever it receives requests from the client. If there is a cache hit, then data is directly sent from the cache to the client, without an intervening blockchain server. If there is a cache miss, then a request is sent to the blockchain server through Gigabit Peripheral Component Interconnect express (PCIe) [27]. A cost-effective secured mobile enabled assisting device (SMEAD) is designed for patients with diabetes. This system consists of wearables, such as a wristband, footwear, neckband, and MEDIBOX (consisting of insulin and other medications). Ethereum blockchain is used to keep data secure through encryption and manage access rights through a smart contract. If an abnormal blood sugar level is detected, this is communicated to the doctor using a mobile app, as well as to friends/relatives, using a social network [28].

A blockchain-based framework for identity management in case of medical insurance claims is described in [29]. The insurance company, policyholder, and hospital create their accounts in an Ethereum virtual machine. The insurance company takes care of client registration and verifies policyholder details using a session key. The client/hospital initiates the processing of the claim. Documents related to the insurance claim amount will be signed by the policyholder, using a private key. The amount is credited to the Ethereum wallet of the client/hospital. Vibranthealthchain is proposed as an interoperable system that combines the capabilities of AI, big data

A Framework for a Secure e-Health Care System 259

analytics, and blockchain in the IoT, to meet the need for participatory, personalized, proactive, preventive, predictive, and precision medication in health care [30].

A decentralized system for RPM with patient agent (PA) is installed on smart phones and fog and cloud servers that guarantee secure communication, as detailed in [31]. The use of a lightweight PoS consensus with a fuzzy inference system (FIS) to assess the fitness of a node is an additional advantage, as this reduces energy consumption and the time needed for block generation. A fine-grained access control scheme using blockchain to preserve patient privacy is proposed in [32]. This is named Healthchain. It uses two blockchains: Userchain, for patients, which is a public blockchain, and DocChain, for doctors, which is a consortium blockchain. PoW is used for consensus in Userchain, and PBFT in DocChain.

The use of a private blockchain owned by an individual is suggested for the health care domain in [33]. The patient alone manages the cross-institutional sharing of information. Fast Healthcare Interoperability Resources (FHIR) are used to provide standard data formats for medical records, and API is used to facilitate the exchange of EHR [34]. This standard format also helps distinguish between data sent by the patient or a clinician. It also helps store data in chronological order. A patient-centric system based on blockchain for health care 4.0 is proposed by in [35]. Data accessibility among the patient, lab, clinician, and system admin is seamlessly possible, using an access control mechanism. A caliper tool is used to verify the performance of the system in terms of latency, throughput, round trip time (RTT), and network security.

A patient-centric blockchain model is designed to have decentralized Ethereum swarm cloud storage for medical records. In this model, multi-signature smart contracts are used, which requires the signatures of both patient and hospital for verification and storage. A swarm hash is generated every time data is accessed or changed, which avoids data breaching [36].

Recently (2020), in [37], the use of a blockchain-based app was suggested, to curb the coronavirus disease–2019 (COVID-19) pandemic. Two apps have been developed based on blockchain: CIVITAS [38] and MiPasa [39]. Deloitte has presented a new model for a health information exchange (HIE). Employing blockchain in health care reduces transaction costs and provides nearly real-time processing. The Office of the National Coordinator (ONC) can provide guidelines, policies, and standards for nationwide use and offer guidance as to which data, and what data size, is to be stored on-chain and off-chain [40].

12.3 DESIGN

Trilevel architecture consists of distributed computing, processing at fog nodes, and sensors working together. Sensors may be embedded in wearable or not wearable devices that are kept close to patients. The applications utilized to check the health of a patient will have parts executed in the edge devices orchestrated in the fog computing layer, or wearable sensors, or on the cloud. Information will stream over this three-layer structure.

Figure 12.1 shows the procedure of collecting data from sensors at the application end. Through the gateway, the data will be moved onto the fog layer, where the

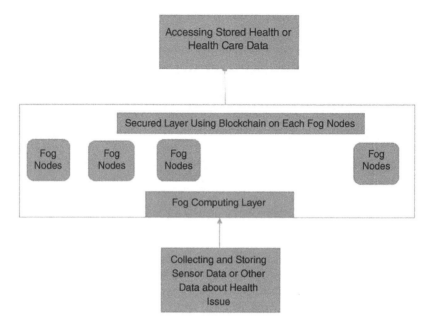

FIGURE 12.1 System Architecture Describing Different Layers.

application of data analytics is been carried out, and the security of those records will be enhanced using blockchain technology. The data can be stored on the cloud for further analysis by the doctors.

The information gathered by the e-Health Shield is sent to a Raspberry Pi. Therefore, this device sends the data to the fog. This layer gives a representational state transfer (REST) interface, permitting different types of users to get to that information, giving the syntactic interoperability of the proposed arrangement. At last, the patient's portable application and the medical caretaker's web application decipher this data and demonstrate it to their clients. The data collection process can be accomplished by interacting with the sensors, and even the manual process of collecting data through a query process with the clients. A scheduler sends it to the gateway, which in turns forwards it to fog nodes.

12.3.1 Information Collector Using Sensors (IoT Techniques)

The IoT is being incorporated in various sorts of enterprises. One of the applications relevant to the topic of this hypothesis is the medical services field. There are portable apps and wearable devices that gather real-time data. Moreover, with respect to how predominant human services are and will become, not only are there financial rewards in the human services field that are implementable using the IoT, but additionally, there are security factors with less pressure on doctors and human health care suppliers to oversee their patients. In the medical services domain, it is urgent for human health caretakers to improve their practice with new technologies to enhance the feasibility of treatment. In the medical services field, the IoT enables

A Framework for a Secure e-Health Care System 261

treatment and, in our investigation, patient data can be screened and analyzed from remote areas. One of the ways to realize this is by using wearable devices, with their ability to interface with the system. The fog-driven cloud can provide complex processing, in light of the fact that it has the capacity for enormous information stockpiling. For our situation, we chose to have a reinforcement of information stockpiling on the fog PC, alongside the capacity provided by putting it on the database. This capacity of the cloud can likewise be used to perform long-haul examination, including pattern recognition and AI. The gateway module is a product part that gets the information created on the sensors and sends it to the fog layer. Gateway segments have two archives: the exchanges archive, which stores all the exchanges to which the module is related, and the data descriptors, which include the whole arrangement of descriptors of the information types put away in a clinical record. A work conveyance task is performed by means of the smart door utilizing a scheduler.

12.3.2 PROCESSING OF HEALTH CARE RECORDS IN FOG NODES

The processing of health care records in fog nodes includes the data preprocessing of medical reports, effective analysis using machine learning algorithms, and, finally, securing the records using blockchain.

12.3.2.1 Preprocessing of Electronic Health Records (EHR)

Most current patient records undergo mining (for example grouping, forecast) and depend on a standard portrayal as organized records with numerical and additional implied qualities. The noteworthy advances in the prepreparation, design acknowledgment, and understanding of clinical pictures, messages, and signals can, and should, be combined with other information-mining and information-disclosure strategies. This incorporation is relied upon to significantly improve the aftereffects of patient records mining, specifically when applied to a thorough arrangement of information that incorporates patient history and status.

12.3.2.2 Extract Database Entries

Recognizing all of the fundamental EHR information components and questioning databases to recover all passages of intrigue, normally utilizing occasion identifiers; extraction delivers a lot of tables.

12.3.2.3 Define Features

Through a precise methodology, every clinical idea contained in EHR information is distinguished and characterized by highlights passed on for every idea, including its sort (numerical).

12.3.2.4 Process Data

The list of capabilities is controlled, to improve homogeneity and maintain a strategic distance from information scattering by moderating repetition (ideas addressed by various assignments) and granularity (a clinical idea is communicated with various degrees of detail), which are handled by joining various highlights that allude to the same clinical idea into a solitary element.

12.3.2.5 Assess Feature Values

The estimation of each clinical component (variable) for each dataset example is decided by questioning the extricated database sections, as indicated by the element types and recording instruments.

12.3.2.6 Integrate Data Elements

Linking frameworks are delivered from each EHR information component by the coordinating lines of each occasion utilizing identifiers, in this manner blending networks side to side; the component remembers coordinating each example with the relating line for the mark lattice (lines speaking to occurrences, and sections speaking to clear-cut or numerical value).

12.3.3 Effective Analysis Using Machine Learning

The general attributes of EMR information are to create, investigate, and test the information. Some of them are as follows:

a. High-dimensionality: EMR information regularly comprises countless clinical highlights, for example, various clinical tests, prescriptions, conclusions, and strategies.
b. Irregularity in time: The irregularity of EMR data is achieved by the way that every patient's clinical features are recorded when they visit the crisis facility. Thus, every patient's records, which can be addressed as a transient progression, have different stretches between each pair of events and also consistently have different lengths.
c. A large portion of missing data and data sparsity: EMR data regularly encounters a significant degree of missing data. This can result either from data combination issues (i.e., patients are simply checked for certain clinical considerations) or from documentation issues. Beside this, data sparsity is another general quality of EMR. Sparsity is unavoidable, since most patients visit the crisis center only multiple times and for the most part take only a small subset of clinical evaluations and medicines [41]. Information distribution centers store gigantic quantities of information created from different sources.

12.3.4 Securing of Health Records Using Blockchain

Using blockchain to secure health records includes initialization, the creation of a new block, verifying the block, and appending the next block. The next iteration again starts with the creation of a new block.

12.3.4.1 Initialization

At the instatement stage, every supplier $PI, I = 1,2,\ldots, n$, will be related with a noteworthiness S_I, dependent on the amount and estimation of the health records in its own database. We have the amount and the estimation of the records characterized as: assume PI has $m\,I$ bits of records in its database, the hugeness Si of supplier PI is characterized as follows:

$$Si = \sum vt \tag{12.1}$$

A Framework for a Secure e-Health Care System | 263

where vt alludes to the estimation of each record Rt for a client Ut. The translation of the estimation of a record may fluctuate for various partners. Here, we mean the estimation of a record dependent on two standards: completeness and redundancy [42].

12.3.4.2 Generating a New Block

All the suppliers that have refreshed records and that need to be included in producing the new block will communicate a tuple (provider's ID, job) in the system, where "job" is a two-bit string showing whether the supplier has any refreshed records and also whether the supplier has to make the new block. Every supplier in the system will initially gather this data and later communicate to the system what it has gathered. The association of the communicated assortments will be the last status acknowledged by all suppliers.

12.3.4.3 Verifying the New Block

Checking the new block has two techniques:

- Each included supplier P_1 checks its logs in the new block and sends its marked confirmation to P_j;
- At the point when P_j has gotten all the marked confirmations, it refreshes its s_j with a motivator c and advises all suppliers to attach the new block.

12.3.4.4 Appending the New Block

After fruitful confirmation of the new block, all suppliers broaden their own blockchain by annexing the new block, where the status and criticalness in each RRC (record relationship contract) are refreshed. The SCs (summary contract) will be refreshed with another timestamp of the last alteration.

12.4 IMPLEMENTATION AND RESULTS

The proposed model works in the layers: the sensor tier, the fog computing tier, and the cloud computing tier. This is quite common; but the kind of processing done in the fog computing layer, and the use of blockchain for securing preprocessed EHRs, are different from other architectures.

12.4.1 SENSORS LAYER

Sensors are the devices that assemble data from patients. They assemble both extraneous and natural qualities. Outward attributes are the temperature, area, etc. Inherent qualities are the pulse, blood glucose level, heartbeat, etc that are gathered by the patient's wearable sensors. The patient can likewise enter information into their advanced cell, and this information will at that point be made accessible for preparing. The activity of the sensors is to gather this information and send it to the fog processing layer.

12.4.1.1 Fog Computing Layer

This layer achieves information examination and total analysis of the information. The information and data gathered by the edge devices are broke down in this layer. This layer carries on as the server. The fog layer at that point disseminates the preparation work to different edge devices associated with the fog layer, and subsequently, gigantic measures of information are broken down.

A. Work Distribution: This errand is performed by means of a smart gateway, utilizing a scheduler.

B. Data Aggregation: When assignments are dispersed, the information must be amassed. Information collection consists of three primary parts: composition planning, copied location, and information combination. Diagram planning guarantees that the information is accumulated so that it bodes well, and there is a stream to the information. Copied identification guarantees that there will not be excess information. Information combination is the last phase of information collection, wherein the last data is assembled as one element [43]. At the fog nodes, different operations, such as data preprocessing, data analysis, and the securing of the data, are performed using blockchain technologies.

12.4.1.2 Cloud Computing Layer

Patients' secured health records are stored in the cloud layer and accessed by the authenticated doctors for further analysis and diagnosis.

12.4.1.3 Data Preprocessing

Data preprocessing is that movement where data is changed, or encoded, to convey it to such an express, that now the machine can parse it efficiently. In a manner of speaking, the features of the data would now have the option to be successfully deciphered by the estimation.

There are significant steps in data preprocessing, as follows:

1. Acquire the dataset: This dataset will include information assembled from different and dissimilar sources, which are then joined in a legitimate organization to shape a dataset.
2. Import all the necessary libraries
3. Import the dataset
4. Extract the independent variables
5. Recognize and deal with the missing attributes
6. Encode the unmitigated information
7. Feature scaling: Scaling is a technique to normalize the factors of a dataset inside a particular range.

12.4.1.4 Effective Analytics Using Machine Learning

We collected data from sensors, creating big data containing antenatal care. Even the diabetic datasets from patients were retrieved for analysis. Results were obtained using different machine learning algorithms. The final analysis was fed into a blockchain module, to secure the data. Premature birth is the closure of pregnancy

A Framework for a Secure e-Health Care System

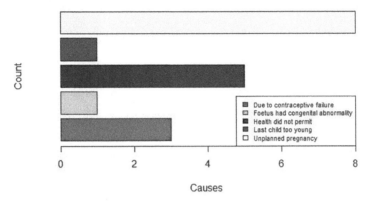

FIGURE 12.2 The Causes of Abortions.

due to evacuating an incipient organism or baby before it is due outside the uterus. At the point when intentional advances are taken to end a pregnancy, it is called an instigated fetus removal. Figure 12.2 is a pictorial representation of count versus causes of abortion.

Diabetic data collected over sensors from different hospitals are analyzed. The patients' test results and other records are collected. A linear regression is carried out, which is a procedure for exhibiting the association between a scalar variable y and at least one informative factor signified as X. The occurrence of one illustrative variable is called an essential straight relapse. Figure 12.3 shows the result of the linear regression.

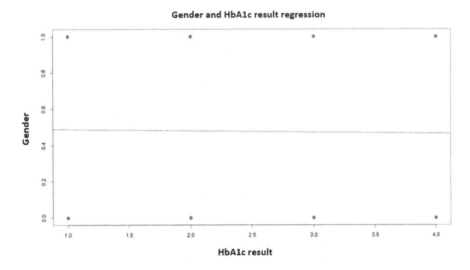

FIGURE 12.3 A Linear Regression Showing the Relation between Gender and an HbA1c Test.

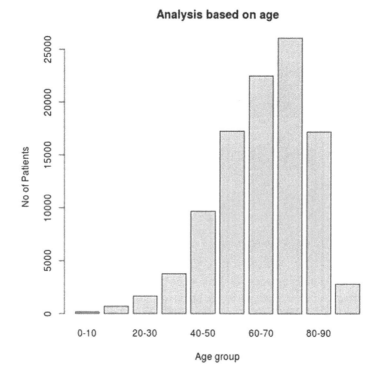

FIGURE 12.4 Distribution of Number of Patients Based on Age Group.

In insights, calculated relapse, or logit relapse, or logit model is a relapse model where the dependent variable (DV) is flat. The instance of paired factors—i.e., the place it can take are only two qualities, for example, pass or fail, win or lose, alive or dead, or solid or unhealthy. Cases with multiple classifications are alluded to as multinomial strategic relapse, or, if the numerous classifications are wished, as ordinal calculated relapse.

We also analyzed the data based on age groups, with an interval of 10 years. We initially made a hypothesis stating that people of the age group 60–80 are most affected by diabetes. We used a graphical method to analyze this hypothesis. The bar graph shown in Figure 12.4 was plotted having age groups on the x-axis and the number of patients on the y-axis. A pie chart representations of an analysis based on gender is shown in Figure 12.5, and an analysis based on the administration of insulin is presented in Figure 12.6.

12.4.1.5 Building a Blockchain and Cryptocurrency

This section discusses the detailed steps to build a blockchain and create cryptocurrency.

12.4.1.5.1 SHA-256 Algorithm

The Secure Hash Algorithm-256-bit is an algorithm that generates a cryptographic hash value for a text or file or transaction. This is used in blockchain to ensure the immutability of data. This algorithm creates a hash of length 256-bit that is unique

A Framework for a Secure e-Health Care System

Analysis Based on Adminstration of Insulin

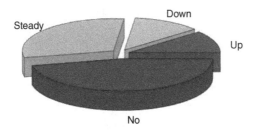

FIGURE 12.5 A 3D Pie Chart Showing Distribution Based on Gender.

Analysis Based on Gender

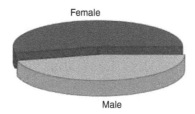

FIGURE 12.6 A 3D Pie Chart Representing Analysis of Administration of Insulin.

for a particular content. It is not an encryption; instead, it transforms input data to hash. It is one-way, i.e., information cannot be obtained from its hash. This algorithm generates the same hash for the same data and different hash for even a small change in the data. This hash can further be signed by encryption, using a private key. This digital signature is authenticated by other participants of the blockchain, by decryption of the signature, using their public key. The hash value generated is compared with original hash, a process that ensures that the data has not been altered.

12.4.1.5.2 P2P

A peer-to-peer network or system is a distributed and decentralized architecture. The workload is shared among peer nodes. Resources such as power, bandwidth, storage space, and access to other peers are equipped within each peer. A peer is capable of both transmitting and receiving data. This kind of network has gained importance, as it has the capability to share resources and also look up peers with the required amount of resources to perform a specific task.

12.4.1.5.3 Byzantine Fault Tolerance (BFT)

It is a challenging task to design a reliable system with a secure consensus, in the presence of faulty nodes and processes in the network. Since blockchain is a distributed system, it is crucial to find malicious nodes or nodes that fail to operate. A byzantine-fault-tolerant system is capable of operating efficiently in the presence of faulty nodes. The Byzantine Generals' Problem can be solved in several ways. One promising solution is the appropriate choice of consensus mechanism. A consensus mechanism should be able to deal with byzantine fault issues. BFT implies that two hubs can convey information securely over a network, understanding that they are showing same information.

12.4.1.5.4 Proof of Work (PoW) Consensus

A PoW is a computationally intensive task executed by miners to add new block into the blockchain. Miners receive rewards for generating hash, performing the task of adding valid blocks,the and linking hash of a new block to the previous block. All verified transactions are arranged in blocks. Every new block is validated, and consensus among the majority of nodes is obtained before a new block is added into the blockchain. The previous hash value for the genesis block is zero, and from then on, the hash value of block 1 becomes the previous hash value for the second block. This is indicated in Figure 12.7.

12.4.1.5.5 Creating Smart Contracts

A smart contract is code written to contain terms and conditions of exchange, agreed to by the sender and receiver of files in a blockchain network. A smart contract is stored in the blockchain with a unique address. It completely eradicates the need for a middleman and makes the exchange more trustworthy and irreversible. It is an agreement between the nodes, which is a legal framework that is automatically executed. It is implemented using solidity language in an Ethereum virtual machine.

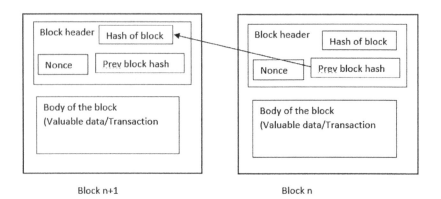

FIGURE 12.7 Details of Block in Blockchain.

A Framework for a Secure e-Health Care System 269

12.4.1.5.6 *Truffle Ganache*

This is a virtual Ethereum blockchain, which is used to set up addresses, smart contracts, and use functions to run and test the workings of a blockchain network. This can be used to control the functioning of a planned blockchain.

We create a cryptocurrency for interacting with the blockchain and postman on the three nodes we have in the network. We run each node on a different console, which is equivalent to having the computers connected to different servers. The steps performed are as follows:

- Creation of chain of blocks: Perform a GET request on node-1 using the get_ chain function. This will check to see if the genesis block was well created. Similarly, perform a get_chain request on node-2. Similar to node-1 and node-2, perform a get_chain request on node-3, to verify that the genesis block is created.
- Set the requests to POST and call the connect_node function to create a connection between the nodes. Node-2 is connected to node-1 and node-3 is calling the connect_node POST request. Similar to node-2, node-3 is also connected to the other nodes, and therefore, all the nodes are connected to one another in the network.
- Mine a block in the first node by calling mine_block GET request. This will mine a block in the blockchain. Once the block is mined, we perform get_chain request on the node to obtain all the details of that node. We can see that two blocks are there in the chain. One is the genesis block and the other is the block that was just mined.
- Perform replace_chain( ) and GET request on other nodes, to replace the chain with the longest chain in the network, thereby following consensus among all the nodes. The replace_chain request can then be verified by the get_chain request. We observe that the chain is now replaced with the longest chain in the network. The add_transaction POST request is a request that is made to add a particular transaction, i.e., EHR to the chain. Here, we specify the amount to be transacted between the sender and the receiver. Once the transaction is added, perform the get_chain request to view the blocks in the chain. This helps to verify if the transaction made is being added to the chain or not.

12.5 CONCLUSION

In this chapter, we propose a blockchain-based IoT framework using fog and cloud infrastructures for the secure and efficient storage and maintenance of electronic health records (EHRs). Our proposed framework is an amalgamation of record storage that is secure, along with the access rules for these records at the granular level. It creates a system that is easy for the users to understand and use. Fog processing encourages medical practitioners to make smart choices during a crisis, for time-critical health care applications. It additionally serves to secure sensitive information with diminished postponement, in contrast to the independent cloud-based application. Since some IoT-empowered frameworks, for example, health checking and utility assistance metering, manage sensitive information, the structure applies verification for information security and blockchain for information integrity. Blockchain innovation is set to change the manner in which human services are overseen, for eternity. Blockchain permits clinicians and patients to be sure that clinical records are exact and cutting-edge, as the appropriated record ensures an ethical, trustless form of the information that can be relied on. The utilization of blockchain in requesting and storing such a quantity of information will eventually prompt productivity upgrades for medical services staff, which implies that they will invest more energy in the individuals who matter: the patients. It would be exceptionally far-fetched to conceal all data completely and keep up an available and interoperable framework; yet, by utilizing smart agreements to isolate data, the structure proposed, despite everything, offers noteworthy protection conservation and information trustworthiness. To move information across less secure systems, encryption methods are applied. We additionally presume that the total encryption of records and maintaining usability can't be handled separately, and there is a trade-off that should be taken consideration by the system administrator. Contrasted with existing structures, the product parts of this system are lightweight, sufficiently responsive, and fit for tackling both edge and remote assets.

REFERENCES

1. Ozair, F. F., Jamshed, N., Sharma, A., and Aggarwal, P., 2015. Ethical issues in electronic health records: A general overview. *Perspectives in Clinical Research,* 6(2), pp. 73–76. doi:10.4103/2229-3485.153997.
2. Casino, F., Dasaklis, T.K., and Patsakis, C., 2019. A systematic literature review of blockchain-based applications: Current status, classification and open issues. *Telematics and Informatics*, 36, pp. 55–81. ISSN 0736-5853. doi:10.1016/j.tele.2018.11.006.
3. Zheng, Z., Xie, S., Dai, H., Chen, X., and Wang, H., 2017. *An Overview of Blockchain Technology: Architecture, Consensus, and Future Trends.* In *2017 IEEE International Congress on Big Data (BigData Congress)* (pp. 557–564). Honolulu, HI: IEEE. doi:10.1109/BigDataCongress.2017.85.
4. Sarwar, S., and Hector, M. G., 2019, April. Assessing blockchain consensus and security mechanisms against the 51% attack. *Applied Sciences*, 9(9), p. 1788.
5. Karim, S., Umar, R., and Rubina, L., 2018. *Conceptualizing blockchains: Characteristics & applications.* In *11th IADIS International Conference Information Systems 2018* (pp. 49–57). hhttps://arxiv.org/abs/1806.03693.
6. Deloitte, 2017. Key characteristics of the blockchain. https://www2.deloitte.com/content/dam/Deloitte/in/Documents/industries/in-convergence-blockchain-key-characteristics-noexp.pdf. Accessed on August 28, 2020.
7. Irfan, M., and Ahmad, N., 2018. *Internet of medical things: Architectural model, motivational factors and impediments.* In *2018 15th Learning and Technology Conference (L& T)* (pp. 6–13) Jeddah: IEEE. doi:10.1109/LT.2018.8368495.
8. Darwish, S., Nouretdinov, I., and Wolthusen, S. D., 2017. Towards composable threat assessment for medical IoT (MIoT). *Procedia Computer Science*, 113, pp. 627–632.
9. Hatzivasilis, G., Soultatos, O., Ioannidis, S., Verikoukis, C., Demetriou, G., and Tsatsoulis, C., 2019. *Review of security and privacy for the internet of medical things (IoMT).* In *2019 15th International Conference on Distributed Computing in Sensor Systems (DCOSS)* (pp. 457–464). Santorini Island, Greece: IEEE. doi:10.1109/DCOSS.2019.00091.

10. Alsubaei, F., Abuhussein, A., and Shiva, S., 2017. *Security and Privacy in the Internet of Medical Things: Taxonomy and Risk Assessment*. In *2017 IEEE 42nd Conference on Local Computer Networks Workshops (LCN Workshops)* (pp. 112–120). Singapore: IEEE. doi:10.1109/LCN.Workshops.2017.72.

11. Koosha, M. H., Mohammad, E., Tooska, D., and Ahmad, K., 2019. *2019 IEEE Canadian Conference of Electrical and Computer Engineering (CCECE)* (pp. 1–4). Canada: IEEE. doi:10.1109/CCECE.2019.8861857.

12. Banerjee, A., Mohanta, B. K., Panda, S. S., Jena, D., and Sobhanayak, S., 2020. *A secure IoT-fog enabled smart decision making system using machine learning for intensive care unit*. In *2020 International Conference on Artificial Intelligence and Signal Processing (AISP)* (pp. 1–6). Amaravati, India: IEEE. doi:10.1109/AISP48273.2020.9073062.

13. Patil, R. M., and Kulkarni, R., 2020. *Universal storage and analytical framework of health records using blockchain data from wearable data devices*. In *2020 2nd International Conference on Innovative Mechanisms for Industry Applications (ICIMIA)* (pp. 311–317). Bangalore, India: IEEE. doi:10.1109/ICIMIA48430.2020.9074909.

14. Jabbar, R., Fetais, N., Krichen, M., and Barkaoui, K., 2020. *Blockchain technology for health care: Enhancing shared electronic health record interoperability and integrity*. In *2020 IEEE International Conference on Informatics, IoT, and Enabling Technologies (ICIoT)* (pp. 310–317). Doha, Qatar: IEEE. doi:10.1109/ICIoT48696.2020.9089570.

15. Sabir, A., and Fetais, N., 2020. *A practical universal consortium blockchain paradigm for patient data portability on the cloud utilizing delegated identity management*. In *2020 IEEE International Conference on Informatics, IoT, and Enabling Technologies (ICIoT)* (pp. 484–489). Doha, Qatar: IEEE. doi:10.1109/ICIoT48696.2020.9089583.

16. Uddin, M. A., Stranieri, A., Gondal, I., and Balasubramanian, V., 2018. Continuous patient monitoring with a patient centric agent: A block architecture. *IEEE Access*, 6, pp. 32700–32726. doi:10.1109/access.2018.2846779.

17. Rahman, M. A., Hassanain, E., Rashid, M. M., Barnes, S. J., and Hossain, M. S., 2018. Spatial blockchain-based secure mass screening framework for children with dyslexia. *IEEE Access*, 6, pp. 61876–61885. doi:10.1109/access.2018.2875242.

18. Abdur, R., Shamim, H. M., George, L., Elham, H., Syed, R., Mohammed, A., and Mohsen, G., 2018. Blockchain-based mobile edge computing framework for secure therapy applications. *IEEE Access*, 6, pp. 72469–72478. doi:10.1109/access.2018.2881246.

19. Abou-Nassar, E. M., Iliyasu, A. M., El-Kafrawy, P. M., Song, O., Bashir, A. K., and El-Latif, A. A. A., 2020. DITrust chain: Towards blockchain-based trust models for sustainable health care IoT systems. *IEEE Access*, 8, pp. 111223–111238. doi:10.1109/access.2020.2999468.

20. Attia, O., Khoufi, I., Laouiti, A., and Adjih, C., 2019. *An IoT-blockchain architecture based on Hyperledger Framework for health care monitoring application*. In *2019 10th IFIP International Conference on New Technologies, Mobility and Security (NTMS)* (pp. 1–5). Canary Islands, Spain: IEEE. doi:10.1109/NTMS.2019.8763849.

21. Azbeg, K., Ouchetto, O., Andaloussi, S. J., Fetjah, L., and Sekkaki, A., 2018. *Blockchain and IoT for security and privacy: A platform for diabetes self-management*. In *2018 4th International Conference on Cloud Computing Technologies and Applications (Cloudtech)* (pp. 1–5). Brussels, Belgium: IEEE. doi:10.1109/CloudTech.2018.8713343.

22. Chakraborty, S., Aich, S., and Kim, H., 2019. *A secure health care system design framework using blockchain technology*. In *2019 21st International Conference on Advanced Communication Technology (ICACT)* (pp. 260–264). PyeongChang Kwangwoon_Do, Korea (South): IEEE. doi:10.23919/ICACT.2019.8701983.

23. Dwivedi, A. D., Malina, L., Dzurenda, P., and Srivastava, G., 2019. *Optimized block-chain model for Internet of Things based health care applications*. In *2019 42nd International Conference on Telecommunications and Signal Processing (TSP)* (pp. 135–139). Budapest, Hungary: IEEE. doi:10.1109/TSP.2019.8769060.

24. Islam, A., and Shin, S. Y., 2019. *BHMUS: Blockchain based secure outdoor health monitoring scheme using UAV in smart city*. In *2019 7th International Conference on Information and Communication Technology (ICoICT)* (pp. 1–6). Kuala Lumpur, Malaysiae: IEEE. doi:10.1109/ICoICT.2019.8835373.

25. Pham, H. L., Tran, T. H., and Nakashima, Y., 2018. *A secure remote health care system for hospital using blockchain smart contract*. In *2018 IEEE Globecom Workshops (GC Wkshps)* (pp. 1–6). Abu Dhabi, UAE: IEEE. doi:10.1109/GLOCOMW.2018.8644164.

26. Ray, P. P., Kumar, N., and Dash, D., 2020, March. BLWN: Blockchain-based light-weight simplified payment verification in IoT-assisted e-health care. *IEEE Systems Journal*, pp. 1–12. doi:10.1109/JSYST.2020.2968614.

27. Sanka, A. I., and Cheung, R. C. C., 2018. *Efficient high performance FPGA based NoSQL caching system for blockchain scalability and throughput improvement*. In *2018 26th International Conference on Systems Engineering (ICSEng)* (pp. 1–8). Sydney, Australia: IEEE. doi:10.1109/ICSENG.2018.8638204.

28. Saravanan, M., Shubha, R., Marks, A. M., and Iyer, V., 2017. *SMEAD: A secured mobile enabled assisting device for diabetics monitoring*. In *2017 IEEE International Conference on Advanced Networks and Telecommunications Systems (ANTS)* (pp. 1–6). Bhubaneswar: IEEE. doi:10.1109/ANTS.2017.8384099.

29. Shobana, G., and Suguna, M., 2019. *Block chain technology towards identity management in health care application*. In *2019 Third International conference on I-SMAC (IoT in Social, Mobile, Analytics and Cloud) (I-SMAC)* (pp. 531–535). Palladam, India: IEEE. doi:10.1109/I-SMAC47947.2019.9032472.

30. Talukder, A. K., Chaitanya, M., Arnold, D., and Sakurai, K., 2018. *Proof of disease: A blockchain consensus protocol for accurate medical decisions and reducing the disease burden*. In *2018 IEEE SmartWorld, Ubiquitous Intelligence & Computing, Advanced & Trusted Computing, Scalable Computing & Communications, Cloud & Big Data Computing, Internet of People and Smart City Innovation (SmartWorld/SCALCOM/UIC/ATC/CBDCom/IOP/SCI)* (pp. 257–262). Guangzhou, China: IEEE. doi:10.1109/SmartWorld.2018.00079.

31. Uddin, M. A., Stranieri, A., Gondal, I., and Balasubramanian, V., 2019. *A decentralized patient agent controlled blockchain for remote patient monitoring*. In *2019 International Conference on Wireless and Mobile Computing, Networking and Communications (WiMob)* (pp. 1–8). Barcelona, Spain: IEEE. doi:10.1109/WiMOB.2019.8923209.

32. Xu, J., Xue, K., Li, S., Tian, H., Hong, J., Hong, P., and Yu, N., 2019. Healthchain: A blockchain-based privacy preserving scheme for large-scale health data. *IEEE Internet of Things Journal*, 6(5), pp. 8770–8781. doi:10.1109/JIOT.2019.2923525.

33. Bhuiyan, M. D., Aliuz, Z., Tian, W., Guojun, W., Hai, T., and Mohammad, H., 2018. *Blockchain and big data to transform the health care*. In *ICDPA 2018: Proceedings of the International Conference on Data Processing and Applications* (pp. 62–68). Guangzhou, China: ACM Digital Library. doi:10.1145/3224207.3224220.

34. Fast Healthcare Interoperability Resources (FHIR), 2019. FHIR overview. https://www.hl7.org/fhir/overview.html. Accessed on August 28, 2020.

35. Sudeep, T., Karan, P., Richard, E., 2020. Blockchain-based electronic health care record system for health care 4.0 applications. *Journal of Information Security and Applications*, 50, pp. 102407. doi:10.1016/j.jisa.2019.102407.

A Framework for a Secure e-Health Care System 273

36. Chen, H. S., Jarrell, J. Y., Carpenter, K. A., Cohen, D. S., and Huang, X., 2019. Blockchain in health care: A patient-centered model. *Biomedical Journal of Scientific & Technical Research*, 20(3), pp. 15017–15022. Epub 2019 Aug 8. PMID: 31565696; PMCID: PMC6764776.

37. Chamola, V., Hassija, V., Gupta, V., and Guizani, M., 2020. A comprehensive review of the COVID-19 pandemic and the role of IoT, drones, AI, Blockchain, and 5G in managing its impact. *IEEE Access*, 8, pp. 90225–90265. doi:10.1109/access.2020.2992341.

38. Wright, T., 2020, April. Blockchain app used to track COVID-19 cases in Latin America. https://cointelegraph.com/news/blockchain-app-used-to-track-covid-19-ca%ses-in-latin-america. Accessed on August 27, 2020.

39. IBM, 2020, March. Mipasa project and IBM blockchain team on open data platform to support Covid-19 response. https://www.ibm.com/blogs/blockchain/2020/03/mipasa-project-and-ibm-blo%ckchain-team-on-open-data-platform-to-support-covid-19-response/. Accessed on August 27, 2020.

40. I... 2016. *Blockchain: A new model for health information exchanges.*

41. Guo, R., Fujiwara, T., Li, Y., Lima, K. M., Sen, S., Tran, N. K., Ma, K.-L., 2020. Comparative visual analytics for assessing medical records with sequence embedding. *Visual Informatics*, 4(2), pp. 72–85. ISSN 2468-502X. doi:10.1016/j.visinf.2020.04.001. http://www.sciencedirect.com/science/article/pii/S2468502X20300139.

42. Yang, G., and Li, C., 2018. *A design of blockchain-based architecture for the security of electronic health record (EHR) systems.* In *2018 IEEE International Conference on Cloud Computing Technology and Science (CloudCom)* (pp. 261–265). Nicosia: IEEE. doi:10.1109/CloudCom2018.2018.00058.

43. Paul, A., Pinjari, H., Hong, W.-H., Cheol Seo, H., Rho, S., 2018. Fog computing-based IoT for health monitoring system. *Journal of Sensors*, 2018(2), pp. 1–7. doi:10.1155/2018/1386470.

13 Blockchain in EHR
A Comprehensive Review and Implementation Using Hyperledger Fabrics

Ravinder Kumar
Shri Vishwakarma Skill University, India

CONTENTS

13.1 Introduction 276
    13.1.1 Blockchain Technology Stack and Protocol 278
        13.1.1.1 The User Experience (UX)...............................278
        13.1.1.2 Application Layer..278
        13.1.1.3 The Blockchain Protocol...................................279
        13.1.1.4 Internet 279
    13.1.1 Smart Contracts 280
    13.1.2 Blockchain Protocol Projects....................................280
    13.1.3 Blockchain Ecosystem...280
    13.1.4 Motivation and Contribution.....................................281
    13.1.5 Organization of Chapter..281
13.2 Review of Privacy Preservation in Electronic Health Records....................281
13.3 How Blockchain Solutions Can Address Health Care Challenges..............284
13.4 Blockchain Implementation Frameworks..285
    13.4.1 Blockchain Development Platform and APIs 285
    13.4.2 Ethereum Platform..286
    13.4.3 HyperLedger Platform...288
13.5 Challenges of Blockchain..289
    13.5.1 Technological Challenges 289
    13.5.2 Business Model Challenges 290
    13.5.3 Scandals and Public Perception 290
    13.5.4 Government Regulation..290
    13.5.5 Primary Challenges for Personal Records 290
13.6 Conclusion..291
References..291

13.1 INTRODUCTION

Because of its built-in security features, blockchain is astonishingly popular nowadays. The following obvious questions arise for users and designers of centralized information systems:

- What do we stand to gain by using blockchain?
- How does it work?
- What issues does it resolve?
- How do we use blockchain?

The term "blockchain" signifies the chain of numerous blocks, as shown in Figure 13.1, which consists of transaction information. This blockchain was originally proposed in 1991 by a group of researchers, who wanted to create tamper-proof digitally signed documents. However, Bitcoin, the digital currency created in 2009 by Satoshi Nakamoto, has been exchanged over blockchain [1]. Blockchain is also referred to as a distributed ledger that is available for everyone to see. One interesting characteristic of blockchain is that once data is written into it, it cannot be altered. Changing data stored in a blockchain is very, very difficult, because to do so, one has to change the all block hashes. This can be seen in Figure 13.1, which explains the working of blockchain. Figure 13.1 consists of a block that contains stored data, the hash code of this block, and the previous block's hash code of the chain.

The type of the blockchain determines the type of the data recorded inside it. For an example, a Bitcoin blockchain is used to store the details of data about the transactions, i.e., the information about the sender, the details of the receivers, and the number of Bitcoins being transacted. The hash of each block is unique, just like a fingerprint, and can be used to identify the block and the contents of that block. The block's hash code is created at the time of creation of the block. Whenever someone wants to make changes inside the block, the hash of the block needs to be changed, and thus the changes in the block can be easily detected. Once a change in the hash of a block is detected, the block no longer remains part of the chain. The previous hash of the block determines the sequence of blocks in the chain. The chain of the block, known as blockchain, is shown in Figure 13.2.

The use of hashes in blockchain adds security. For example, Figure 13.2 has four blocks, wherein each block of the chain contains a hash code and the hash code of the

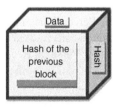

FIGURE 13.1 Block of a Blockchain. [This figure shows the block of a blockchain. Each block of the blockchain consists of three types of information: the hash of the block, the hash of the previous block, and the transaction data contained inside the block.]

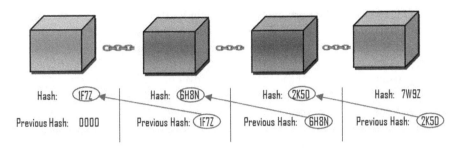

FIGURE 13.2 Chain of Blocks in a Blockchain. [This figure shows the chain of blocks in a blockchain. Each block of the blockchain consists of three types of information: the hash of the block, the hash of the previous block, and the transaction data contained inside the block. The block's hash code is created when the block is created. Whenever someone wants to make changes inside the block, the hash of the block needs to be changed, and thus, the change in the block can be easily detected. Once a change in the hash of a block is detected, the block no longer remains part of the chain. The previous hash of the block determines the sequence of blocks in the chain.]

preceding block. As we can see, the hash of previous block value of block number 4 contains the hash value of block number 3, and so on. The block whose previous hash value is "0000" is the first block of the chain, and this block is called the "genesis" block.

To clarify this point, let us assume that the second block has been tampered with and hence, its hash was changed. Therefore, the previous hash value contained by the third block is no longer valid. The hash has to be recalculated, to form a valid blockchain again. On average, the computation of each hash will take, say, 10 minutes. Therefore, on tampering, the hash of each block has to be recalculated, and it takes lots of time to make the chain valid again. This is referred to hashing or proof of work, and it makes blockchain secure from tampering.

Blockchains are evolving at a rapid rate, and one recent development of the blockchain is the smart contact [2]. These contracts are short pieces of code, which are used to exchange Bitcoin coins between blockchains, as and when some specific condition is met. The new uses of blockchain technology have recently attracted the interest of many people. It has quickly been realized that this technology could be used for many types of transactions, such as preserving medical histories, providing a digital notary, collecting taxes, etc.

The Hyperledger is defined as a set of tools and frameworks for the implementation of open source blockchain. Developed by the Linux Foundation in 2015, it is supported by IBM, Intel, and SAP Ariba [3]. It provides support for the development of blockchain-based distributed ledgers using collaborative development. For the convenience of the organization blockchain technologies are divided into three main categories: Blockchain V1.0, Blockchain V2.0, Blockchain V3.0, and Blockchain V4.0 [4]. Blockchain V1.0 is related to cryptocurrencies. The cryptocurrencies are used in the application associated with digital payment systems and the cash transfer of currency. Blockchain V2.0 is the process of coding a new application that has to run on the new set of protocols called the Blockchain V2.0 protocol. However, the relationship between Blockchain V1.0 and Blockchain V2.0 can be demonstrated in terms of the layers in their protocols. Blockchain V1.0 can be viewed as the TCP/IP Transport layer, whereas the Blockchain V2.0 can be seen as FTP, HTTP, and SMTP. Hence, Blockchain V2.0 applications would be similar to file sharing, social networks, and browser services [5,6]. Blockchain V3.0 offers safety and security for the user's personal identification and better control over the sharing of personal information over the internet [7]. In the Industry 4.0'-era, Blockchain V4.0 proposes a hierarchical framework that comprises four concrete layers designed to consolidate interorganizational value networks, manufacturing factories, the engineering value chain, smart factories, smart industry, etc. [8]. The following subsection discusses the underlying technologies of blockchain based distributed ledger development.

13.1.1 Blockchain Technology Stack and Protocol

Bitcoin and blockchain can become confused, because Bitcoin denotes the digital currency that is the underlying technology platform of blockchain, which is responsible to transfer the assets over blockchain technology. The technology stack over which blockchain operates includes the following components, as shown in Figure 13.3. A detailed discussion of these components is provided in the following subsection [9].

13.1.1.1 The User Experience (UX)

User experience is the highest layer of the blockchain technology stack. It integrates all of the underlying technological developments to support the user application development for day-to-day activities. Specialized Dapps (Decentralized applications) are being continuously developed, in order to access specialized Dapps browsers on the user devices. Blockchain applications are continuously growing, as users are becoming conversant with these applications.

13.1.1.2 Application Layer

Lots of blockchain applications have been developed in recent years. However, Bitcoin, is the intial and most important application, devised using indigenous blockchain protocol, except for the recent changes that have been introduced. The Bitcoin network is mainly used in the centralized financial sector, banks, resolution agencies, and other financial intermediaries. The decentralized autonomous organization (DAO) was developed with to decentralize the transfer of value between

| Laptops, Smartphones, Tablets, Desktops |
| User Experience (UX) |
| Cryptocurrency: Bitcoin, Litecoin |
| Smart Contracts |
| Application Layers |
| Distributed Ledger, P2P Network |
| Blockchain Protocol |
| Internet Infrastructure TCP/IP |

FIGURE 13.3 Layers of Blockchain Stack and Protocols. [This figure shows the layers of a blockchain stack, and protocols such as the user experience layer, the cryptocurrency layer, the distributed ledger, and the internet infrastructure (TCP/IP.]

agencies. In recent years, the development of more complex blockchain-based applications has been needed, to support transaction and data storage and computation.

Ethereum was developed to create an alternate protocol to Bitcoin, to build a decentralized public application by executing smart contracts. Smart contract applications run as programs on a peer-to peer-network, which allows for improvements in security, time, and scaling. Ethereum tokens have been moved up in the technological stack, along with distributed application (dApp) or DAO tokens. This provides a framework to support the development of complex applications using smart contacts, which are critical to the development of new applications based on blockchain.

13.1.1.3 The Blockchain Protocol

Blockchain is implemented on a peer-to-peer network of computing devices (also called "nodes"), which are connected through the internet (TCP/IP) and operate on blockchain protocols. Each node consists of identical copies of the distributed ledger of completed transactions. A cryptographic consensus algorithm is used to provide security for transactions completed on each node of the P2P. The blockchain protocol is based on the trusted public and an open, shared ledger of transactions, which is not under the control of any single entity in the network. Once recorded on the blockchain, these transactions cannot be changed. These protocols also take care of the financial benefits to the authors/owners and the miners (operators) of devices, in order to retain the cryptographically transactions registered into that distributed ledger. These financial benefits are known as tokens.

New transactions are validated using these tokens, which are generated for that transaction well before these transactions have been appended to the blockchain ledger.

13.1.1.4 Internet

This is the lowest and most basic technological layer. The internet is a network infrastructure that allows all the devices to talk to other devices using the most popular internetwork protocol, TCP/IP. The TCI/IP became popular after the invention in 1989 of the World Wide Web (WWW). The first version of the WWW is referred as Web version 1 (Web1), which provided the facility of the internet to all of its users. Web 1 has provided the facility of web resources such as multimedia, documents, images, services, etc. Each service on the internet has been assigned a URL (uniform resource locater), to access it.

In the early 2000s, technology advanced with the upgradate to the initial version of WWW (i.e., Web 1), which this was referred as Web version 2 (Web2). That innovation gave rise to services such as ecommerce, transactions over the internet, and social networking. There was further growth of the WWW, known as Web version 3.0 (Web3): the Internet of Value. Initially, point-to-point transactions were completed with the help of intermediaries; but with the invention of blockchain protocol, P2P interactions were performed without the help of intermediaries. The absolute first and most significant blockchain was Bitcoin, which allows value (as digital currency) to be moved between people without requiring any of the budgetary delegates of banks. These are situations where blockchain innovation could likewise be utilized for the advancement of web-based life applications without requiring

280 Blockchain Technology for Data Privacy Management

Facebook and Twitter, the improvement of ride-sharing applications without requiring Uber and Grab, convenience sharing without Airbnb, and interactive media sharing without YouTube. This will prompt a genuinely decentralized web.

13.1.1 SMART CONTRACTS

Blockchain might also be used in other applications being developed beyond sell/buy currency transactions, and could also carry more focused instructions that can be embedded into a smart contract blockchain. In other words, a smart contract is a methodology using Bitcoin to form agreements between people by using blockchain technology.

There are basically three elements in a smart contract, namely autonomy, decentralization, and self-sufficiency. In this context, autonomy means that once a smart contract has been launched and run, no further contract with its owner is required. The smart contract in itself is sufficient to raise the funding to provide the services for issuing equity and speedy than the needed resources like storage or processing capacity.

Smart contracts are likewise decentralized, such that they don't remain alive on a solitary brought together worker; they are circulated and self-executable over the shared system [2]. Agreements don't make anything conceivable which was already conceivable, rather they permit general issues to be settled, which limits the requirement for trust. Negligible trust frequently makes situations more advantageous by removing human judgment from the conditions, which permits total mechanization. One case of smart contracts is the setting up of programmed installments for wagering. A code or smart agreement can be composed that delivers an installment when a specific estimation of a specific trade of products is set off, or when something unfolds in reality.

13.1.2 BLOCKCHAIN PROTOCOL PROJECTS

There are many projects being designed and developed that loosely use blockchain technology. Some common projects are discussed here. Ripple is one such category. It is a gateway, a payment system which exchange, and a remittance network, and it contains a smart contract system called Codius. Another type of project counterparty which is an overlay protocol for exchange system and currency issuance. Ethereum is a well-known and widely used blockchain protocol project. It is a general-purpose complete Turing cryptocurrency platform. NXT is an alternative coin mined with a proof-of-state consensus model. Many more such projects are available in the market, such as Open Transactions, BitShare, Open Assets and colored coins.

13.1.3 BLOCKCHAIN ECOSYSTEM

Blockchain is a decentralized record that is utilized to store transactions in an enormous calculating framework that might also include other useful functions to help with correspondence, stockpiling, document preparation, and filing. In the event of lack of capacity, the greatest need is for the secure, decentralized, off-chain stockpiling of documents, for example, electronic health records, or the genome, or

Blockchain in EHR

straightforward Microsoft Word reports. The blockchain transaction that enlists the advantages of enhanced security, greater transparency, improved traceability, increased efficiency and speed, reduced costs, etc., can incorporate a pointer and access control strategy and benefits from off-chain record storage. In the case of file processing, the InterPlanetary File System (IPFS) is used. The IPFS is a worldwide system of shared frameworks. It is a framework that is utilized for requesting and serving a record from various locations where the document may exist. Blockchain record exchange isn't just made for safeguarding, in addition it requires a method for recovering and controlling recently recorded blockchain resources of past dates, which may be had with specific algorithms.

13.1.4 MOTIVATION AND CONTRIBUTION

In today's ICT Era, user/customers expect to access their health information (and any other useful information, for example, related to news, banking, test results) instantly. Therefore, the information provider must make the information available as early as possible, by adopting the use of technology. Based on the literature, the health care industry is lagging behind, compared to other industries, such as food, retail, banks, insurance, etc. Also, the sharing of health data among different organizations is very difficult, due to variations in formats and standards. Therefore, centralized storage is needed, to ensure the integrity, privacy, and security of data. The contribution of the proposed work is to:

1. Propose an implementation of EHR on the Hyperledger platform.
2. Make available the full health history of a patient at any hospital.
3. Design methods to deal with the security and privacy of data while it is being transmitted and stored.
4. Store the data centrally, to ensure its integrity.
5. Design a system or process on the cloud, to achieve the desired goals.
6. Propose the analysis of the implementation, and propose the validation of and suggestions for blockhain-based EHR implementation.

13.1.5 ORGANIZATION OF CHAPTER

The chapter is organized as follows: Section 13.2 presents and discusses a review of privacy preservation in the EHR. Section 13.3 presents how blockchain solutions can address health care challenges. Section 13.4 presents the application of blockchain beyond digital currency (Bitcoin), with special mention of EHRs. Section 13.5 presents the limitations and challenges of blockchain. Finally, the conclusion is presented in Section 13.6.

13.2 REVIEW OF PRIVACY PRESERVATION IN ELECTRONIC HEALTH RECORDS

Various researchers have proposed their work on protecting the privacy of EHRs. Some trends are discussed in this section.

So as to repeat the patients' medical services information, the authors in [10] have proposed a procedure utilizing blockchain innovation. As blockchain is computationally costly and requires extra computational force, the authors have attempted to analyze these issues while utilizing blockchain innovation for IoT devices. In this work, the authors have proposed a novel structure of an adjusted blockchain model, which is reasonable for IoT devices that rely upon efficient processing. They give protection and security calculations by presenting cryptography calculations.

Another calculation has been proposed, utilizing de-recognized exploration datasets of patients' information [11]. This technique depends on Bayesian displaying of binarized conclusion codes, and it gives a back the likelihood of coordinating for every patient pair. They have done a reenactment study and contrasted it and two genuinely utilized cases connecting patients' EHRs from an enormous tertiary care network. The author has suggested that it is possible to link de-identified datasets without any personal health identification using only diagnosis codes, provided sufficient information is stored between data sources.

In Yang et al. [12], a work privacy system for smart IoT-based health care big data is introduced. In this work, data generated from the system is encrypted by the algorithms and sent to a big data storage system [13]. These encrypted files are shared with the other health care system. This secure system is compared with existing methods, and the result demonstrates the efficiency of the system.

Dubovitskaya et al. [13] proposed a patient's health care data management system by using blockchain technology in order to maintain the privacy of user data [14]. They used cryptographic functions in order to encrypt the user's data. The data was analyzed also used the cost effectiveness of the smart contracts.

In order to meet the security needs in system provenance, information was used by the authors in [14]. They have proposed a domain-independent provenance model. This model was based on symmetric web technology and an open provenance model. The aim of the proposed model is to provide the privacy of data, by integrating the provenance and security concepts.

The authors in [15] present a new type of attack on 1:M records that one user has many records, with multiple sensitive attributes. For these attacks, formal modeling and analysis have been performed in order to provide a privacy preservation technique, which has been proposed for 1:M with MSAs' data.

The work done in [16] describes the legal aspect of privacy in health care data. The two major threats discussed are: (1) a lack of understanding of the different policies and regulations that are in place, and how that affects the handling of patients' data, and, (2) the threat of data being hacked.

In [17], threats and challenges posed by IoT devices and a personal area network (PAN) on a patient's data base are discussed. After reviewing the literature, they describe various challenges of the health care sector that are related to IoT devices and networks.

The authors in [18] discuss the method of formulating privacy issues. They also discuss the importance of equity, consent, and patient governance in data collection, limitations on data uses, and issues related to the handling of data breaches.

In [19], the significance and utility of maintaining data security are identified with the sharing of individual health information for various protections and k-obscurity. From the findings of this research, a health proposal framework was recommended.

The work done in [20] depends on information association in electronic health services. Information association manages the expelling or covering up of individual identifier information, similar to an ID, SSN, name, and so forth, from the health datasets and which are not to be recognized by the beneficiary of the information. Various procedures and models for association are utilized in this work.

In another paper, security protecting a CP_ABKS framework with concealed access strategy in a shared multi-proprietor setting is introduced [21]. They also describe an improved methodology for tracing malicious users.

Data is protected in the health care system by a shared key system, in the work proposed by the authors in [22]. Here, data from the treatment process are encrypted and stored in a blockchain system, with the help of a shared key. This work meets the security requirements for the integrity, privacy, and usability of medical data.

The research done in [23] describe the application of the IoT in the health care field. They identify the challenges, vulnerabilities, and risk factors resulting from the heterogeneity and diversity of communication in the IoT. They also have performed a privacy risk assessment.

Another work [24] has exhibited a security-protecting framework that is dependent on the cloud for constant health observations, by identifying changes in different crucial health indications of smart network individuals. Indispensable signs utilized in this framework are acquired from the information received from IoT-empowered wearable devices. This work predominantly centers on the structure and improvement of a prescient model utilized for the smart network. The actualized model exhibits the productivity and precision of the model that will be utilized for building a smart network.

Another work [25] has proposed a secure network that offers clinical choice and emotional support. It uses privacy preserving random forest algorithms to diagnose symptoms without disclosing the vital information of patients. The simulation result shows that the proposed method performed better, as compared with other algorithms.

Another survey has been done [26] in order to show existing research work in the field of blockchain technology that was applied in electronic health-related fields. The study exposes possible research trends on which this technology may be focused.

A new hybrid reasoning-based methodology for predicting diseases has been introduced [27]. The authors use the combination of IC–nearest neighbor, fuzzy set theory, and case-based reasoning, to enhance the prediction's results. This proposed model is evaluated with the spatial evaluated metric; results obtained from the experimental study reveal the improved performance of privacy-aware disease prediction support systems.

A method for rank-swapping capable of protecting nominal data from the prospect of semantic are presented in this work [28]. For real-time clinical records, empirical experimentation has been carried out using standard medical ontology. The result shows that this method is able to preserve the semantic features of data significantly.

In another work [29], blockchain technology is used in health-related applications. The main reason to use blockchain technology is to provide a strong structure for storing health-related data. This also helps in the analysis of data, without affecting the privacy of sensitive health-related data.

A lightweight strong key management model is proposed by the authors [30]. Their system requires a few computations of keys and provides a null rekeying mechanism, for backward and forward services. This provides a secure and privacy-preserving mechanism for electronic health systems.

In order to protect privacy of users, a new model is proposed [31] with t-safe (i, k) diversity, based on the generalization and segmentation of recourse. This model ensures that for each record, the signature maintains consistency, in order to prevent interaction in all receivers.

An efficient and privacy-preserving priority classification model for an e-health care system has been proposed [32]. The authors have also designed an algorithm for a noninteractive privacy-preservation priority classification. Detailed experimentation and analysis show that the proposed model achieved the priority classification and packets relay, without affecting the privacy of the user's electronic health system.

An end-to-end private deep-learning framework has been proposed for privacy preservation in e-health records [33]. The implementation results demonstrated that the proposed model could maintain high performance and also provide robust privacy guarantees against the leaking of information due to outside attacks and data transmission.

Another work [34] proposes a balanced p+ sensitivity, K-anonymity model for preserving the privacy of electronic health records for user health data. The proposed model is then analyzed using high-level Petri Nets, and the results are verified using SMT-lab and Z3 solutions.

A cloud EHR model has been proposed [35]. It uses an attribute-based access control mechanism using extensive access control mark-up language. This model is mainly focused on security. It performs partial encryption and uses electronic signatures when a user's document is sent to document requesters.

Other research [36] demonstrates that by using extensible access control markup languages, along with semantic capabilities, the risk of disclosing data can be controlled.

For HER, new privacy preserving access control mechanisms have been proposed [12]. They use an attribute-based sign crypts method to sign crypt data that was based on the access policy. This method protects the owner's privacy information in EHR.

In other research [37], a famous privacy presentation framework is analyzed. In this study, the framework's loopholes are identified. The authors also proposed the enhancement of the model, after careful analysis.

A novel framework called Model chain [38] to adopt blockchain technology for protecting the privacy of patients in an electronic health care system is proposed. In this approach, each participating site contributes to the parameters of the model, without revealing any patient's sensitive data.

13.3 HOW BLOCKCHAIN SOLUTIONS CAN ADDRESS HEALTH CARE CHALLENGES

Because of its inherent transparency and security features, blockchain technologycan reimagine every aspect of health care services, including drug traceability, the accessibility of medical services, insurance processing, payment and settlement, and medical financial services.

Blockchain in EHR

TABLE 13.1
Blockchain Application in Health Care Systems

Blockchain Application in Medical Devices Supply Chain	• Unique identifiers for clinical devices or resources on the blockchain. • Autonomous observing and preventive upkeep of clinical devices. • Secure tracking and management of medeical devices and medical assets with unique identifiers on the blockchain. • Encryption and permanent storage of device-generated health data with access control and smart contact highlights.
Blockchain for Health Care Insurance	• Automating claims, the executives to dispose of non-value-producing procedures and mediators. • Automating endorsing, strategy protection, and other BIR exercises. • Improving petitioner and recipient KYC process.
Blockchain for Electronic Health Care Records	• Longitudinal Electronic Health Records (EHRs). • Improving health data trade across different human services suppliers. • Ensuring information consistency and security.
Blockchain for Drug Supply Chain	• Automating the serialization and geo-labeling process across value chain exercises, for example, creation, improvement, and testing by manufacturing offices. • Improving track-and-follow frameworks and guarantee consistence with US FDA, EU FMD. • Checking drug duplication.
Blockchain for Clinical Trials	• Clinical exploration and information sharing. • Managing IP and RandD resource exchanges on the blockchain.

Blockchain is a dispersed framework that records shared exchanges, tracks progressions across systems, and stores and trades information for cryptographies. Blockchain innovation can possibly change health insurance, putting the patient at the center of the human services biological system and upgrading the security, protection, honesty, and interoperability of health information. This innovation will offer another, more successful and secure model for health data trades (HIE) as shown in Table 13.1 [39].

13.4 BLOCKCHAIN IMPLEMENTATION FRAMEWORKS

13.4.1 Blockchain Development Platform and APIs

As a part of the Blockchain V2.0 protocol projects, several blockchain developer platform and project companies offer various tools and APIs, to facilitate blockchain application development. Blockchain offers a variety of APIs specific to the type of application, which are discussed in the following sections and shown in Figure 13.4.

Receive Payments API: Version 2 of this interface has been accessible since January 1, 2016. This is the least complex path for an organization or business to start to acknowledge mechanized installments in Bitcoin. The API depends on HTTP GET demands and is accountable for making a solitary location for every client and for each receipt given in each Bitcoin exchange, a fundamental condition for good praxis.

Blockchain Wallet API: Since January 1, 2016, to utilize this API, it has been important to introduce a nearby worker to deal with the virtual wallet. The specialized strategy utilized depends on HTTP POST or GET calls. The procedure for

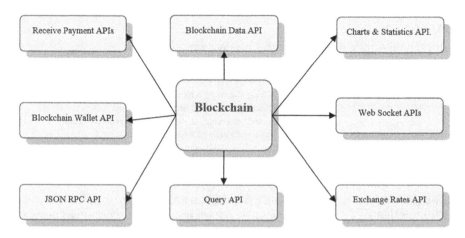

FIGURE 13.4 Blockchain APIs. [This figure shows the Blockchain Development Platform and APIs, such as blockchain Data API, Charts and Statistics API, Web Socket API[s], Exchange Rates API, Query API, JSON RPC API, blockchain Wallet API, and Receive Payment APIs.]

making a virtual wallet is known as create_wallet from this url: http://localhost:3000/programming interface/v2/make. Every wallet is related to a secret key with a base length of 10 characters, a validation code for the API, a private code for every client, the organizer where the wallet is made, and an email address.

JSON RPC API: Since March 2016, the general suggestion for Bitcoin clients is to utilize the new Blockchain Wallet API, despite the fact that the interface dependent on RPC calls keeps on being perfect with the old Bitcoin RPC, to communicate with virtual monetary forms. It may very well be introduced and utilized from libraries in various programming dialects: linguistic structure, for example, Python, Ruby, PHP, Node.js, and .NET.

Blockchain Data API: This can be utilized to counsel the information on the exchanges and activities inside the blockchain in JSON design.

Inquiry API: Plain content API for questioning blockchain information.

WebSocket API: This application programming interface gives software engineers access to continuous notices on exchanges and squares.

Trade Rates API: This deals with the data on Bitcoin trade rates and universal monetary forms continuously and in JSON.

Graphs and Statistics API: The Blockchain Charts and Statistics API gives a straightforward interface to automatically collaborate with the graphs and insights shown on blockchain.info.

13.4.2 ETHEREUM PLATFORM

Ethereum is the most well-known stage for creating and distributing a conveyed application utilizing programming dialects. Ethereum was developed to overcome the issues of non-Turing completeness associated with Bitcoin 1.0. Creating applications utilizing Bitcoin content expects engineers to fork the Bitcoin center code-base

Blockchain in EHR

and include the rationale for their own utilization cases. The forking is tedious and hard to keep up. Hence, to address these difficulties, Ethereum was created. Ethereum gives a stage to software engineers to assemble applications on the head of the blockchain, called an Ethereum blockchain. It was first proposed in late 2013, by a Bitcoin software engineer named Vitalik Buterin, in the white paper, *Ethereum: A Next-Generation Smart Contract and Decentralized Application Platform.* This theory proposes Turing-complete programming language for composing contents (smart agreements) and an Ethereum Virtual Machine (EVM) to execute the smart agreements and exchanges. An Ethereum client can make smart contracts and transfer them to the Ethereum blockchain for a small fee. Other Ethereum clients can get to these agreements by far-off system calls given by the Ethereum Application Program Interface (API). The agreements can store information, send exchanges, and interface with different agreements. The agreements are executed in bytecode. When agreements are transferred to the blockchain, they are put away, executed, and deciphered by the EVM. The EVM requires a modest quantity of expenses to execute exchanges. These charges are called gas, and the measure of gas relies upon the size of guidance. The more extended the agreement directions, the more gas is 33 required. Also, Ethereum has it own cryptographic money, called ether, and it is introduced by the shortened form, ETH. Ether is a sort of token that powers applications on the decentralized Ethereum network. The smallest unit of ether is the Wei. One ether is equivalent to 1018 Wei. Clients can utilize the Ethereum trade to change the physical or typical cash to ether. As of May 13, 2017, the trade estimation of 1 ether was $86.59.

Ethereum is an open-source stage that empowers designers to construct and convey decentralized applications. Similar to Bitcoin, Ethereum is a comprehensive platform with support for smart contracts and a complete programming language. Solidity helps in creating customized contract agreements that are executed when the stipulated events occur. The programming potential is enormous, and a large number of token-based applications are being built on this platform. Since it was the first complete Turing machine, and evolved out of Bitcoin, it also supports cryptocurrency, called Ether. Ether can easily be converted into other cryptocurrencies using an exchange. For consensus, Ethereum uses a proof-of-work (PoW) protocol, but is planning to update to proof of stake (PoS).

It has defined the first industry pseudo-standard for tokens, ERC20, which is in use for most common token development. In fact, all tokenized application uses this format for exchange and transferability. Of course, there are new standards, ERC223 (merged transfer), ERC621 (token supply), ERC721 (non-fungible), ERC998 (non-fungible and composable), ERC827 (token approval) for applications that are specific and need a more robust way of handling tokens.

Another interesting aspect of Ethereum is the transaction cost in gas units. A transaction in an Ethereum smart contract can invoke data reads and writes, carry out other high-end computations, such as using cryptographic primitives, make calls or send messages to other contracts, etc. Each of these operations has a cost, which is measured in gas. The gas unit consumed by a transaction must be paid for in Ether, the native cryptocurrency of Ethereum. The originator of the transaction usually pays this cost to the node that is successful in completing the transaction, based on the consensus.

Comparatively, Ethereum has long industry standing to its credit; it is time-tested and proven for many successful tokenization projects. Ether is a highly traded cryptocurrency and has been gaining new users on a daily basis. The Ethereum environment is rich and comes complete with a wallet, command line tools, a testing environment, and a host of GUI apps. A solid user community and a continuously evolving set of open source tools are big positives for this platform.

13.4.3 Hyperledger Platform

Hyperledger is not a specific technology, but rather a group of blockchain- and DLT-based projects under the Linux Foundation banner, for collaborative development. There are multiple frameworks under Hyperledger, each of which has slightly different characteristics. It also comes with a host of tools that help in development.

Hyperledger Fabric: a permissioned blockchain giving a secluded design help with the execution of smart sontracts and configurable accord and participation administrations (MSP). Texture organize has peer hubs that execute smart agreements written in chaincode. Texture underpins smart agreement execution in golang, Javascript and Java, and is conceivably more adaptable than a static smart agreement language.

- **Hyperledger Sawtooth** – a secluded stage for building, sending, and running DLTs; utilizes a Proof of Elapsed Time (PoET) agreement, which targets huge circulated validator populaces with insignificant asset utilization.
- **Hyperledger Iroha** – an appropriated record venture that was intended to be straightforward and simple to fuse into infrastructural ventures requiring conveyed record innovation.
- **Hyperledger Indy** – provides instruments, libraries, and reusable segments for interoperable computerized personalities established on blockchains or other conveyed records.
- **Hyperledger Burrow** – provides a modular blockchain client with a permission smart contract interpreter partially developed to the specification of the Ethereum Vital Machine (EVM).

Hyperledger Tools:

- **Hyperledger Caliper** – a benchmark instrument for the execution estimation of explicit blockchain usage, utilizing a lot of predefined use cases.
- **Hyperledger Explorer** – view, summon, send, or question squares, exchanges, and related information.
- **Hyperledger Cello** – a deployment apparatus for the blockchain biological system, to lessen the exertion required for making, overseeing, and ending blockchains.
- **Hyperledger Composer** – a collaboration apparatus for quickening the improvement of smart agreements and their sending over a dispersed record.

Blockchain in EHR

- **Hyperledger Quilt** – offers interoperability between record frameworks by actualizing ILP, which is essentially an installments convention and is intended to move an incentive across dispersed records and non-conveyed records.

Hyperledger supports CouchDB for storing world state and for full data-rich inquiries. Enrollment Service Providers (MSPs) permit associations to characterize altered personality, jobs, and validation.

Given its open source modular framework, Hyperledger can be used to build specific blockchains, as it enables a mix-and-match approach to the platform features.

13.5   CHALLENGES OF BLOCKCHAIN

As discussed earlier, the blockchain industry is still in the initial phase of development and implementation, and there are many kinds of challenges. These challenges may be both internal and external. They are related to the technology, government regulations, public perception, and industry readiness.

13.5.1   TECHNOLOGICAL CHALLENGES

There are many technological issues related to blockchain; the most common issues are discussed here as clarity of blockchain implementation among developers [40]. Different developers may suggest different solutions for any given problem, as there is no common coding standard for potential problems. Specialists, additionally, have diverse degrees of certainty with regard to whether and how these issues can be defeated, to advance into the subsequent periods of improvement in the blockchain industry. One of the most common tests for the underlying Bitcoin innovation is scaling up from the current greatest constraint of seven exchanges per second, particularly if there were to be standard reception of Bitcoin. A portion of the different issues includes expanding the block size, tending to blockchain swell, countering weakness to 51 percent mining assaults, and executing hard forks to code, as summed up here:

Throughput: the Bitcoin network has a significant issue with throughput, as it is handling just a single exchange for every second (tps), with the dramatic greatest pace of 7 tps.

Dormancy: at present, each Bitcoin exchange block takes 10 minutes to process/affirmed. For satisfactory security, one should stand by for over an hour, for the exchange of larger sums.

Size and Bandwidth: the size of the blockchain is very large: 25 GB, and has increased from 14 GB since 2019, and hence it might take approximately one day to download.

Security: Bitcoin blockchain still suffers from various potential security issues. At present, it is vulnerable to 51 percent attacks, which are attacks on blockchain by a group of miners.

Squandered Resources: Bitcoin mining expends an immense measure of energy, every last bit of it squandered. The previous gauge referred to was $15 million every day, significantly more than the resources usually used.

Ease of use: the API for working with Bitcoin is far less easy to use than the current principles of other simple-to-utilize present-day APIs, for example, the generally utilized REST APIs.

Forming hard forks, various chains: At present, such a large number of blockchain issues are identified with the framework. One issue is the multiplication of blockchain. Another issue is that when chains are used partly for authoritative or forming purposes, it is difficult to merger or cross-execute on forked chains.

The most significant technical challenge and requirement is that a full ecosystem of plug-and-play solutions be developed to provide the entire value chain of service delivery.

13.5.2 Business Model Challenges

Another significant test is identified with practical and specialized plans of action. Conventional plans of action probably will not appear to be material to Bitcoin, since the general purpose of decentralized peer-to-peer models is that there are no encouraging mediators to take a cut/exchange expense.

13.5.3 Scandals and Public Perception

Public perception of the adoption of Bitcoin is seen as the biggest challenge, as it is seen as a venue for the dark net's money-laundering, drug-related, and other illicit activity, e.g., illegal online marketplaces, such as Silk Road. Bitcoin and the blockchain are themselves neutral, as any technology, and are "dual use"; that is, they can be used for good or evil.

13.5.4 Government Regulation

A lack of all-around planned government guidelines is likewise one of the biggest variables and dangers, as to whether the blockchain business will thrive into a developed budgetary administrations industry [41]. One such issue is the likely down-to-earth difficulties of carrying out tax assessment with the current techniques. The second issue is that blockchain technology raises with regard to government regulation is the value preposition offered by governments and their business models. In this era of big data, the government is unable to keep its data and archive information in easily accessible modes.

13.5.5 Primary Challenges for Personal Records

There are many issues related to individuals, for keeping their personal records on a decentralized manner with a pointer and possible access via the blockchain [42]. The potential threat is that their data is online, and their secret key could be

Blockchain in EHR

stolen or exposed. In the current scenario, there are ample chances that a password might be leaked, stolen, or hacked.

13.6 CONCLUSION

In this chapter, the author has tried to explore all above applications, usages, limitations, and others concepts associated with Bitcoin and blockchain. The chapter divided blockchain into categories 1.0, 2.0, 3.0, and 4.0. Blockchain could be implemented in both centralized and decentralized models and offers a longer all and layer flat and crypto currencies. The application domain of Blockchain is growing day by day, in all the domains of security and financial transactions, and also for security and privacy. The basic challenges related to the adoption of Bitcoin blockchain were also discussed. Smart contracts, Ethereum, and Hypeledger are good choices for the implementation of privacy and security. Still, blockchain has to go through more than one revolution, to be adopted large-scale at the governmental and individual level.

REFERENCES

1. Nakamoto, S., 2009, February 11. *Bitcoin open source implementation of P2P currency.* http://p2pfoundation.ning.com/forum/topics/bitcoin-open-source. Accessed on August 19, 2019.
2. Christidis, K., and Devetsikiotis, M., 2016. Blockchains and smart contracts for the Internet of Things. *IEEE Access*, 4, pp. 2292–2303.
3. Weinman, J., 2015. *Digital disciplines: attaining market leadership via the cloud, big data, social, mobile, and the internet of things.* New York: John Wiley and Sons.
4. Yoon, H. J., 2019. Blockchain technology and health care. *Health Care Informatics Research*, 25(2), pp. 59–60.
5. Ulieru, M., 2016. Blockchain 2.0 and beyond: Adhocracies. In Tasca, P., Aste, T., Pelizzon, L., and Perony, N. (eds.) *Banking Beyond Banks and Money*. Cham: Springer, pp. 297–303.
6. Nakamoto, S., 2019. Bitcoin: A peer-to-peer electronic cash system. *Manubot*. https://bitcoin.org/bitcoin.pdf. Accessed on November 29, 2020.
7. Manohar, A., and Briggs, J., 2018. *Identity management in the age of Blockchain 3.0.* Paper presented at *2018 ACM Conference on Human Factors in Computing Systems*, Montréal, Canada.
8. Lin, C., He, D., Huang, X., Choo, K. K. R., and Vasilakos, A. V., 2018. BSeIn: A blockchain-based secure mutual authentication with fine-grained access control system for industry 4.0. *Journal of Network and Computer Applications*, 116, pp. 42–52.
9. Corbet, S., Larkin, C., Lucey, B., Meegan, A., and Yarovaya, L., 2020. Cryptocurrency reaction to FOMC announcements: Evidence of heterogeneity based on blockchain stack position. *Journal of Financial Stability*, 46, p. 100706.
10. Esposito, C., De Santis, A., Tortora, G., Chang, H., and Choo, K. K. R., 2018. Blockchain: A panacea for health care cloud-based data security and privacy? *IEEE Cloud Computing*, 5(1), pp. 31–37.
11. Hejblum, B., Weber, G., Liao, K. P., Palmer, N. P., Churchill, S., Shadick, N. A., Szolovits, P., Murphy, S. N., Kohane, I. S., and Cai, T., 2019. Probabilistic record linkage of de-identified research datasets with discrepancies using diagnosis codes. *Science Data*, 6, p. 180298. doi:10.1038/sdata.2018.298.

12. Yang, Y., Zheng, X., Guo, W., Liu, X., and Chang, V., 2019. Privacy-preserving smart IoT-based health care big data storage and self-adaptive access control system. *Information Sciences*, 479, pp. 567–592.
13. Dubovitskaya, A., Xu, Z., Ryu, S., Schumacher, M., and Wang, F., 2017. *Secure and trustable electronic medical records sharing using blockchain.* In *AMIA Annual Symposium Proceedings* (Vol. 2017, p. 650). Washington, D.C.: American Medical Informatics Association.
14. Can, O., and Yilmazer, D., 2020. Improving privacy in health care with an ontology-based provenance management system. *Expert Systems*, 37(1), e12427.
15. Kanwal, T., Shaukat, S. A. A., Anjum, A., Choo, K. K. R., Khan, A., Ahmad, N., ... and Khan, S. U., 2019. Privacy-preserving model and generalization correlation attacks for 1:M data with multiple sensitive attributes. *Information Sciences*, 488, pp. 238–256.
16. Hsieh, C. Y., Su, C. C., Shao, S. C., Sung, S. F., Lin, S. J., Yang, Y. H. K., and Lai, E. C. C., 2019. Taiwan's national health insurance research database: Past and future. *Clinical Epidemiology*, 11, p. 349.
17. Ahmed, S. M., and Rajput, A., 2020. Threats to patients' privacy in smart health care environment. In Lytras, M. and Sarirete, A. (eds.) *Innovation in Health Informatics.* London: Academic Press, Elsevier, pp. 375–393.
18. Price, W. N., and Cohen, I. G., 2019. Privacy in the age of medical big data. *Nature Medicine*, 25(1), pp. 37–43.
19. Kumar, A., and Kumar, R., 2020. Privacy preservation of electronic health record: Current status and future direction. In Gupta, B., Perez, G. M., Agrawal, D. P., and Gupta, D. (eds.) *Handbook of Computer Networks and Cyber Security.* Cham: Springer, pp. 715–739.
20. Bellovin, S. M., Dutta, P. K., and Reitinger, N., 2019. Privacy and synthetic datasets. *The Stanford Technology Law Review*, 22, p. 1.
21. Miao, Y., Liu, X., Choo, K. K. R., Deng, R. H., Li, J., Li, H., and Ma, J., 2019. Privacy-preserving attribute-based keyword search in shared multi-owner setting. *IEEE Transactions on Dependable and Secure Computing.* http://www.ieeeprojectmadurai. in/2019%20IEEE%20PROJECT%20BASEPAPERS/Privacy-Preserving%20Attribute-Based%20Keyword%20Search%20in%20Shared%20Multi-owner%20Setting.pdf.
22. Tian, H., He, J., and Ding, Y., 2019. Medical data management on blockchain with privacy. *Journal of Medical Systems*, 43(2), p. 26.
23. Habibzadeh, H., Dinesh, K., Shishvan, O. R., Boggio-Dandry, A., Sharma, G., and Soyata, T., 2019. A survey of health care Internet of Things (HIoT): A clinical perspective. *IEEE Internet of Things Journal*, 7(1), pp. 53–71.
24. Li, M., Zhu, L., and Lin, X., 2019. Privacy-preserving traffic monitoring with false report filtering via fog-assisted vehicular crowdsensing. *IEEE Transactions on Services Computing.* doi:10.1109/TSC.2019.2903060.
25. Alabdulkarim, A., Al-Rodhaan, M., Tian, Y., and Al-Dhelaan, A., 2019. A privacy-preserving algorithm for clinical decision-support systems using random forest. *CMC Computers, Materials and Continua*, 58, pp. 585–601.
26. Li, X., Jiang, P., Chen, T., Luo, X., and Wen, Q., 2020. A survey on the security of blockchain systems. *Future Generation Computer Systems*, 107, pp. 841–853.
27. Malathi, D., Logesh, R., Subramaniyaswamy, V., Vijayakumar, V., and Sangaiah, A. K., 2019. Hybrid reasoning-based privacy-aware disease prediction support system. *Computers and Electrical Engineering*, 73, pp. 114–127.
28. Kabou, S., Benslimane, S. M., and Kabou, A., 2020, February. *Toward a new way of minimizing the loss of information quality in the dynamic anonymization.* In *2020 2nd International Conference on Mathematics and Information Technology (ICMIT)* (pp. 186–189). Adrar, Algeria: IEEE.

Blockchain in EHR

29. Hussien, H. M., Yasin, S. M., Udzir, S. N. I., Zaidan, A. A., and Zaidan, B. B., 2019. A systematic review for enabling of develop a blockchain technology in health care application: Taxonomy, substantially analysis, motivations, challenges, recommendations and future direction. *Journal of Medical Systems*, 43(10), p. 320.

30. Iqbal, S., Kiah, M. L. M., Zaidan, A. A., Zaidan, B. B., Albahri, O. S., Albahri, A. S., and Alsalem, M. A., 2019. Real-time-based E-health systems: Design and implementation of a lightweight key management protocol for securing sensitive information of patients. *Health and Technology*, 9(2), pp. 93–111.

31. Zigomitros, A., Casino, F., Solanas, A., and Patsakis, C., 2020. A survey on privacy properties for data publishing of relational data. *IEEE Access*, 8, pp. 51071–51099.

32. Wang, G., Lu, R., and Guan, Y. L., 2019. Achieve privacy-preserving priority classification on patient health data in remote eHealth care system. *IEEE Access*, 7, pp. 33565–33576.

33. Mamdouh, M., Awad, A. I., Hamed, H. F., and Khalaf, A. A., 2020, April. *Outlook on security and privacy in IoHT: Key challenges and future vision*. In *Joint European-US Workshop on Applications of Invariance in Computer Vision* (pp. 721–730). Cham: Springer.

34. Aggarwal, S., Kumar, A., and Kumar, R., 2019, March. *The privacy preservation of patients' health records using soft computing in python*. In *2019 6th International Conference on Computing for Sustainable Global Development (INDIACom)* (pp. 156–160). New Delhi: IEEE.

35. Al-Sharhan, S., Omran, E., and Lari, K., 2019. An integrated holistic model for an eHealth system: A national implementation approach and a new cloud-based security model. *International Journal of Information Management*, 47, pp. 121–130.

36. Drozdowicz, M., Ganzha, M., and Paprzycki, M., 2020. Semantic access control for privacy management of personal sensing in smart cities. *IEEE Transactions on Emerging Topics in Computing*, PP(99), p. 1. doi:10.1109/TETC.2020.2996974.

37. Aloufi, R., Haddadi, H., and Boyle, D., 2019. Emotionless: Privacy-preserving speech analysis for voice assistants. arXiv preprint arXiv: 1908.03632.

38. Hussien, H. M., Yasin, S. M., Udzir, S. N. I., Zaidan, A. A., and Zaidan, B. B., 2019. A systematic review for enabling of develops a blockchain technology in health care application: Taxonomy, substantially analysis, motivations, challenges, recommendations and future direction. *Journal of Medical Systems*, 43(10), p. 320.

39. Ayer, T., Ayvaci, M. U., Karaca, Z., and Vlachy, J., 2019. The impact of health information exchanges on emergency department length of stay. *Production and Operations Management*, 28(3), pp. 740–758.

40. Bhushan, B., Sahoo, C., Sinha, P., and Khamparia, A., 2020. Unification of Blockchain and Internet of Things (BIoT): Requirements, working model, challenges and future directions. *Wireless Networks*. doi:10.1007/s11276-020-02445-6.

41. Bhushan, B., Khamparia, A., Sagayam, K. M., Sharma, S. K., Ahad, M. A., and Debnath, N. C., 2020. Blockchain for smart cities: A review of architectures, integration trends and future research directions. *Sustainable Cities and Society*, 61, p. 102360. doi:10.1016/j.scs.2020.102360.

42. Khamparia, A., Singh, P. K., Rani, P., Samanta, D., Khanna, A., and Bhushan, B., 2020. An internet of health things-driven deep learning framework for detection and classification of skin cancer using transfer learning. *Transactions on Emerging Telecommunications Technologies*. doi:10.1002/ett.3963.

14 Attacks, Vulnerabilities, and Blockchain-Based Countermeasures in Internet of Things (IoT) Systems

Nayanika Shukla
HMR Institute of Technology and Management, India

Bharat Bhushan
School of Engineering and Technology,
Sharda University, India

CONTENTS

14.1 Introduction 296
14.2 Background of Blockchain...297
    14.2.1 History of Blockchain..298
    14.2.2 Ingredients of Blockchain..298
        14.2.2.1 Ledger 298
    14.2.3 Cryptography 299
        14.2.3.1 Peer-to-Peer Network..299
        14.2.3.2 Assets 299
        14.2.3.3 Merkle Trees...299
        14.2.3.4 Consensus Algorithms...300
    14.2.4 Classification of Blockchain 300
        14.2.4.1 Public versus Private 300
        14.2.4.2 Permissioned versus Permissionless 300
    14.2.5 Challenges in Blockchain 300
    14.2.6 Security Analysis of Blockchain..301
14.3 Attacks in the IoT 302
    14.3.1 Physical Attacks...302
        14.3.1.1 Tampering...302
        14.3.1.2 RF Interface/Jamming 302
        14.3.1.3 Fake Node Injection 303
        14.3.1.4 Sleep Denial 303
        14.3.1.5 Network Attacks...303
        14.3.1.6 Traffic Analysis Attack..303

295

		14.3.1.7 RFID Spoofing	303
		14.3.1.8 Routing Information Attacks	304
		14.3.1.9 Man-in-the-Middle Attack	304
	14.3.2	Software Attacks	304
	14.3.3	Viruses, Worms, Trojan Horses, Spyware, and Adware	304
		14.3.3.1 Malware	305
	14.3.4	Data Attacks	305
		14.3.4.1 Data Inconsistency	305
		14.3.4.2 Unauthorized Access	305
		14.3.4.3 Data Breach	306
14.4	Blockchain Applications for Smart Industrial Automation		307
	14.4.1	Industries 4.0	307
	14.4.2	Autonomous Vehicles (VANET)	307
	14.4.3	Smart Home	308
	14.4.4	Smart City	308
	14.4.5	Health Care 4.0	309
	14.4.6	Unmanned Aerial Vehicles	309
	14.4.7	Smart Grid	310
14.5	Challenges of Blockchain in the IoT		310
	14.5.1	Computation	310
	14.5.2	Storage	311
	14.5.3	Communication	311
	14.5.4	Energy	311
	14.5.5	Mobility and Partition of the IoT	312
	14.5.6	Latency and Capacity	312
14.6	Conclusion		312
References			313

14.1 INTRODUCTION

The Internet of Things (IoT) is a web that consists of physical objects, or "things," that facilitate the exchange of information over the internet to create new things, by using sensors that record data and provide a fixed network connection [1]. By "things," we do not point to a particular object; instead, this is a broad term for calculating devices, automatic machines, objects, and other electronic devices. As the IoT has some anticipated advantages, innovation leaders have seen a rapid growth in its usage. It has sufficient potential to achieve a great economic impact worldwide in the coming years. As a decentralized system, the IoT helps in the analysis, observation, and control of various procedures, which revamps various activities and helps achieve better performance and efficiency [2]. The IoT's latest techniques help in maintenance purposes, by examining recurring patterns and contributing to their analysis, which provides many benefits. Traditional procedures and services could be improved with the help of this accurate information. Though the IoT provides many advantages, there are also disadvantages related to it, such as the lack of privacy, an overdependence on technology, as well as the loss of jobs [3]. The cost of sensors, and other expenses, are also a major cause of worry.

Attacks, Vulnerabilities, and Blockchain-Based Countermeasures 297

A proper linkage between the IoT and blockchain could be made in order to avoid the aforementioned problems. As a decentralized, dispersed, enduring, and collective database ledger, blockchain helps to store all the transactions taking place and contains a hash that points to the previous block [4]. A complete history of all transactions is provided by the blockchain. The information-sharing component of the IoT could be enhanced by blockchain technology. Essential information can be shared safely by IoT devices, by using blockchain's methods [5]. Blockchain technology can enable the processing of transactions and incorporating billion of connected devices, which is extremely important in tracking those devices [6]. Therefore, blockchain technology enables IoT industry manufacturers to facilitate notable savings [7]. Once disseminated, this technique would abolish any chance of failure and provide a pliable environment for the smooth functioning of devices. In order to make customer data more secure, blockchain technology significantly uses cryptographic algorithms [8]. Various sectors, such as health care, banking, industry, etc. are applying the IoT to gain various advantages [9]. The IoT could use blockchain technology to address security issues.

Our work brings an integrated approach to the applications of blockchain technology and its contribution in the operation of IoT devices. The major contribution of this work is an elaborated discussion of blockchain technology, its applications in IoT systems, and the challenges it poses. This work also presents in-depth analysis of the distribution, access, and growth of blockchain. The chapter demonstrates how blockchain has evolved and provided a diverse range of processes to ensure privacy and the smooth retrieval of data. With the help of blockchain and the IoT, advanced technology could be introduced to the world, with elements such as smart homes and smart cities. The concept of smart health care and industrial systems, now considered "health care 4.0" and "industry 4.0," respectively, has greatly benefited from the advent of blockchain technology [10,11]. Given the challenges faced by IoT devices using blockchain, including communication between devices, the quantity of storage needed, and cost, it is difficicult to integrate the two technologies properly [12].

The remainder of the chapter is organized as follows. Section 14.2 starts off with the history and background of blockchain technology. It highlights the fundamental components of blockchain, along with its types and associated security issues. Section 14.3 explores common security attacks on the IoT and how blockchain protects IoT devices from these attacks. Section 14.4 discusses the most useful applications of blockchain technology. Finally, Section 14.5 highlights the major challenges associated with blockchain technology, followed by a conclusion in Section 14.6.

14.2 BACKGROUND OF BLOCKCHAIN

Blockchain technology introduced the term "trustlessness," to ensure a secure transmission of data. In the absence of blockchain, there was no such authority who could take account of all the data entries and transactions taking place. Many events led to the evolution of blockchain and made it popular around the globe. In this section, we discuss the history and the complete architecture of blockchain technology, along with its features.

14.2.1 History of Blockchain

Satoshi Nakamoto et al. [13] proposed the idea of blockchain as the fundamental technology in Bitcoin. While blockchain has acquired an adequate amount of consideration outside the domain of cryptocurrencies, it all started with Bitcoin. With the benefit of public-key cryptography and cryptographic hashing, Bitcoin enables its users to be tremendously unique. A digital wallet generally consists of the user's Bitcoin and the private key for an account, which is helpful in signing all transactions from that particular account. The network will verify any transaction introduced by the account, with the help of the corresponding public key for that account [14], whereas the blockchain platform does not require anonymity [15].

Blockchain can be summarized as a transparent process in the following way. In the first place, before implementing multiple witnesses, an announcement is created. The details of the announcement for each participant are documented in their own distinctive duplicate of the ledger, in the form of "blocks." On the network, each participant routinely tries to contrast their current block with other participants' blocks. If a majority of the population has access to common version of the current block, then that version is regarded as the truth. Governments and various established companies, e.g., American Express and Microsoft, are exploring blockchain technology. The most essential and beneficial incidents regarding breakthroughs in the evolution of the blockchain architecture are summarized in Table 14.1.

14.2.2 Ingredients of Blockchain

Various components work together to implement blockchain technology. These components include:

14.2.2.1 Ledger

At its most fundamental level, blockchain is an entrenched record, similar to a traditional ledger, often used to control and track asset ownership. Blockchain is just an innovative coalescence of various old ideas, approaches, and methodologies, though

---

TABLE 14.1

Essential Events of the Blockchain Architecture Regarding the Breakthroughs in Its Evolution

Year	Breakthrough
2009	Creation of first Bitcoin block.
2010	December marks the disappearance of Satoshi.
2012	e-Estonia: Estonian blockchain technology.
2015	Hyperledger and Ethereum go live.
2018	Blockchain demand increases.
2019	Produce suppliers demanded by Walmart [16].
2020–2021	Dubai initiative [17].

Attacks, Vulnerabilities, and Blockchain-Based Countermeasures 299

it is usually expressed as a new and innovative technology. Ledgers, cryptography, group consensus, and immutability are some of these components. The ledger is a record-keeping infrastructure located at the core of blockchain, which enables the users of a ledger to review past transactions. The ownership in the past the ownership itself was the usual element of this story, although any type of data can be recorded with the help of ledgers [18].

14.2.3 CRYPTOGRAPHY

The second most fundamental component of blockchain technology, cryptography confidentially specifies the way of communication. In Bitcoin technology, cryptography is used to vindicate claims made by people against the assets being controlled on the blockchain, which provides ledger immutability and preserves anonymity. A "cryptographic hash" is a particular function that implements all the data in a block, in order to chain those blocks. A hash or ID is generated that does not match the true value documented on the upcoming block in the chain when one tries to convert the information in the block. A completely new hash is engendered when we convert the information in any block. As a result, this will break the blockchain and refute all the blocks associated to where the conversion was created, as the newly generated hash will not match the hash in the upcoming block header.

14.2.3.1 Peer-to-Peer Network

In blockchain, peer-to-peer network (P2P) infrastructure makes extensive use of existing computer network technology. Blockchain operates as a principal element of our contemporary internet, which also uses this networking technology. Redundancy and fault tolerance are increased with the removal of a single point of flaw, using P2P network architecture, generally available in client/server network infrastructure.

14.2.3.2 Assets

All blockchain solutions have assets as one of their essential component. Assets are those items that are considered in the context of a given solution. Items that require a record of ownership are known as assets. They contain general information, like health records, event tickets, or a patent, and can be monetary as well as nonmonetary. The transfer of Bitcoin and other cryptocurrencies is recorded with the help of blockchain, which initially begin as a record keeping system. Blockchain was born to convert a documentation of digital ownership. Blockchain supplements the internet of knowledge that we have in our contemporary world with the internet of the value that we are inventing for tomorrow, in various ways.

14.2.3.3 Merkle Trees

Merkle trees in blockchain are used for the quick and methodical validation of data. All the data in a block can be summarized with the help of Merkle trees, by generating a root hash of that particular data. The child nodes of data are paired and hashed repeatedly, until a single node is left, in order to find out the root hash. A Merkle root is defined as the last remaining child node.

14.2.3.4 Consensus Algorithms

In order to validate transactions and accord with the order of the nodes on the network and their existence on the ledger, a consensus is used. This process is censorious in the case of applications like cryptocurrency, as it fails to avoid the invalid data to be documented to the underlying ledger that acts as a database for all the transactions. There are distinct solutions relevant to distinct situations with consensus. One decides on the basis of expenses related to opportunity (e.g., safety, rapidity, etc.) which consensus mechanism is to be used. Proof of work (PoW) and proof of stake (PoS) are the usual consensus algorithms. Private and permissioned scenarios contain other consensus mechanisms, where we do not require computationally intensive consensus mechanisms, for example in Hyperledger. If blockchain is not public, then there are many more alternatives for consensus.

14.2.4 CLASSIFICATION OF BLOCKCHAIN

This section classifies blockchain technology with the help of the following two metrics.

14.2.4.1 Public versus Private

Data can be added to the ledger by the public on their own, with the help of public blockchains. In Bitcoin, we have no regulations or consents concerning who can trade it, thus Bitcoin serves as an example of a public blockchain network. Bitcoin can be sent to anyone, and anyone can buy or sell it. We can determine how charitable donations are used by a nonprofit organization with the help of a blockchain solution, thus it is an example of a private solution. Only designated officers are allowed to share metrics describing the allocations and expenditures of the nonprofit organization.

14.2.4.2 Permissioned versus Permissionless

Solutions where there is little need for public consent are known as permissionless platforms. These platforms neither have the capability to track and control identity nor can they eventually define and impose permissions based on that identity. A prominent way to detect which kind of blockchain is required is to detect whether all candidates are given equal access to permissions or not, for developing a solution. This question will help in deciding either to apply a permissioned or permissionless blockchain technology [19]. In an enterprise blockchain solution, the access to permissions is only given to authorized employees, thus this is an example of permissioned blockchain. On the other hand, we have permissionless blockchain, i.e., digital currency, which could be exchanged and dealt in by everyone.

The major types of blockchain, along with their aforementioned characteristics, are summarized in Table 14.2.

14.2.5 CHALLENGES IN BLOCKCHAIN

- **Storage capability and scalability:** Communication and information are stored as blockchain transactions, in order to secure them [20]. But blockchain cannot store large amounts of data, because of its design. Sadly, the size of the blockchain and the number of transactions increase proportionally with performance and synchronization time.

TABLE 14.2
Types of Blockchain

Type of Blockchain	Characteristics
Public blockchain	• Data can be added to the ledger by the public on their own • Example: Bitcoin
Private blockchain	• Only designated officers are allowed to share metrics • Donations made to a nonprofit organization
Permissionless blockchain	• Little need for the public to consent • Digital currency
Permissioned blockchain	• Access to permissions is only given to authorized users • Enterprise blockchain solution

- **Data privacy:** From the system's initial transaction, every single transaction can be patterned, inspected, and copied in blockchain. The first cryptocurrency that focused on privacy was Dash [21]. Some proposed approaches to increase privacy are mix coin, coin shuffle, coin swap, and blind coin.
- **Legal issues:** Although the emergence of new technologies such as the IoT and blockchain has rejuvenated the traditional system of transactions and benefitted the world, it also has forced countries to revise, control, and update their existing laws and orders. Centralization can be eased with the help of new laws and regulations.
- **No specified software:** As there is no standard software on which blockchain and the IoT can be connected, each company has to make its own software and the entire set of use cases. Also, they have to build the entire architecture from scratch. It takes several years for the proper establishment of the procedure.
- **Rapid change in technology:** Existing blockchains become quickly outmoded because of their immature nature. If the technology becomes outmoded in six months, or does not perform well for the use cases, or has slow validation times with IoT devices, then no organization would want to commit to that technology.

14.2.6 SECURITY ANALYSIS OF BLOCKCHAIN

Blockchain is vulnerable to various security threats:

- **Attacks on consensus protocols:** By gaining an enormous section of the computational capacity of the whole network, assailants could exploit the security of the consensus protocols. The chain can be controlled and rebuilt by such attackers. For example, Bitcoin [22], where a 51 percent attack in PoW could be observed.
- **Eclipse attacks:** When adversaries in a P2P network fabricate each and every connection to the legal nodes, and avert these nodes from connecting to any truthful peers, this is known as an eclipse attack. For example, it was reported that eclipse attacks were introduced to Ethereum via a kademlia P2P protocol it had adapted [23].

- **Vulnerability of smart contracts:** With the irretrievability and openness of blockchain, smart contracts are susceptible. There is transparency between the public and the bugs and frauds. Due to the irreversibility of blockchain, it also becomes extremely difficult to fix viruses and other bugs in established smart contracts. The 2016 attack on the decentralized autonomous organization (DAO) is an excellent example of such a case.
- **Double spending:** With conflicting transactions, the transaction's receivers may become deceived by adversaries. For instance, in Bitcoin, there is an expenditure of similar coins. Dispatching conflicting transactions, and the early extraction of one or more blocks to achieve conflicting transactions, are possible attack methods included in it.

14.3   ATTACKS IN THE IOT

We can broadly classify security attacks in the IoT with the help of the following four domains:

14.3.1   Physical Attacks

When an attacker maintains a physical closeness with the network or other devices of the system, then physical attacks are launched. Physical attacks include:

14.3.1.1   Tampering

When a device (e.g., RFID) or any transmission link is enhanced physically, then tampering is introduced [24]. Confidential and sensitive information can be acquired as the result of tampering. Research has been done to identify the vulnerabilities in popular physical devices (such as smart meters, IP cameras, and Amazon Echo). A camera's password can be acquired by an attacker, regardless of its length and a configuration that is confidential and sensitive to a particular user. A physically unclonable function (PUF) is the proposed countermeasure for this attack. The idea of a PUF was proposed for small-sized IoT devices, in order to exploit the integral instability of integrated circuits (IC). A PUF uses a challenge-response mechanism, where the physical structure at the micro level of the device primarily decides the output of the system. Thus, the PUF is an efficient measure to eliminate attacks such as tampering.

14.3.1.2   RF Interface/Jamming

In order to hinder a communication, instead of sending radio frequency (RF) signals, an attacker creates and transmits noise signals to launch DoS attacks on RFID tags. This is known as RF interfacing/jamming. Hindering or jamming communication is the prominent effect of this attack. A customizable and trustable end device mote (CUTE mote) is the proposed countermeasure used for this attack. With the help of a CUTE mote, in order to achieve productivity and overall performance, a solution of the device is presented in the work [25]. With a hardcore microcontroller unit (MCU) and an IEEE 802.15.4 radio transceiver, we have some essential elements of the architecture, i.e., a reconfigurable computing unit (RCU).

Attacks, Vulnerabilities, and Blockchain-Based Countermeasures 303

14.3.1.3 Fake Node Injection

In order to control data flow, a fake node is dropped by an attacker into the connecting legal nodes of the entire web, and this is known as a fake node injection. Managing the flow of data is the effect of this attack. An attacker can acquire control of the process of any data. Several physical devices are vulnerable to this attack. Pathkey is the proposed countermeasure against this attack. In today's world, distributed IoT applications are a fundamental component of great importance. The establishment of secure links between each and every sensor node and end user is extremely important, in such an environment.

14.3.1.4 Sleep Denial

Those attacks in which battery-powered devices are continuously used, by providing them with incorrect inputs by an attacker, are known as sleep denial attacks. Sleep denial attacks are vulnerable to node shutdown, which is their main effect. A CUTE mote and support vector machine (SVM) are the proposed countermeasures for this attack. A CUTE mote is beneficial to prevent sleep denial attacks, because of its heterogeneous structure. An SVM has been designed that uses medical access patterns from a patient's device. Because of its classification algorithm, an SVM is capable of determining resource exhaustion, which makes it invulnerable to sleep denial attacks.

14.3.1.5 Network Attacks

Those attacks that damage an entire system by orchestrating the IoT network are termed network attacks. Without even being close to the network, these attacks can be launched easily. Common network attacks include:

14.3.1.6 Traffic Analysis Attack

Attackers acquire confidential information even without being close to the network, so that they can obtain information about the network. These attacks are vulnerable to data leakage, i.e., unauthorized access to network information. The efficient and privacy preserving traffic obfuscation framework (EPIC) is the proposed countermeasure for traffic analysis attacks. Jianqing Liu et al. [26] have proposed the EPIC framework, so that smart homes could be protected from traffic analysis. It assures the pseudonymity of the flow of traffic to a specific smart home and also between source and destination. This framework acts as a secure multihop routing protocol, assuring strong privacy prevention.

14.3.1.7 RFID Spoofing

In RFID spoofing, in order to acquire the data stamped on the RFID tag, an RFID signal is first spoofed by the attacker. The attacker then sends its data by posting it as valid, with the help of the original tag ID. This attack is designed to manage and modify data (i.e., reading, writing, and deleting). The physically unclonable function (PUF) based on SRAM is the countermeasure against RFID spoofing. A PUF based on an on-board SRAM has been invented, which fabricates a unique device ID with the help of the unique device footprint. The possibility of the impersonation of an ID

Blockchain Technology for Data Privacy Management

by an adversary can be minimized by using device ID matching, which prevents the risk of spoofing, as well as fraudulent access.

14.3.1.8 Routing Information Attacks

Direct attacks where the attacker creates a nuisance through activities such as creating routing loops, sending error messages and spoofs, or altering routing information, are known as routing information attacks. Routing loops are the prominent targets of these attacks. Hash chain authentication is the proposed countermeasure for routing information attacks. Hash chain authentication is required in order to efficiently deal with routing attacks, by preventing malicious codes from exploiting control messages. Selective forwarding and sinkhole attacks can be effectively reduced by using hash chain authentication and rank threshold in combination.

14.3.1.9 Man-in-the-Middle Attack

In a man-in-the-middle (MitM) attack, an attacker can gain access to the private data of any user by eavesdropping or monitoring the communication between two IoT devices. The violation of the data privacy of any user is the prominent effect of MitM attacks. An IoT system can be seriously impacted by MitM attacks. For example, an attacker can take control of a smart actuator in an industrial IoT setting [27]. They can potentially damage an assembly line, by knocking an industrial robot out of its designed lane and speed limit. MQTT and inter-device authentication are the proposed countermeasures for this attack. MQTT ensures device-to-device (D2D) communication by using key-police (KP) to implement elliptic curve cryptography (ECC), which prevents MitM attacks.

14.3.2 SOFTWARE ATTACKS

When an assailant avails itself of the auxiliary software or safety vulnerabilities provided by an IoT system, software attacks are launched. There are various software attacks, including:

14.3.3 VIRUSES, WORMS, TROJAN HORSES, SPYWARE, AND ADWARE

This dangerous software tampers with data or steals information by adversely infecting the system or even launching DoS. These attacks are vulnerable to resource destruction in any IoT system. A high-level synthesis (HLS) and lightweight framework are the proposed countermeasures for these software attacks. It is important to mitigate these attacks, in order to secure any IoT system. A lightweight framework has been developed that integrates three safety features to avoid Trojans from being introduced to IoT-based devices. Initially, in order to assure a trusted transmission among untrusted nodes, vendor diversity is established. Then, in order to prevent unwanted parties from accessing communications, message encryption is established. Last, legal nodes are allowed to validate the encryption status and content of messages by mutual auditing. The HLS is introduced in order to prevent hardware Trojans. HLS designs a security-enhanced hardware that directly prevents Trojans from being injected into the network.

Attacks, Vulnerabilities, and Blockchain-Based Countermeasures

14.3.3.1 Malware

Malware is software that affects the data present in IoT devices, which further results in the contamination of data or cloud centers [28]. In reality, it is the most common software attack launched in any IoT system. Malware infects the data, which is the main effect of this software attack. A lightweight neural network framework and malware image classification system (MICS) are the proposed countermeasures for this software attack. An MICS is software that classifies malware and represents images globally and locally. Before capturing hybrid local and global malware features to perform malware family classification, the MICS first converts the suspicious program into a gray-scale image. A lightweight conventional neutral network framework is an alternate solution, which can classify malware samples gathered from two distinct households. In this case, program binaries are converted to gray-scale images that detect malware with high accuracy.

14.3.4 DATA ATTACKS

In order to exert pressure on certain computing resources which maintains the features of data collectivity and managing data connection required by IoT devices, the development and evolution of the IoT is extremely important [29]. As the fundamental root of each and everything that the IoT could offer to users, cloud computing steps into the action. With the help of cloud computing, virtual servers, a database example, and the creation of data pipelines that help IoT solutions to run could become effortless. The cloud ensures better safety by supplying firmware and other software techniques, better management, and different procedures, in order to assure data security. Some of the major data attacks include:

14.3.4.1 Data Inconsistency

In the central database, when an attack is made on data ethics that leads to the incompatibility of data and its storage, this is known as data inconsistency. As the name suggests, the inconsistency in data is the main effect of this attack. Blockchain architecture and a chaos-based scheme are the proposed countermeasures for this data attack. Chaos-based schemes, along with a chaos-based privacy preserving cryptographic scheme, also use a message authentication code (MAC), in order to provide the safe transmission of data within a home with smart features. This scheme guarantees data integrity by using a logistic map to create symmetric keys, to provide a safe mean of communication. In the case of remote semi-trusted data storage, blockchain is used, based on a three-level split [30].

14.3.4.2 Unauthorized Access

In the IoT, it is extremely important to prevent unauthorized users from getting access control and to ensure authorized users complete access. Unauthorized users can achieve access to private data if they are given complete access control. The violation of data privacy is the main effect of this data attack. Blockchain-based attribute based encryption (ABE) and privacy-preserving ABE are the proposed countermeasures for this attack. It becomes extremely difficult to acquire access to private data, because of the wireless sensor network's (WSN) open wireless channel, although the WSN plays an important role in the IoT. For that reason, access control is extremely important and must be incorporated, either with or without the use cryptography. A blockchain-based infrastructure, with the help of ABE, has been proposed. The proposed scheme works in such a way that the privacy of the transaction data is protected, supports integrity, and supports nonrepudiation. In order to preserve the IoT system, this scheme provides end-to-end privacy and directs the privacy of shared data by obtruding access control in the blockchain.

14.3.4.3 Data Breach

When an unauthorized user gets access to private or confidential data, this is known as a data breach. It is also known as memory leakage. Data leakage is the main effect of this data attack. Two-factor authentication, dynamic privacy protection (DPP), and improved secure directed diffusion (ISDD) are the proposed countermeasures for this attack. In recent years, the privacy of the user's personal data has been at risk, due to data breaches. A lightweight authentication scheme has been proposed by the work [31]. This acts as a security safeguard to ensure proper transmission between multiple IoT devices. A DPP model has been devised by some authors in other work [32], with the help of dynamic programming. This model enables the resource-constrained devices in order to attain optimal solutions for security-preserving levels. In order to avoid any chance of privacy leakage, content-oriented data pairs (CODP) and optimal data alternatives (ODA) are used. An (ISDD) is a protocol proposed by some authors, which assures data confidentiality in IoT systems.

Table 14.3 summarizes major attacks that may harm the IoT device and its functions.

TABLE 14.3
IoT Devices Are Prone to the Following Attacks

Type	Attacks
Physical attacks [33]	Tampering
	RF interface
	Jamming
	Fake node injection
	Sleep denial
Network attacks	Traffic analysis
	RFID spoofing
	Routing information
	Man-in-the-middle attack
Software attacks	Virus
	Worms
	Trojan horses
	Spyware and adware
	Malware
Data attacks [34]	Data inconsistency
	Unauthorized access
	Data breach

Attacks, Vulnerabilities, and Blockchain-Based Countermeasures 307

14.4 BLOCKCHAIN APPLICATIONS FOR SMART INDUSTRIAL AUTOMATION

Because it is a decentralized system, blockchain technology is transforming our lives, from the way we manage business activities to lifestyle, i.e., smart homes (Section 14.4.3). Blockchain is simplifying our affairs, facilitating procedures, minimizing errors, and saving us from appointing third-party resources. The following are some applications of blockchain that provide great benefits and improve our way of life.

14.4.1 INDUSTRIES 4.0

The robotization of industrial and business operations has become a part of today's world. Industry 4.0 has emerged as the new approach in production, with massive advancements in technology. The integration of blockchain technologies into industrial IoT systems has strengthened the whole system, offering advantages such as data confidentiality, the secure sharing of data, etc. The main aim of industry 4.0 is to combine several technical realms, such as blockchain and cyber physical systems (CPS). In order to succeed in this highly competitive world, business process management (BPM) has computerized and automated business operations, for a notable rise in profits. But malicious and untrusted clients can misuse this system for their own benefits. Moreover, with the addition of the unique agents to these business operations, a significant rise in transaction expenses and the risks related to them could also be seen. Therefore, Aleksandr Kapitonav et al. [35] have proposed the implementation of blockchain technology, in order to build decentralized systems to assure secure communication in a multi-agent system. Additionally, in order to impose a safe, interactive validation system with fine-gained access control, Chao Lin et al. [36] have devised a system known as BSeIn, which is based on the blockchain technology. There are two major challenges of the industrial IoT. These challenges are: achieving a high quality of data and assuring secure communication among mobile terminals (MTs). With the help of blockchain technology, BPM provides various benefits; it builds trust, reduces costs, provides accurate transactions, provides efficiency, increases the agility of modern businesses, and integrates cross-organizational business. In the present era, modern business also includes QoS. The QoS blockchain requires the instantaneous upgrade of information, contrary to Ethereum and Bitcoin, which are dispersed ledgers.

14.4.2 AUTONOMOUS VEHICLES (VANET)

Intelligent transportation systems (ITS) have grown immensely in recent years, with quick technical advancements in observation, communications, examination, and calculations. It has provided modish, secured, and more convenient conveyance facilities. Because of ITS' tendency toward centralization, there is a critical security threat from malicious users. To overcome this problem, a safe and a trusted ecosystem based on a blockchain technology known as B2ITS has been proposed. It aims at optimizing real-world transportation systems, using parallel interactions with their counterparts, and thus acts as the expedient for parallel transportation management systems (PTMS). A wireless communication system has been devised to ensure safe and efficient transportation, by interchanging information among themselves as well as with road-sized base stations. This is known as a vehicular ad hoc network (VANET) [37]. It consists of automobiles furnished with radio interfaces. In order to achieve secure and accurate message delivery, blockchain has been introduced in the VANET network. Othman S. Al-Heety et al. [38] provide a brief review of the infrastructure and operations of a VANET. A brief description of the SDN-VANET is given, along with its applications. This work suggests how VANET technology can help in the growth of IoT-based systems.

14.4.3 SMART HOME

The manifestation of a technically advanced environment, which helps in improving the quality of life, is known as a smart home. A smartphone application is of great value to smart home owners, as it allows them to control settings according to their preferences and thus provide security, convenience, and comfort. An energy-efficient smart home provides consecutive services as per user's choices. Network connectivity, sensor devices, and mobile applications based on IoT technology are the essential elements of a smart home [39]. Smart lighting, smart doors, smart thermostats, video surveillance, and smart parking are some of the advantages offered by a smart home. A smart door locking system prevents prohibited users from getting inside the house and thus acts as an efficient part of any smart house. However, a smart home is vulnerable to any kind of intrusion where unauthorized users gain access to the system. A smart lock system based on blockchain technology has been proposed, to ensure important features such as security, authentication, data integration, and non-abrogation. Because of the quiescence offered by IoT devices, it is difficult to detect any kind of intrusion. 5G wireless technology could be used to address this issue; it provides low latency and block mining of the transactions on the blockchain. Due to the high latency offered in transaction confirmation, the model suggested in the work had a minor drawback. To overcome this drawback, a smart home was proposed, based on blockchain technology, and consisting of three tiers: the smart home, the overlay, and cloud storage. There were many similarities between this overlay network and the P2P network that can be seen in Bitcoin. Weixian Li et al. [40] have proposed a smart energy theft system (SETS) that monitors the energy for a system. This system uses machine language and data science. The system combines various machine learning applications, to provide the power that is consumed in the process.

14.4.4 SMART CITY

The increasing trend toward urbanization has generated complex challenges, such as for cities' overall infrastructure, and to ensure whether its people have access to basic needs such as the water supply, transportation, the power supply, and health facilities. Various factors, including climate change, the rise in mass population, and the insufficiency of resources have resulted in this unprecedented urban growth.

Attacks, Vulnerabilities, and Blockchain-Based Countermeasures

In such a situation, a smart city could work as a proactive response to the above-mentioned problems, by ensuring the use of available resources efficiently and smartly. With a reduction in the overall cost of management, a smart city aims to provide an improved and efficient quality of services to its residents. Saber Talari et al. [41] have proposed its various applications, benefits, and disadvantages, based on a survey conducted on IoT-based smart cities. An advanced parking system, which lessens costs by recruiting relevant staff, is an essential element of any smart city. With the objective of reducing the number of attempts to locate a parking site, Thanh Nam Pham et al. [42] have devised an algorithm to increase the effectiveness of the smart parking systems based on cloud computing. The information provided by IoT sensor devices builds a fundamental structure for a highly efficient smart city.

14.4.5 HEALTH CARE 4.0

It is important to improve the health care facilities of any country, for its complete advancement. The load on the modern health care system has increased, due to an increase in the mass population and, also, in medical conditions. E-health care has come into the picture, with modern technologies and advanced communication systems for the benefit of doctors, patients, policy makers, and others [43]. As a consequence of using blockchain technology, the entire health care setup can be restructured for a nation's complete welfare. The load on the health care system can be assuaged with the help of a 5G-enabled IoT [44,45]. Stephanie B. Baker et al. [46] have identified remote health monitoring as a key element of a health care system based on IoT technology. The collection of a patient's health care information can be digitally controlled, with the help of electronic health controls (EHC) [47]. On the other hand, for an individual patient's information, a personal health record (PHC) is used. EHCs assure security and provide for the real-time sharing of medical information. A dispersed record management system, known as MedRec, was proposed, which uses blockchain technology in order to handle EHRs. EHRs could result in the infringement of the privacy of a user's data. A searchable encryption scheme that hinges on blockchain technology could be used to prevent the sharing of private data. With the help of this scheme, data is fully controllable by its owners. Another work [48] has proposed a blockchain-based platform that is EHC-friendly and assures full privacy and security. Adarsh Kumar et al. [49] have proposed various algorithms that can be applied to enhance the operations of health care 4.0. They show how the applications of blockchain technology can be enormously useful to the health care sector.

14.4.6 UNMANNED AERIAL VEHICLES

An aircraft that can be automatically operated by any user is known as an unmanned aerial vehicle (UAV). It is also called a drone [50]. The pictures acquired from UAVs could assist in various industrial implementations, such as supervision, urban modeling, communications, and agriculture [51]. A Wi-Fi connection in remote areas is extremely essential for the operation of UAVs. However, in the case of line-of-sight

310 Blockchain Technology for Data Privacy Management

(LOS) communication, UAVs may not be suitable. Therefore, in such cases, a 5G network is considered an ideal option that also covers a broad area and provides fast speed, as well as a secure network connection [52]. With the rise in the number of autonomous UAVs, the hazard of obtrusion increases, as well. Therefore, a UAV network becomes vulnerable to attacks, such as the compression of computing networks, a disturbance in networking operations, and the insertion of false data [53]. An allocable, extensible, and safe model has been devised, which is known as a UAV communication network (UAVNet). This network model encourages a secure and real-time communication between different UAVs.

14.4.7 Smart Grid

The entire web of communication lines, transformers, resisters, and other elements constitute the electricity infrastructure. This infrastructure carries electricity from power plants and delivers it to our houses; this is known as the "electric grid." A smart grid, however, amalgamates digital technology with the electric grid. The elements of the electric grid combine with computing devices, in order to fulfill the user's needs properly. The smart grid is a hot topic for researchers and developers, who have been discussing and studying it extensively. With better safety measures for network connectivity, permission access, and data interchange, the smart grid has acquired great popularity in recent times [54]. But some unauthorized and fraudulent markets could intervene in the security standards of smart grid systems [55]. While offering service to smart meters (SMs), the respective service providers (SPs) may become vulnerable to some point of failure. An effective and safe decentralized keyless signature scheme has been introduced to overcome this issue.

14.5 CHALLENGES OF BLOCKCHAIN IN THE IOT

Recently, blockchain systems have been invented that are able to run on a consistent P2P network. Various properties of the IoT avert the direct deployment of blockchain technology. For instance, resource-bounded end devices are used, in contrast to speedy servers. The challenges of using blockchain in IoT devices include:

14.5.1 Computation

With respect to the major expenses, blockchain technology is not favorable for some IoT devices, which, because of their light weight, are unable to afford blockchain activity. In order to preserve privacy, blockchain technology uses advanced cryptographic algorithms, such as attribute-based encryption (ABE) and zero knowledge, which are way too heavy for IoT devices. An entire node verifies and searches for each and every block and transaction in blockchain, which becomes considerably heavy for the IoT devices to manage, as they are usually resource-limited [56]. IoT devices do not support the consensus protocols. The network in Bitcoin conducts around 10^{19} hashes per second [57]. A Raspberry pi 3 [58] is a strong IoT device that can conduct around 10^4 hashes per second.

Attacks, Vulnerabilities, and Blockchain-Based Countermeasures

14.5.2 STORAGE

Blockchain requires enormous amount of storage, which may be repressive for some IoT devices. In Bitcoin, we have about 5×10^5 blocks from the last nine years. The entire Bitcoin blockchain is around 150 gigabytes in size. Ethereum contains about 5×10^6 blocks. The overall size of Ethereum is around 400 gigabytes. The storage of each and every block is extremely important. If this enormous data is not present, then IoT devices cannot validate the transactions generated by others. In order to create a new transaction, a transaction sender requires data about past events, such as the balance index and transaction index. Either by trusting itself or by trusting remote servers, IoT devices can ensure a secure and efficient communication with other reliable servers. In blockchain technology, if IoT devices are used as light nodes, then some of the storage load could be decreased. However, we must not forget that the block headers must be included in this process. As we know, storing information in blockchain is extremely important. In a blockchain network, the overall size of the data may be eruptive, since in an n-node, each and every block would be replicated n times.

14.5.3 COMMUNICATION

Since blockchain runs on a P2P network, all its nodes need persistent transmissions and data exchanges. In order to maintain congruous records, such as for fresh blocks and transactions, blockchain keeps on exchanging its data. Some IoT devices use a wireless communication technique that is less manageable then wired communication technology, because it is vulnerable to various elements, such as interference, shadowing, and fading, which halt the flow of data. Various blockchain projects, such as Bitcoin, use wireless communication technology. The capacity required by the blockchain is much lower than that given by wireless technologies. For instance, Bluetooth (IEEE 802.15.1) can yield 0.72 mbps of data; Zigbee (IEEE 802.15.4) can yield 0.25 mbps of data; ultra-wideband (UWB, IEE 802.15.3) can yield 0.11 mbps of data; Wi-Fi (802.11 a/b/g) can yield 54 mbps of data. The NB-IoT can produce around 100 kbps per signal.

14.5.4 ENERGY

With the help of a battery energy supply, IoT devices can work for a long time. For instance, with the help of a CR2032 battery of capacity 600mWh, an IoT device with power consumption of 0.3mWh per day can work for at least five years. IoT devices adapt various power-saving strategies, such as sleep mode, and highly-efficient communication techniques like the NB-IoT. However, the calculation and transmission compelled by various blockchain activities are usually power-dependent. For instance, SHA-256 needs around 90 nJ/B. The regularized transmission power expense of Bluetooth is around 140 mJ/Mb; for Zigbee, it is around 300 mJ/Mb; for UWB, it is 7 mJ/Mb, and for Wi-Fi it is around 13 mJ/Mb. Thus, with the help of the Zigbee protocol, the aforenamed power budget of 0.3 mWh per day can support 0.5 MB of processing and communication.

312 Blockchain Technology for Data Privacy Management

14.5.5 Mobility and Partition of the IoT

The network of wireless communication can be segregated into two modes, i.e., an infrastructure mode and an ad hoc mode. In an infrastructure mode, with the help of network architecture, all packets are forwarded. By contrast, in an ad hoc mode, without depending on preexisting architectures, the network enables each and every node to forward information for several distinct nodes. The blockchain operation becomes weak, due to the mobility of IoT devices. In a wireless network, the devices based on an infrastructure mode can affect the development of signaling and controlling messages. On the other hand, in an ad hoc network, partitioning segregates the networks into independent routes when mobile nodes shift with multiple patterns. With innovative sensing and transmission abilities, mobile phones and vehicles are well equipped. Different sensors with different technologies are used, which results in more power consumption and is not very sustainable. Hence, it is really important to level and manage these sensors.

14.5.6 Latency and Capacity

In order to ensure stability in its dispersed network, blockchain should have high latency. Various IoT devices may not favor the latency accepted by blockchain. For instance, Bitcoin takes around 10 minutes to confirm a block, which may be too long for some IoT applications, which are delay-sensitive. Vehicle networks are an example of such IoT applications. But based on research, blockchain's high latency may limit the capacity of blockchain. For example, Bitcoin uses 10 MB of data in 10 minutes, which is much lower than what is required by IoT applications. The capacity required by various IoT devices depends on their different applications. For instance, in a smart city, which is an IoT application [59], 4.03 GB is the vehicular traces of 700 cars in 24 hours. The aforementioned data tells us that every car uses around 0.24 MB per hour. In the meantime, 294 KB of data is procured from the parking lot from 55 points in around five months, which makes about 36 B per day per point. Moreover, with the rise in the number of IoT devices, the capacity required by some IoT applications would continually escalate.

14.6 CONCLUSION

With the dispersed network of architecture provided by blockchain, the world has been technologically revolutionized. Blockchain is technologically sound and helps in developing a secure environment for IoT devices. It is self-regulating and self-managing. A combination of the two technologies, i.e., IoT and blockchain, has helped various industries to assemble a massive amount of data properly and efficiently. In this chapter, we have provided a brief insight about how IoT-enabled devices can benefit from using blockchain technology. The deployment of blockchain technology in sectors such as health care, industry, smart cities, etc. offers various advantages, as we have seen. However, certain characteristics of the technology present risks that need to be taken care of. Therefore, the advantages of combining the two different technologies should be observed carefully and must be handled properly. This paper has demonstrated the major challenges that blockchain technology presents to IoT devices. These challenges must be analyzed to make both technologies work together efficiently and successfully.

REFERENCES

1. Ganz, F., Puschmann, D., Barnaghi, P., and Carrez, F., 2015. A practical evaluation of information processing and abstraction techniques for the Internet of Things. *IEEE Internet of Things Journal*, 2(4), pp. 340–354. doi:10.1109/jiot.2015.2411227.
2. Arellanes, D., and Lau, K., 2019. *Decentralized data allows in algebraic service compositions for the scalability of IoT systems.* In *2019 IEEE 5th World Forum on Internet of Things (WF-IoT)* (pp. 668–673). Limerick, Ireland: IEEE. doi:10.1109/wf-iot. 2019.8767238.
3. Arora, A., Kaur, A., Bhushan, B., and Saini, H., 2019. *Security concerns and future trends of Internet of Things.* In *2019 2nd International Conference on Intelligent Computing, Instrumentation and Control Technologies (ICICICT).* (pp. 891–896). Kannur, Kerala, India: IEEE. doi:10.1109/icicict46008.2019.8993222.
4. Liu, Y., Wang, K., Qian, K., Du, M., and Guo, S., 2020. Tornado: Enabling blockchain in heterogeneous Internet of Things through a space-structured approach. *IEEE Internet of Things Journal*, 7(2), pp. 1273–1286.
5. Bhushan, B., Sahoo, C., Sinha, P., and Khamparia, A., 2020. Unification of Blockchain and Internet of Things (BIoT): Requirements, working model, challenges and future directions. *Wireless Networks*. doi:10.1007/s11276-020-02445-6.
6. Zhaofeng, M., Xiaochang, W., Jain, D. K., Khan, H., Hongmin, G., and Zhen, W., 2020. A blockchain-based trusted data management scheme in edge computing. *IEEE Transactions on Industrial Informatics*, 16(3), pp. 2013–2021. doi:10.1109/tii.2019.2933482.
7. Arora, D., Gautham, S., Gupta, H., and Bhushan, B., 2019. *Blockchain-based security solutions to preserve data privacy and integrity.* In *2019 International Conference on Computing, Communication, and Intelligent Systems (ICCCIS)* (pp. 468–472). Greater Noida, India: IEEE. doi:10.1109/icccis48478.2019.8974503.
8. Wang, D., Jiang, Y., Song, H., He, F., Gu, M., and Sun, J., 2017. Verification of implementations of cryptographic hash functions. *IEEE Access*, 5, pp. 7816–7825. doi:10.1109/access.2017.2697918.
9. Varshney, T., Sharma, N., Kaushik, I., and Bhushan, B., 2019. *Architectural model of security threats and their countermeasures in IoT.* In *2019 International Conference on Computing, Communication, and Intelligent Systems (ICCCIS)* (pp. 424–429). Greater Noida, India: IEEE. doi:10.1109/icccis48478.2019.8974544.
10. Bhushan, B., Khamparia, A., Sagayam, K. M., Sharma, S. K., Ahad, M. A., and Debnath, N. C., 2020. Blockchain for smart cities: A review of architectures, integration trends and future research directions. *Sustainable Cities and Society*, 61, p. 102360. doi:10.1016/j.scs.2020.102360.
11. Sharma, T., Satija, S., and Bhushan, B., 2019. *Unifying Blockchain and IoT: Security requirements, challenges, applications and future trends.* In *2019 International Conference on Computing, Communication, and Intelligent Systems (ICCCIS)* (pp. 341–346). Greater Noida, India: IEEE. doi:10.1109/icccis48478.2019.8974552.
12. Memon, R. A., Li, J. P., Nazeer, M. I., Khan, A. N., and Ahmed, J., 2019. DualFog-IoT: Additional Fog Layer for Solving Blockchain Integration Problem in Internet of Things. *IEEE Access*, 7, pp. 169073–169093. doi:10.1109/access.2019.2952472.

13. Nakamoto, S., 2019. Bitcoin: A peer-to-peer electronic cash system. *Manubot.* https://bitcoin.org/bitcoin.pdf. Accessed on November 24, 2020.
14. Motohashi, T., Hirano, T., Okumura, K., Kashiyama, M., Ichikawa, D., and Ueno, T., 2019. Secure and scalable mHealth data management using blockchain combined with client hashchain: System design and validation. *Journal of Medical Internet Research*, 21(5). doi:10.2196/13385.
15. Zyskind, G., Nathan, O., and Pentland, A., 2015. *Decentralizing privacy: Using blockchain to protect personal data.* In *2015 IEEE Security and Privacy Workshops.* San Jose, CA: IEEE. doi:10.1109/spw.2015.27.
16. Al-Jaroodi, J., and Mohamed, N., 2019. Blockchain in Industries: A Survey. *IEEE Access*, 7, pp. 36500–36515. doi:10.1109/access.2019.2903554.
17. Bishr, A. B., 2019. Dubai: A city powered by Blockchain. *Innovations: Technology, Governance, Globalization*, 12(3–4), pp. 4–8. doi:10.1162/inov_a_00271.
18. Pilkington, M., n.d.. Blockchain technology: Principles and applications. In Olleros, F. X., and Zhegu, M. (eds.) *Research Handbook on Digital Transformations.* Cheltenham, UK: Edward Elgar Publishing, pp. 225–253. doi:10.4337/9781784717766.00019.
19. Reyna, A., Martín, C., Chen, J., Soler, E., and Díaz, M., 2018. On blockchain and its integration with IoT: Challenges and opportunities. *Future Generation Computer Systems*, 88, pp. 173–190. doi:10.1016/j.future.2018.05.046.
20. Soni, S., and Bhushan, B., 2019. *A comprehensive survey on Blockchain: Working, security analysis, privacy threats and potential applications.* In *2019 2nd International Conference on Intelligent Computing, Instrumentation and Control Technologies (ICICICT)* (pp. 922–926). Kannur, Kerala, India: IEEE. doi:10.1109/icicict46008.2019.8993210.
21. Dash, 2017. *Your money, your* way. https://www.dash.org/es/. Accessed on October 20, 2019.
22. Bastiaan, M., 2015. *Preventing the 51%-attack: A stochastic analysis of two phase Proof of Work in Bitcoin.* https://fmt.ewi.utwente.nl/media/175.pdf. Accessed on November 24, 2020.
23. Atzei, N., Bartoletti, M., and Cimoli, T., 2017. A survey of attacks on Ethereum smart contracts (SoK). *Lecture Notes in Computer Science Principles of Security and Trust*, 164–186. https://eprint.iacr.org/2016/1007.pdf. Accessed on November 24, 2020.
24. Andrea, I., Chrysostomou, C., and Hadjichristofi, G., 2015. *Internet of Things: Security vulnerabilities and challenges.* In *2015 IEEE Symposium on Computers and Communication (ISCC)* (pp. 180–187). Larnaca: IEEE. doi:10.1109/iscc.2015.7405513.
25. Gomes, T., Salgado, F., Tavares, A., and Cabral, J., 2017. CUTE mote, a customizable and trustable end-device for the Internet of Things. *IEEE Sensors Journal*, 17(20), pp. 6816–6824. doi:10.1109/jsen.2017.2743460.
26. Liu, J., Zhang, C., and Fang, Y., 2018. EPIC: A differential privacy framework to defend smart homes against internet traffic analysis. *IEEE Internet of Things Journal*, 5(2), pp. 1206–1217. doi:10.1109/jiot.2018.2799820.
27. Eom, J. H., 2015. Security threats recognition and countermeasures on smart battlefield environment based on IoT. *International Journal of Security and Its Applications*, 9(7), pp. 347–356. doi:10.14257/ijsia.2015.9.7.32.
28. Rambus, n.d. *Industrial Iot: Threats and countermeasures.* https://www.rambus.com/iot/industrial-iot/.1230. Accessed on November 24, 2020.
29. Varga, P., Plosz, S., Soos, G., and Hegedus, C., 2017. *Security threats and issues in automation IoT.* In *2017 IEEE 13th International Workshop on Factory Communication Systems (WFCS)* (pp. 1–6). Trondheim: IEEE. doi:10.1109/wfcs.2017.7991968.

30. Machado, C., and Frohlich, A. A., 2018. *IoT data Integrity verification for cyber-physical systems using blockchain.* In *2018 IEEE 21st International Symposium on Real-Time Distributed Computing (ISORC)* (pp. 83–90). Singapore: IEEE. doi:10.1109/isorc.2018.00019.
31. Gope, P., and Sikdar, B., 2019. Lightweight and privacy-preserving two-factor authentication scheme for IoT devices. *IEEE Internet of Things Journal*, 6(1), pp. 580–589. doi:10.1109/jiot.2018.2846299.
32. Gai, K., Choo, K. R., Qiu, M., and Zhu, L., 2018. Privacy-preserving content-oriented wireless communication in Internet-of-Things. *IEEE Internet of Things Journal*, 5(4), pp. 3059–3067. doi:10.1109/jiot.2018.2830340.
33. Li, F., Shi, Y., Shinde, A., Ye, J., and Song, W., 2019. Enhanced cyber-physical security in Internet of Things through energy auditing. *IEEE Internet of Things Journal*, 6(3), pp. 5224–5231. doi:10.1109/jiot.2019.2899492.
34. Xie, S., Yang, J., Xie, K., Liu, Y., and He, Z., 2017. Low-sparsity unobservable attacks against smart grid: Attack exposure analysis and a data-driven attack scheme. *IEEE Access*, 5, pp. 8183–8193. doi:10.1109/access.2017.2680463.
35. Kapitonov, A., Lonshakov, S., Krupenkin, A., and Berman, I., 2017. *Blockchain-based protocol of autonomous business activity for multi-agent systems consisting of UAVs.* In *2017 Workshop on Research, Education and Development of Unmanned Aerial Systems (RED-UAS)* (pp. 84–89). Linköping, Sweden: IEEE. doi:10.1109/red-uas.2017.8101648.
36. Lin, C., He, D., Huang, X., Choo, K. R., and Vasilakos, A. V., 2018. BSeIn: A blockchain-based secure mutual authentication with fine-grained access control system for Industry 4.0. *Journal of Network and Computer Applications*, 116, pp. 42–52. doi:10.1016/j.jnca.2018.05.005.
37. Kang, J., Xiong, Z., Niyato, D., Ye, D., Kim, D. I., and Zhao, J., 2019. Toward secure blockchain-enabled Internet of Vehicles: Optimizing consensus management using reputation and contract theory. *IEEE Transactions on Vehicular Technology*, 68(3), pp. 2906–2920. doi:10.1109/tvt.2019.2894944.
38. Al-Heety, O. S., Zakaria, Z., Ismail, M., Shakir, M. M., Alani, S., and Alsariera, H., 2020. A comprehensive survey: Benefits, services, recent works, challenges, security, and use cases for SDN-VANET. *IEEE Access*, 8, pp. 91028–91047. doi:10.1109/access.2020.2992580.
39. Roshan, R., and Ray, A. K., 2016. Challenges and risk to implement IOT in smart homes: An Indian perspective. *International Journal of Computer Applications*, 153(3), pp. 16–19. doi:10.5120/ijca2016911982.
40. Li, W., Logenthiran, T., Phan, V., and Woo, W. L., 2019. A novel smart energy theft system (SETS) for IoT-based smart home. *IEEE Internet of Things Journal*, 6(3), pp. 5531–5539. doi:10.1109/jiot.2019.2903281.
41. Talari, S., Shafie-Khah, M., Siano, P., Loia, V., Tommasetti, A., and Catalão, J., 2017. A review of smart cities based on the Internet of Things concept. *Energies*, 10(4), p. 421. doi:10.3390/en10040421.
42. Pham, T. N., Tsai, M., Nguyen, D. B., Dow, C., and Deng, D., 2015. A cloud-based smart-parking system based on Internet-of-Things technologies. *IEEE Access*, 3, pp. 1581–1591. doi:10.1109/access.2015.2477299.
43. Chen, L., Lee, W.-K., Chang, C.-C., Choo, K.-K. R., and Zhang, N., 2019. Blockchain based searchable encryption for electronic health record sharing. *Future Generation Computer Systems*, 95, pp. 420–429.

44. Jain, R., Gupta, M., Nayyar, A., and Sharma, N., 2020. Adoption of fog computing in health care 4.0. In Tanwar, S. (eds.) *Fog Computing for Health Care 4.0 Environments*. Signals and Communication Technology Series. Cham: Springer, pp. 3–36. doi:10.1007/978-3-030-46197-3_1.

45. Islam, S. M., Kwak, D., Kabir, M. H., Hossain, M., and Kwak, K., 2015. The Internet of Things for health care: A comprehensive survey. *IEEE Access*, 3, pp. 678–708. doi:10.1109/access.2015.2437951.

46. Baker, S. B., Xiang, W., and Atkinson, I., 2017. Internet of Things for smart health care: Technologies, challenges, and opportunities. *IEEE Access*, 5, pp. 26521–26544. doi:10.1109/access.2017.2775180.

47. Hathaliya, J. J., Tanwar, S., Tyagi, S., and Kumar, N., 2019. Securing electronics health care records in health care 4.0: A biometric-based approach. *Computers and Electrical Engineering*, 76, pp. 398–410. doi:10.1016/j.compeleceng.2019.04.017.

48. Omar, A. A., Bhuiyan, M. Z., Basu, A., Kiyomoto, S., and Rahman, M. S., 2019. Privacy-friendly platform for health care data in cloud based on blockchain environment. *Future Generation Computer Systems*, 95, pp. 511–521. doi:10.1016/j.future.2018.12.044.

49. Kumar, A., Krishnamurthi, R., Nayyar, A., Sharma, K., Grover, V., and Hossain, E., 2020. A novel smart health care design, simulation, and implementation using health care 4.0 processes. *IEEE Access*, 8, pp. 118433–118471. doi:10.1109/access.2020.3004790.

50. Government of India Department of Space, 2020. Applications of unmanned aerial vehicle (UAV) based remote sensing in NE region: ISRO. https://www.isro.gov.in/applications-of-unmanned-aerial-vehicle-uav-based-remote-sensing-ne-region. Accessed on November 25, 2020.

51. Chandrasekharan, S., Gomez, K., Al-Hourani, A., Kandeepan, S., Rasheed, T., Goratti, L.,... Allsopp, S., 2016. Designing and implementing future aerial communication networks. *IEEE Communications Magazine*, 54(5), pp. 26–34. doi:10.1109/mcom.2016.7470932.

52. Lin, X., Yajnanarayana, V., Muruganathan, S. D., Gao, S., Asplund, H., Maattanen, H., ...Wang, Y. E., 2018. The sky is not the limit: LTE for unmanned aerial vehicles. *IEEE Communications Magazine*, 56(4), pp. 204–210. doi:10.1109/mcom.2018.1700643.

53. Banerjee, M., Lee, J., and Choo, K. R., 2018. A blockchain future for Internet of Things security: A position paper. *Digital Communications and Networks*, 4(3), pp. 149–160. doi:10.1016/j.dcan.2017.10.006.

54. Forbes, 2018. How blockchain can help increase the security of smart grids. https://www.forbes.com/sites/andrewarnold/2018/04/16/how-blockchain-can-help-increase-the-security-of-smart-grids/#1b59ad95b489. Accessed on November 25, 2020.

55. Bhushan, B., and Sahoo, G., 2020. Requirements, protocols, and security challenges in wireless sensor networks: An industrial perspective. In Gupta, B., Pérez, G. M., Agrawal, D. P., and Gupta, D. (eds.) *Handbook of Computer Networks and Cyber Security*. Cham: Springer, pp. 683–713. doi:10.1007/978-3-030-22277-2_27.

56. Protocol rules, 2016. https://en.bitcoin.it/wiki/Protocol_rules. Accessed on November, 25, 2020.

57. Blockchain, 2017. https://blockchain.info. Accessed on November 25, 2020.

58. Raspberry pi, 2017. https://www.raspberrypi.org. Accessed on November 25, 2020.

59. Rathore, M. R., Paul, A. P., and Ahmad, A. A., n.d.. IoT and big data: Application for urban planning and building smart cities. In Huang, J., and Hua, K. (eds.) *Managing the Internet of Things: Architectures, Theories and Applications*. London: The Institution of Engineering and Technology, pp. 155–183. doi:10.1049/pbte067e_ch9.

Index

A

authentication, 2, 5, 8–10, 15, 28, 32, 50, 85–95, 98, 100, 101, 104, 149, 167–171, 174, 187, 192, 193, 199, 202, 203, 211, 214, 217, 218, 225, 246, 250, 254, 257, 304–306, 308
authorization, 9, 15, 23, 28, 85, 88–92, 98, 100, 103, 120, 170, 172, 174, 175, 214, 218, 219, 246

B

big data, 127, 169, 178, 205, 211, 256, 258, 264, 282, 290
Bitcoin, 14, 110–127, 139, 143, 146, 152, 162, 164, 177, 186–188, 194, 199–205, 210, 217, 218, 221, 222, 224, 242, 258, 276–282, 286, 287, 289–291, 298–302, 308, 310–312
blockchain, 3, 14–16, 28, 29, 83, 112–127, 135–154, 160–178, 185–205, 210–235, 239–250, 254–270, 275–291, 296–312
Bluetooth, 7, 8, 13, 38, 169, 257, 311

C

cloud computing, 7, 31, 33, 38, 45, 214, 254, 256, 263, 264, 305, 309
consensus, 112, 113, 116, 117, 120, 123–125, 138, 140, 162–168, 171, 172, 175, 176, 187, 188, 199, 210–214, 217, 220, 222–225, 232–234, 254–259, 268, 269, 279, 280, 287, 299–301, 310
cybersecurity, 22, 27, 75–78, 83, 86, 98–100, 103, 110, 111, 113, 115, 120, 122, 124, 126, 148, 176

D

data integrity, 14, 95, 120, 172, 218, 233, 305
data security, 44, 83, 114, 120, 173, 188, 204, 205, 211, 282, 305
decentralized architecture, 160, 165, 210, 267
deep learning, 38, 44, 68–72, 223, 255
digital currency, 111, 140, 145, 151, 163, 177, 276, 278, 279, 281, 300, 301

E

electronic health record, 161, 254, 269, 280, 281, 284, 285
encryption, 34, 87, 95, 96, 98, 110, 115, 119, 121, 126, 175, 186–192, 197, 214, 224, 234, 246, 258, 267, 270, 284, 285, 304, 305, 309, 310
Ethereum, 14, 15, 110, 112, 113, 116, 117, 124, 125, 140, 149, 166, 177, 223, 246, 249, 258, 259, 268, 269, 279, 286–288, 299, 302, 307, 311

F

fog computing, 28, 210, 211, 214, 254, 255, 259, 260, 263, 264

H

hashing, 102, 116, 174, 188, 191, 193, 198, 217, 225, 243, 277, 298
healthcare, 110, 245, 259
Hyperledger, 113, 165–168, 212, 213, 219, 223, 224, 228, 230, 232–234, 257, 277, 281, 288–300

I

industrial automation, 97, 296, 307
industries, 7, 11, 22, 82, 83, 89, 97, 106, 110, 122, 124, 126, 151, 281, 307, 312
intelligent transportation, 82, 105, 307
Internet of Things, 6, 7, 14, 16, 21, 23, 25, 38, 65, 82, 83, 86, 110, 122, 124, 148, 161, 186, 213, 215, 296

M

medical records, 34, 147, 254, 259

distributed ledger, 110–116, 120, 122, 126, 127, 138, 162, 173–175, 187, 203, 210, 276–279

317

318 Index

P

privacy, 2, 23–27, 30, 31, 34, 87–89, 95, 127, 170, 173–175, 188, 213, 217, 218, 221, 224, 241, 246, 256, 257, 259, 281–286, 291, 301, 303, 305, 306, 309

S

security, 1, 5, 8–10, 14, 15, 22–34, 44, 45, 48–50, 75–78, 83–105, 110–122, 125, 126, 142–148, 161–163, 169–178, 186–191, 204, 205, 211–219, 224–235, 240, 241, 246–249, 254–260, 270, 278–285, 289, 298, 301–310
smart cities, 11, 14, 38, 160, 170, 175, 297, 309, 312
smart contracts, 117, 122, 124, 127, 147, 150, 151, 164, 171–178, 210, 212, 213, 217, 220–222, 227, 228, 240, 254, 255, 258, 259, 268, 269, 278, 280, 282, 287, 291, 302

smart grid, 160, 175, 212, 214, 226, 228, 310
smart home, 6, 8, 11, 13, 82, 97, 161, 177, 214, 239–250, 297, 303, 307, 308

T

traceability, 14, 15, 117, 211, 281, 284
transactions, 13–15, 92, 110–126, 135–147, 150, 152, 160–166, 173, 186–188, 192, 199, 202–205, 211–213, 217, 220, 223, 224, 230–235, 240–243, 275–280, 291, 297, 301–303, 311
transparency, 25, 110–114, 117, 119, 121, 123, 125, 138, 143, 149, 151, 154, 160, 162, 166, 172, 175–177, 203, 212, 222, 241, 254, 281, 284, 302

W

wireless, 3, 12, 13, 21, 26, 30, 38–47, 53, 62, 65, 75, 86, 87, 169, 172, 210, 214, 255, 305, 308, 311, 312